Advances in Intelligent Systems and Computing

Volume 246

Series editor

Janusz Kacprzyk, Polish Academy of Sciences, Warsaw, Poland
e-mail: kacprzyk@ibspan.waw.pl

For further volumes:
http://www.springer.com/series/11156

About this Series

The series "Advances in Intelligent Systems and Computing" contains publications on theory, applications, and design methods of Intelligent Systems and Intelligent Computing. Virtually all disciplines such as engineering, natural sciences, computer and information science, ICT, economics, business, e-commerce, environment, healthcare, life science are covered. The list of topics spans all the areas of modern intelligent systems and computing.

The publications within "Advances in Intelligent Systems and Computing" are primarily textbooks and proceedings of important conferences, symposia and congresses. They cover significant recent developments in the field, both of a foundational and applicable character. An important characteristic feature of the series is the short publication time and world-wide distribution. This permits a rapid and broad dissemination of research results.

Advisory Board

G. Sai Sundara Krishnan · R. Anitha
R. S. Lekshmi · M. Senthil Kumar
Anthony Bonato · Manuel Graña
Editors

Computational Intelligence, Cyber Security and Computational Models

Proceedings of ICC^3, 2013

Springer

Editors
G. Sai Sundara Krishnan
R. Anitha
R. S. Lekshmi
M. Senthil Kumar
Applied Mathematics and Computational Sciences
PSG College of Technology
Coimbatore, Tamil Nadu
India

Manuel Graña
School of Computing
University of Basque Country
Paseo Manuel De Lardizalbal 1
San Sebastian
Spain

Anthony Bonato
Department of Mathematics
Ryerson University
Toronto, ON
Canada

ISSN 2194-5357 ISSN 2194-5365 (electronic)
ISBN 978-81-322-1679-7 ISBN 978-81-322-1680-3 (eBook)
DOI 10.1007/978-81-322-1680-3
Springer New Delhi Heidelberg New York Dordrecht London

Library of Congress Control Number: 2013954037

Printed on acid-free paper

Springer is part of Springer Science+Business Media (www.springer.com)

Dedicated
To

Dr. G. R. Damodaran
Founder Principal
PSG College of Technology
Coimbatore–641004
India

Preface

The rapid development of network technologies and computing machines has broadened the scope of research and development in computer science and allied areas. To provide a broad interdisciplinary research forum, the International Conference on Computational Intelligence, Cyber Security, and Computational Models (ICC^3-2013) has been organized by the Department of Applied Mathematics and Computational Sciences of PSG College of Technology, during 19–21 December, 2013. We are proud to place on record that this International Conference is a part of the centenary year celebrations of Dr. G. R. Damodaran, Founder Principal, PSG College of Technology, Coimbatore, India.

The primary objective of this conference is to present state-of-the-art scientific results, explore cutting-edge technologies, and promote collaborative research in the areas of revolutionary ideas using computational intelligence, cyber security, and computational models. The conference aims to serve as a platform to establish research relations worldwide.

Computational Intelligence (CI), as a branch of science is applicable in many fields of research, including engineering, data analytics, forecasting, and biomedicine. CI systems are inspired by imitable aspects of living systems. They are used in image and sound processing, signal processing, multidimensional data visualization, steering of objects, expert systems, and many practical implementations. The common feature of CI systems is that it processes information by symbolic representation of knowledge. CI systems have the capability to reconstruct behaviors observed in learning sequences, form rules of inference, and generalize knowledge in situations where they are expected to make predictions or classify objects based on previously observed categories. The CI track comprises research articles which exhibit potential practical applications.

With worldwide escalation in the number of cyber threats, there is a need for comprehensive security analysis, assessment, and action to protect critical infrastructure and sensitive information. Large-scale cyber attacks in various countries threaten information security which, could pose a threat to national security and requires effective crisis management. Such information security risks are becoming increasingly diversified, advanced, and complex and conventional means of security fail to ensure information safety. The cyber security track in this conference aims to bring together researchers, practitioners, developers, and users to arrive at a common understanding of the challenges and build a global framework for security and trust.

Fields such as theory of computation, data analytics, high performance computing, quantum computing, weather forecasting, and flight simulation need computational models like stochastic models, graph models, and neural networks to make predictions about performance of complicated systems. Solutions to many technical problems require extensive mathematical concepts to model the problem and to understand the behavior of associated complex systems by computer simulations. With the advent of efficient computations, solutions to complex problems can be found using computational modeling and research.

ICC^3-2013 received a total of 117 technical submissions out of which only 33 full papers and five short papers were selected for presentation and publication in the proceedings. This selection was done through a stringent blind peer review process. Besides these, research papers by invited speakers have also been included in this proceedings.

The organizers of ICC^3-2013 wholeheartedly appreciate the peer reviewers for their support and valuable comments for ensuring the quality of this proceeding. We also extend our warmest gratitude to Springer for their support in bringing out the proceedings volume in time and with excellent production quality.

We would like to thank all invited speakers, international advisory committee members, and the chairpersons for their excellent contributions. We hope that all the participants of the conference will be benefited academically from this event and wish them success in their research career.

Organization

Patron	Shri L. Gopalakrishnan, Managing Trustee, PSG and Sons Charities Trust, Coimbatore, India
Chairman	Dr. R. Rudramoorthy, Principal, PSG College of Technology, Coimbatore, India
Organizing Chair	Dr. R. Nadarajan, Professor and Head, Department of Applied Mathematics and Computational Sciences, PSG College of Technology, Coimbatore, India
Program Chair	Dr. G. Sai Sundara Krishnan, Associate Professor, Department of Applied Mathematics and Computational Sciences, PSG College of Technology, Coimbatore, India
Computational Intelligence Track Chair	Dr. M. Senthil Kumar, Associate Professor, Department of Applied Mathematics and Computational Sciences, PSG College of Technology, Coimbatore, India
Cyber Security Track Chair	Dr. R. Anitha, Associate Professor, Department of Applied Mathematics and Computational Sciences, PSG College of Technology, Coimbatore, India
Computational Models Track Chair	Dr. R. S. Lekshmi, Associate Professor, Department of Applied Mathematics and Computational Sciences, PSG College of Technology, Coimbatore, India

Advisory Committee Members

Prof. Anthony Bonato	Department of Mathematics, Ryerson University, Canada
Prof. Dan Kannan	Department of Mathematics, University of Georgia, Georgia, USA

Contents

Part V Short Papers

Part I
Keynote Address

The Robber Strikes Back

Anthony Bonato, Stephen Finbow, Przemysław Gordinowicz, Ali Haidar, William B. Kinnersley, Dieter Mitsche, Paweł Prałat and Ladislav Stacho

Abstract We consider the new game of Cops and Attacking Robbers, which is identical to the usual Cops and Robbers game except that if the robber moves to a vertex containing a single cop, then that cop is removed from the game. We study the minimum number of cops needed to capture a robber on a graph G, written $\mathrm{cc}(G)$. We give bounds on $\mathrm{cc}(G)$ in terms of the cop number of G in the classes of bipartite graphs and diameter two, $K_{1,m}$-free graphs.

Keywords Cops and robbers · Cop number · Bipartite graphs · Claw-free graphs

1 Introduction

Cops and Robbers is a vertex-pursuit game played on graphs, which has been the focus of much recent attention. Throughout, we only consider finite, connected, and simple undirected graphs. There are two players consisting of a set of *cops* and

A. Bonato (✉) · W. B. Kinnersley · P. Prałat
Ryerson University, Toronto, Canada
e-mail: abonato@ryerson.ca

S. Finbow
St. Francis Xavier University, Antigonish, Canada

P. Gordinowicz
Technical University of Lodz, Lodz, Poland

A. Haidar
Carleton University, Ottawa, Canada

D. Mitsche
University of Nice Sophia-Antipolis, Nice, France

L. Stacho
Simon Fraser University, Burnaby, Canada

G. S. S. Krishnan et al. (eds.), *Computational Intelligence, Cyber Security and Computational Models*, Advances in Intelligent Systems and Computing 246,
DOI: 10.1007/978-81-322-1680-3_1, © Springer India 2014

a single *robber*. The game is played over a sequence of discrete time steps or *rounds*, with the cops going first in the first round and then playing on alternate time steps. The cops and robber occupy vertices, and more than one cop may occupy a vertex. When a player is ready to move in a round, they may move to a neighbouring vertex or *pass* by remaining on their own vertex. Observe that any subset of cops may move in a given round. The cops win if after some finite number of rounds, one of them can occupy the same vertex as the robber. This is called a *capture*. The robber wins if he can avoid capture indefinitely. A *winning strategy for the cops* is a set of rules that if followed result in a win for the cops, and a *winning strategy for the robber* is defined analogously.

If we place a cop at each vertex, then the cops are guaranteed to win. Therefore, the minimum number of cops required to win in a graph G is a well-defined positive integer, named the *cop number* of the graph G. We write $c(G)$ for the cop number of a graph G. For example, the Petersen graph has cop number 3. Nowakowski and Winkler [14], and independently Quilliot [19], considered the game with one cop only; the introduction of the cop number came in [1]. Many papers have now been written on cop number since these three early works; see the book [8] for additional references and background on the cop number. See also the surveys [2, 4, 5].

Many variants of Cops and Robbers have been studied. For example, we may allow a cop to capture the robber from a distance k, where k is a non-negative integer [7], play on edges [12], allow one or both players to move with different speeds or teleport, or allow the robber to be invisible. See Chap. 8 of [8] for a non-comprehensive survey of variants of Cops and Robber.

We consider a new variant of the game of Cops and Robbers, where the robber is able to essentially strike back against the cops. We say that the robber *attacks* a cop if he chooses to move to a vertex on which a cop is present and eliminates her from the game. In the game of *Cops and Attacking Robbers*, the robber may attack a cop, but cannot start the game by moving to a vertex occupied by a cop; all other rules of the game are the same as in the classic Cops and Robbers. We note that if two cops are on a vertex u and the robber moves to u, then only one cop on u is eliminated; the remaining cop then captures the robber, and the game ends. We write $\mathrm{cc}(G)$ for the minimum number of cops needed to capture the robber. Note that $\mathrm{cc}(G)$ is the analogue of the cop number in the game of Cops and Attacking Robbers; our choice of notation will be made more transparent once we state Theorem 1. We refer to $\mathrm{cc}(G)$ as the *cc-number of* G. Since placing a cop on each vertex of G results in a win for the cops, the parameter $\mathrm{cc}(G)$ is well defined.

To illustrate that $\mathrm{cc}(G)$ can take different values from the cop number, consider that for the cycle C_n with n vertices, we have the following equalities (which are easily verified):

$$\mathrm{cc}(C_n) = \begin{cases} 1 & \text{if } n = 3, \\ 2 & \text{if } 4 \leq n \leq 6, \\ 3 & \text{else.} \end{cases}$$

We outline some basic results and bounds for the cc-number in Sect. 2. We consider bounds on cc(G) in terms of $c(G)$ in Sect. 3. In Sect. 4, we give the bound of $\mathrm{cc}(G) \leq c(G) + 2$ in the case that G is bipartite; see Theorem 9. In the final section, we supply in Theorem 10 an upper bound for cc(G) for $K_{1,m}$-free, diameter two graphs.

For background on graph theory, see [20]. For a vertex u, we let $N(u)$ denote the neighbour set of u, and let $N[u] = N(u) \cup \{u\}$ denote the closed neighbour set of u. The set of vertices of distance 2 to u is denoted by $N_2(u)$. We denote by (G) the minimum degree in G. In a graph G, a set S of vertices is a *dominating set* if every vertex not in S has a neighbour in S. The *domination number* of G, written $\gamma(G)$, is the minimum cardinality of a dominating set. The *girth* of a graph is the length of the shortest cycle contained in that graph and is ∞ if the graph contains no cycles.

2 Basic Results

In this section, we collect together some basic results for the cc-number. As the proofs are either elementary or minor variations of the analogous proofs for the cop number, they are omitted. The first result on the game of Cops and Attacking Robbers is the following theorem; note that the second inequality naturally inspires the notation cc(G). We use the notation $\bar{c}(G)$ for the edge cop number, which is a variant where the cops and robber move on edges; see [12].

Theorem 1 *If G is a graph, then*

$$c(G) \leq \mathrm{cc}(G) \leq \min\{2c(G), 2\bar{c}(G), \gamma(G)\}.$$

The following theorem is foundational in the theory of the cop number.

Theorem 2 [1] *If G has girth at least* 5, *then*

$$c(G) \geq \delta(G).$$

The following theorem extends this result to the cc-number.

Theorem 3 *If G has girth at least* 5, *then*

$$\mathrm{cc}(G) \geq \delta(G) + 1.$$

Isometric paths play an important role in several key theorems in the game of Cops and Robbers, such as the cop number of planar graphs (see Chap. 4 of [8]). We call a path P in a graph G *isometric* if the shortest distance between any two vertices is equal in the graph induced by P and in G. For a fixed integer $k \geq 1$, an induced subgraph H of G is *k-guardable* if, after finitely many moves, k cops can move only in the vertices of H in such a way that if the robber moves into H at round t, then he will be captured at round $t + 1$ by a cop in H. For example, a clique in a graph is 1-guardable.

Aigner and Fromme [1] proved the following result.

Theorem 4 [1] *An isometric path is* 1*-guardable.*

We have an analogue of Theorem 4 for the cc-number.

Theorem 5 *An isometric path is* 2*-guardable in the game of Cops and Attacking Robbers, but need not be* 1*-guardable.*

See Fig. 1 for an example where the robber can freely move onto an isometric path without being captured by a sole cop.

A graph G is called *planar* if it can be embedded in a plane without two of its edges crossing. It was shown first in [1] that planar graphs require at most three cops to catch the robber; see [8] for an alternative proof of this fact. Given the results above, we may conjecture that the cc-number of a planar graph is at most 4 or even 5, but either bound remains unproven.

Outerplanar graphs are those that can be embedded in the plane without crossings in such a way that all of the vertices belong to the unbounded face of the embedding. Clarke proved the following theorem in her doctoral thesis.

Theorem 6 [11] *If G is outerplanar, then $c(G) \leq 2$.*

The counterpart to Theorem 6 is the following.

Theorem 7 *If G is outerplanar, then* $\mathrm{cc}(G) \leq 3$.

Meyniel's conjecture—first communicated by Frankl [13]—is one of the most important open problems surrounding the game of Cops and Robbers. The conjecture states that $c(n) = O(\sqrt{n})$, where $c(n)$ is the maximum of $c(G)$ over all n-vertex, connected graphs. Cops and Robbers has been studied extensively for random graphs (see for example, [3, 9, 15, 16]), partly owing to a search for counterexamples to Meyniel's conjecture. However, it was recently shown that Meyniel's conjecture holds asymptotically almost surely (that is, with probability tending to 1 as the number of vertices tends to infinity) for both binomial random graphs $G(n, p)$ [17] as well as random d-regular graphs [18].

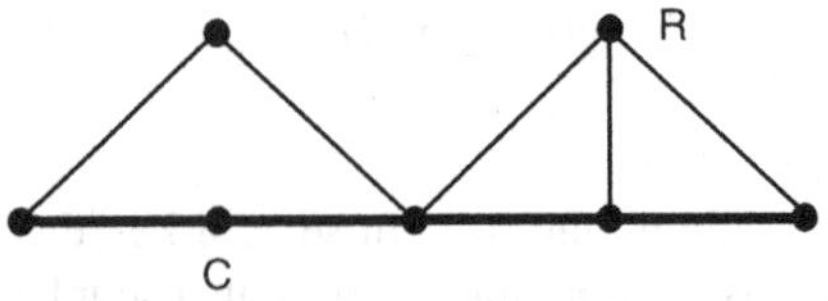

Fig. 1 One cop cannot guard the isometric path (*depicted in bold*). We assume that the robber has just arrived at their vertex, and it is the cop's turn to move

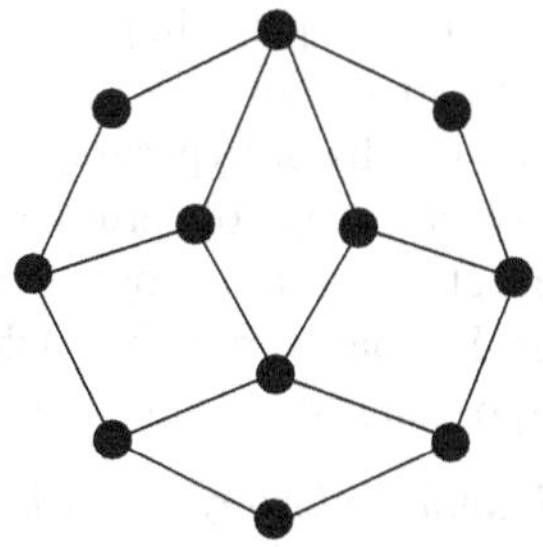

Fig. 2 A graph G with $c(G) = \mathrm{cc}(G) = 2$ and $\gamma(G) = 3$

In [9], it was shown that for dense random graphs, where $p = n^{-o(1)}$ and $p < 1 - \epsilon$ for some $\epsilon > 0$, asymptotically almost surely we have that

$$c(G(n,p)) = (1 + o(1))\gamma(G(n,p)) = (1 + o(1)) \log_{1/(1-p)} n. \tag{1}$$

Note that (1) implies that $c(G(n, p)) = (1 + o(1))\mathrm{cc}(G(n, p))$ for the stated range of p; in particular, applying (1) to the $p = 1/2$ case (which corresponds to the uniform probability space of all labelled graphs on n vertices), we have that for every $\epsilon > 0$, almost all graphs satisfy $\mathrm{cc}(G)/c(G) \in [1, 1 + \epsilon]$. Unfortunately, the asymptotic value of the cop number is not known for sparser graphs. However, it may be provable that $c(G(n,p)) = (1 + o(1))\mathrm{cc}(G(n,p))$ for sparse graphs, without finding an asymptotic value.

We finish the section by noting that graphs with $\mathrm{cc}(G) = 1$ are precisely those with a universal vertex. However, characterizing those graphs G with $\mathrm{cc}(G) = 2$ is an open problem. Graphs with $\mathrm{cc}(G) = 2$ include cop-win graphs without universal vertices and graphs which are not cop win but have domination number 2. Before the reader conjectures this gives a characterization, note that the graph in Fig. 2 with cc-number equalling 2 is in neither class.

3 How Large Can the cc-Number Be?

One of the main unanswered questions on the game of Cops and Attacking Robbers is how large the cc-number can be relative to the cop number. Many of the results from the last section might lead one to (mistakenly) conjecture that

$$\mathrm{cc}(G) \leq c(G) + 1$$

for all graphs, and this was the thinking of the authors and others for some time. We provide a counterexample below.

By Theorem 1, we know that $\mathrm{cc}(G)$ is bounded above by $2c(G)$. For example, this is a tight bound for a path of length at least 3. However, we do not know an improved bound which applies to general graphs, nor do we possess graphs G with $c(G) > 2$ whose cc-number equals $2c(G)$. In this section, we outline one approach which may ultimately yield such examples. Improved bounds for several graph classes are outlined in the next two sections.

Our construction utilizes line graphs of hypergraphs. For a positive integer k, a *k-uniform hypergraph* has every hyperedge of cardinality k. A hypergraph is *linear* if any two hyperedges intersect in at most one vertex. The *line graph* of a hypergraph H, written as $L(H)$, has one vertex for each hyperedge of H, with two vertices adjacent if the corresponding hyperedges intersect.

Lemma 8 *Let H be a linear k-uniform hypergraph with minimum degree at least* 3 *and girth at least* 5. *If $L(H)$ has domination number at least* $2k$, *then* $\mathrm{cc}(L(H)) > 2k$.

Proof Suppose there are at most $2k - 1$ cops. Since the domination number of $L(H)$ is at least $2k$, the robber can choose an initial position that lets him survive the cops' first move. To show that $2k - 1$ cops cannot catch the robber in the game of Cops and Attacking Robbers on $L(H)$, suppose otherwise, and consider the state of the game on the robber's final turn (that is, just before he is to be captured). Let v be the robber's current vertex, E_v the corresponding edge of H, and $w_1, w_2, \ldots, w_k$ the elements of E_v. The neighbours of v in $L(H)$ are precisely those vertices corresponding to the edges of H that intersect E_v; denote by S_{w_i} the set of vertices (other than v) corresponding to edges containing w_i. Each S_{w_i} is a clique; moreover, since H has minimum degree at least 3, each contains at least two vertices. By hypotheses for H, it follows that the S_{w_i} are disjoint and that no vertex outside S_{w_i} dominates more than one vertex inside. Finally, since H has girth at least 5, no vertex in G dominates vertices in two different S_{w_i} (that is, the neighbourhoods $N[S_{w_i}]$ only have v in common).

Consider the cops' current positions. The cops must dominate all of $N[v]$, since otherwise the robber would be able to survive for one more round (by moving to an undominated vertex). Since the $N[S_{w_i}]$ only have v in common, for some j, we have at most one cop in $N[S_{w_j}]$. If in fact there are no cops in $N[S_{w_j}]$, then no vertices of S_{w_j} are dominated, a contradiction. Thus, S_{w_j} contains exactly one cop. Since each vertex outside S_{w_j} dominates at most one vertex inside and S_{w_j} contains at least two vertices, the cop must actually stand within S_{w_j}. However, since she is the only cop within $N[S_{w_j}]$, the robber may attack the cop without leaving himself open to capture on the next turn. Thus, the robber always has a means to avoid capture on the cops' next turn. Hence, at least $2k$ cops are needed to capture the robber, as claimed. □

We aim to find, for all k, graphs G such that $c(G) = k$ and $\mathrm{cc}(G) = 2k$. This, however, remains open for all $k \geq 3$.

As an application of the lemma, take H to be the Petersen graph. It is easily verified that $c(L(H)) = 2$; see also [12]. Lemma 8 with $k = 2$ shows that $\mathrm{cc}(L(H)) \geq 4$; hence, Theorem 1 then implies that $\mathrm{cc}(L(H)) = 4$. See Fig. 3 for a drawing of the line graph of the Petersen graph.

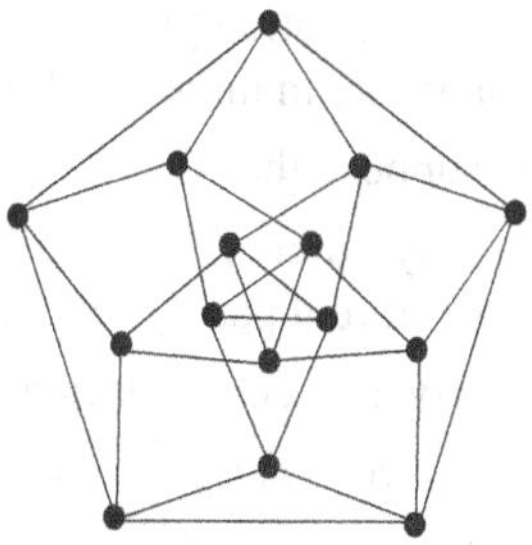

Fig. 3 The line graph of the Petersen graph

4 Bipartite Graphs

For bipartite graphs, we derive the following upper bound.

Theorem 9 *For every connected bipartite graph G, we have that* $\mathrm{cc}(G) \leq c(G) + 2$.

Proof Fix a connected bipartite graph G. Let $k = c(G)$; we give a strategy for $k + 2$ cops to win the game of Cops and Attacking Robbers on G. Label the cops $C_1, C_2, \ldots, C_k, C_1^*, C_2^*$. Intuitively, cops C_1, C_2, …, C_k attempt to follow a winning strategy for the ordinary Cops and Robber game on G; since they must avoid being killed by the robber, they may not be able to follow this strategy exactly, but can follow it "closely enough". Cops C_1^* and C_2^* play a different role: They occupy a common vertex throughout the game, and in each round, they simply move closer to the robber. This has the effect of eventually forcing the robber to move on every turn. (Since the cops move together, the robber cannot safely attack either one.) Further, when the robber passes, the cops $C_1, C_2, \ldots, C_k$ pass. Therefore, we may suppose throughout that the robber moves to a new vertex on each turn.

It remains to formally specify the movements of $C_1, C_2, \ldots, C_k$. To each cop C_i, we associate a *shadow* S_i. Throughout the game, the shadows follow a winning strategy for the ordinary game on G. Let $C_i^{(t)}, S_i^{(t)}$, and $R^{(t)}$ denote the positions of C_i, S_i, and the robber, respectively, at the end of round t. We maintain the following invariants for $1 \leq i \leq k$ and all t:

1. $S_i^{(t)} \in N\left[C_i^{(t)}\right]$ (that is, each cop remains on or adjacent to her shadow).
2. if $C_i^{(t+1)} \neq S_i^{(t+1)}$, then $S_i^{(t+1)}$ and $R^{(t)}$ belong to different partite sets of G.
3. $C_i^{(t+1)}$ is not adjacent to $R^{(t)}$ (that is, the robber never has the opportunity to attack any cop).

On round $t + 1$, each cop C_i moves as follows:

(a) If $C_i^{(t)} \neq S_i^{(t)}$, then C_i moves to $S_i^{(t)}$.
(b) If $C_i^{(t)} = S_i^{(t)}$, and $S_i^{(t+1)}$ is not adjacent to $R^{(t)}$, then C_i moves to $S_i^{(t+1)}$.
(c) Otherwise, C_i remains at her current vertex.

By invariant (1), this is clearly a legal strategy.

We claim that all three invariants are maintained. Invariant (1) is straightforward to verify. For invariant (2), first suppose that $C_i^{(t)} = S_i^{(t)}$, but $C_i^{(t+1)} \neq S_i^{(t+1)}$. By the cops' strategy, this can happen only when $S_i^{(t+1)}$ is adjacent to $R^{(t)}$, in which case, the shadow and robber belong to different partite sets, as desired. Now, suppose that $C_i^{(t)} \neq S_i^{(t)}$ and $C_i^{(t+1)} \neq S_i^{(t+1)}$. By the cops' strategy, we have $C_i^{(t+1)} = S_i^{(t)}$. It follows that $C_i^{(t+1)} \neq C_i^{(t)}$, $S_i^{(t+1)} \neq S_i^{(t)}$, and $R^{(t-1)} \neq R^{(t)}$. Thus, if $S_i^{(t)}$ and $R^{(t-1)}$ belong to different partite sets, then so must $S_i^{(t+1)}$ and $R^{(t)}$; that is, the invariant is maintained. For invariant (3), if $S_i^{(t+1)}$ is adjacent to $R^{(t)}$, then we may suppose that $S_i^{(t+1)} \neq S_i^{(t)}$, since otherwise the shadow would have captured the robber in round $t + 1$. By the cops' strategy, we now have that $C_i^{(t+1)} \neq S_i^{(t+1)}$. But now, the cop and her shadow are in different partite sets by invariant (1), and the shadow and robber are in different partite sets by invariant (2), so the cop and robber are in the same partite set, contradicting adjacency of the cop and the robber.

Since the shadows follow a winning strategy, eventually some shadow S_i captures the robber; that is, for some t, we have that either $S_i^{(t)} = R^{(t)}$ or $S_i^{(t+1)} = R^{(t)}$. In the former case, invariant (3) implies that $C_i^{(t)} \neq S_i^{(t)}$ and invariant (1) implies that C_i captures the robber in round $t + 1$. Now, consider the case when $S_i^{(t+1)} = R^{(t)}$. By invariant (2), since $S_i^{(t+1)}$ is not adjacent to $R^{(t)}$, we in fact have that $C_i^{(t+1)} = S_i^{(t+1)} = R^{(t)}$ so the cops have won. □

5 $K_{1,m}$-Free, Diameter 2 Graphs

We provide one more result giving an upper bound on the cc-number for a set of graph classes.

Theorem 10 *Let G be a $K_{1,m}$-free, diameter* 2 *graph, where* $m \geq 3$. *Then,*

$$\mathrm{cc}(G) \leq c(G) + 2m - 2.$$

When $m = 3$, Theorem 10 applies to claw-free graphs; see [10] for a characterization of these graphs. The cop number of diameter 2 graphs was studied in [6].

Proof of Theorem 10 A cop C is *backup* to a cop C' if C is in $N[C']$, note that a cop with a backup cannot be attacked without the robber being captured in the next round.

Now, let $c(G) = r$, and consider $c(G)$ cops labelled $C_1, C_2, \ldots, C_r$. We refer to these r-many cops as *squad 1*. Label an additional $2m - 2$ cops as $\widehat{C_{i,1}}$ and $\widehat{C_{i,2}}$, where $1 \leq i \leq m - 1$; these cops form *squad 2*. The intuition behind the proof is that the cops in squad 2 act as backup for those in squad 1, who play their usual strategy on G. Further, the cops $\widehat{C_{i,j}}$ are positioned in such a way that the cops C_k need only restrict their movements to the second neighbourhood of some fixed vertex.

More explicitly, fix a vertex x of G. Move squad 2 so that they are contained in $N[x]$. Next, position each of the cops $\widehat{C_{i,1}}$ on x. Hence, R must remain in $N_2(x)$ or he will lose in the next round (in particular, no squad 2 cop is ever attacked). Throughout the game, we will always maintain the property that there are $m - 1$ cops on x.

We note that the squad 2 cops in $N(x)$ can move there essentially as if that subgraph were a clique, and in addition, preserve the property that $m - 1$ cops remain on x. To see this, if $\widehat{C_{i,2}}$ were on $y \in N(x)$ and the cops would like to move to $z \in N(x)$, then move $\widehat{C_{i,2}}$ to x, and move some squad 2 cop from x to z. In particular, a cop from squad 2 can arrange things so that she is adjacent to a cop in squad 1 after at most one move. We refer to this movement of the squad two cops as a *hop*, as the cops appear to jump from one vertex of $N(x)$ to another (although what is really happening is that the cops are cycling through x). Note that hops maintain $m - 1$ cops on x.

We now describe a strategy $\mathcal{S}$ for the cops, and then show that it is winning. The cops in squad 1 play exactly as in the usual game of Cops and Robbers; note that the squad 1 cops may leave $N_2(x)$ depending on their strategy, but R will never leave $N_2(x)$. The squad 2 cops play as follows. Squad 2 cops do not move unless the following occurs: a squad 1 cop C_k moves to a neighbour of R, and C_k has no backup from a squad 1 cop. In that case, some squad 2 cop $\widehat{C_{i,j}}$ hops to a vertex of $N(x)$ which is adjacent to C_k. There are a sufficient number of squad 2 cops to ensure this property, since if m (or more) squad 1 cops move to neighbours of R, then some of these cops must be adjacent to each other as G is $K_{1,m}$-free (in particular, the cops in $N(R)$ play the role of backups to each other).

Hence, the squad 1 cops may apply their winning strategy in the usual game and ensure that whenever they move to a neighbour of R, some squad 2 cop serves as backup. In particular, R will never attack a squad 1 cop for the duration of the game. Thus, $\mathcal{S}$ is a winning strategy in the game of Cops and Attacking Robbers. □

Acknowledgments The authors were supported by grants from NSERC and Ryerson University.

References

1. M. Aigner, M. Fromme, A game of cops and robbers, *Discrete Applied Mathematics* **8** (1984) 1–12.
2. W. Baird, A. Bonato, Meyniel's conjecture on the cop number: a survey, *Journal of Combinatorics* **3** (2012) 225–238.
3. B. Bollobas, G. Kun, I. Leader, Cops and robbers in a random graph, *Journal of Combinatorial Theory Series B* **103** (2013) 226–236.

4. A. Bonato, WHAT IS... Cop Number? *Notices of the American Mathematical Society* ***59*** *(2012) 1100–1101.*
5. A. Bonato, Catch me if you can: Cops and Robbers on graphs, In: *Proceedings of the 6th International Conference on Mathematical and Computational Models* (ICMCM'11), *2011.*
6. A. Bonato, A. Burgess, Cops and Robbers on graphs based on designs, *Journal of Combinatorial Designs* **21** (2013) 404–418.
7. A. Bonato, E. Chiniforooshan, P. Pralat, Cops and Robbers from a distance, *Theoretical Computer Science* **411** (2010) 3834–3844.
8. A. Bonato, R.J. Nowakowski, *The Game of Cops and Robbers on Graphs,* American Mathematical Society, Providence, Rhode Island, 2011.
9. A. Bonato, P. Pralat, C. Wang, Network security in models of complex networks, *Internet Mathematics* **4** (2009) 419–436.
10. M. Chudnovsky, P. Seymour, Clawfree Graphs IV - Decomposition theorem, *Journal of Combinatorial Theory. Ser B* **98** (2008) 839–938.
11. N.E. Clarke, *Constrained Cops and Robber,* Ph.D. Thesis, Dalhousie University, 2002.
12. A. Dudek, P. Gordinowicz, P. Pralat, Cops and Robbers playing on edges, preprint, 2013.
13. P. Frankl, Cops and robbers in graphs with large girth and Cayley graphs, *Discrete Applied Mathematics* **17** (1987) 301–305.
14. R.J. Nowakowski, P. Winkler, Vertex-to-vertex pursuit in a graph, *Discrete Mathematics* **43** (1983) 235–239.
15. T. Luczak, P. Pralat, Chasing robbers on random graphs: zigzag theorem, *Random Structures and Algorithms* **37** (2010) 516–524.
16. P. Pralat, When does a random graph have constant cop number?, *Australasian Journal of Combinatorics* **46** (2010) 285–296.
17. P. Pralat, N.C. Wormald, Meyniel's conjecture holds for random graphs, preprint, 2013.
18. P. Pralat, N.C. Wormald, Meyniel's conjecture holds for random *d*-regular graphs, preprint, 2013.
19. A. Quilliot, Jeux et pointes fixes sur les graphes, These de 3eme cycle, Universite de Paris VI, 1978, 131–145.
20. D.B. West, *Introduction to Graph Theory,* 2nd edition, Prentice Hall, 2001.

Some Applications of Collective Learning

Balaraman Ravindran

Abstract Much of the real-world data have complex dependencies between the individual tuples. For example, the chance that a patient has a particular disease depends on the prevalence of the disease in the immediate neighborhood. One approach to handling such linked data is "collective learning." In collective learning, one deals with a set of data points taken at a time. The dependencies between the data points are modeled as a graph, with the nodes representing the tuples and the edges between them representing the influence of the tuples on one another. A variety of domains lend themselves naturally to such graph-based modeling. There have been a variety of collective learning and inferencing approaches that have been proposed in the literature. In this talk, I will give a brief introduction to collective learning and describe two applications.

keywords Sentiment analysis · Functional site prediction · Interaction network · Label misclassification

The first of these is a sentiment analysis task. Sentiment analysis is the task of identifying the sentiment expressed in the given piece of text about the target entity under discussion. In this work, we look at the problem of analyzing sentiments at different granularities. For example, we want to analyze sentiment about a movie as whole as well as about the acting and directing. Models built for such multigrain sentiment analysis assume fully labeled corpus at fine-grained level or coarse-grained level or both. Huge amount of online reviews are not fully labeled at any of the levels, but are partially labeled at both the levels. We propose a multigrain collective classification framework to not only exploit the information available at all the levels but also use intra dependencies at each level and interdependencies between the levels. We demonstrate empirically that the proposed framework enables better performance at both the levels compared to

B. Ravindran (✉)
Computer Science and Engineering, Indian Institute of Technology Madras,
Chennai 600036, India
e-mail: ravi@cse.iitm.ac.in

G. S. S. Krishnan et al. (eds.), *Computational Intelligence, Cyber Security and Computational Models*, Advances in Intelligent Systems and Computing 246,
DOI: 10.1007/978-81-322-1680-3_2,

baseline approaches. Part of this work was reported in ECAI 2010, and it is a joint work with S. Shivashankar and Shamshu Dharwez.

The second task is that of functional site prediction in proteins. Functional site prediction is an important problem in the structural genomics era where we have a large number of experimentally determined protein structures with unknown function. The functional sites provide useful insights into protein function. In this paper, we propose a method for prediction of functional residues in a given protein from its three-dimensional (3D) structure. Our method exploits correlation between labels of interacting residues to obtain significant performance improvements over the existing methods on the benchmark dataset. We represent each protein as a weighted undirected residue interaction network, where spatially proximal residues in terms of their van der Waal's radii are connected by an edge. The edge weight captures correlation between the labels of interacting residues. The correlation is estimated based on the features of interacting residues. We then obtain a label assignment by minimizing combined cost of residue-wise label misclassification and violation of label correlation constraints. We solve this problem in two stages, where the first stage minimizes residue-wise label misclassification cost followed by an iterative collective inference scheme that adjusts the labels predicted in the first stage so as to minimize the correlation constraint violations. Our approach significantly outperforms state-of-the-art methods on standard benchmark dataset. This work was reported in ACM BCB 2012, and it is a joint work with Ashish V. Tendulkar, Saradindu Kar, and Deepak Vijayakeerthi.

Subconscious Social Computational Intelligence

M. Graña

Abstract The success of social network Web services mediating social interactions, as well as the increasing observation capabilities of human interactions in real life, has prompted the emergence of new computational paradigms, namely social computing, computational social science, and social intelligence. Subconscious social intelligence appears when the social network service is able to provide solutions, generated by a hidden intelligent layer, to problems posed by the social player. This paper discusses some features of subconscious social intelligence and ensuing challenges for machine learning systems implementing the hidden intelligent layer.

Keywords Social computing · Social intelligence · Subconscious reasoning · Learning systems

1 Introduction

This paper discusses the requirements for machine learning systems contributing to the development of a nascent computational field, which can be identified by the name of subconscious social computing. Reviewing the related fields of social computing and computational social science will help to clarify the subtle distinctive features of this new class of systems. We describe the general scheme of subconscious social computing highlighting its contrast with the previous ones, including the description of one instance, the EU-financed SandS project [1, 3, 4, 6]. Then, we identify requirements posed by this kind of systems on learning subsystems implementing the subconscious intelligent layer, discussing how computational intelligence approaches can cope with them.

M. Graña (✉)
Grupo de Inteligencia Computacional, UPV/EHU, Erandio, Spain
e-mail: manuel.grana@ehu.es
URL: www.ehu.es/ccwintco

G. S. S. Krishnan et al. (eds.), *Computational Intelligence, Cyber Security and Computational Models*, Advances in Intelligent Systems and Computing 246,
DOI: 10.1007/978-81-322-1680-3_3, © Springer India 2014

2 Social Computing Paradigms

Computational social science [2] aims to understand the dynamics of social systems form of data that can be extracted from all existing sources of human behavior observation, ranging from surveillance cameras, mobile identification tags to social Web services or electronic commercial transactions. From computational social science point of view, the social players are subjects of observation and experimentation, searching for answers to questions such as:

- which interaction pattern leads to economic success?
- how social interaction influences contagious sickness diffusion?
- what is the best way to promote a product?
- what social hints can be useful to predict the fate of a stock asset?

To this end, computational social science deals with intelligent and efficient hardware/software systems able to process huge amounts of data coming from all kinds of observational sources within some real-time constraints. Big networking data are subject to statistical and data mining analysis, providing answers to the institutional or corporate costumer.

Social computing [5, 8] concerns the development of software for the enhanced interaction between social players and to develop simulation scenarios to forecast the effects of policies and forces, such as technological innovation, on societies. Examples of these systems are entertainment/therapeutic social games involving autonomous intelligent agents, negotiation, recommender and reputation systems, security applications for the detection of criminal social activities, and artificial societies of agents designed to provide adaptation to changing environments (i.e., traffic) through competition, platforms for scientific collaboration offering information about the current state of the research community and research effort planning. Social computing is developing into a productive model where rewarding mechanisms are required to control the desired output of the system [7].

Social intelligence is the emergence of problem-solving behavior out of social interactions from the point of view of the social player. In other words, the social player expects to obtain solutions to his/her problems from the pool of intelligence available from a social network and the computational resources that may be at work behind the social service. We may further distinguish between conscious and subconscious intelligent computing. In the former, social players contribute information and the intelligence to create/discover solutions. In the latter, an underlying intelligent layer is able to provide innovative solutions to old and new problems, following an autonomous process that is not directly controlled by the social players. Figures 1 and 2 illustrate the differences between paradigms. The social interaction layer in both figures includes all means of sharing information between users, but does not contemplate any information transformation.

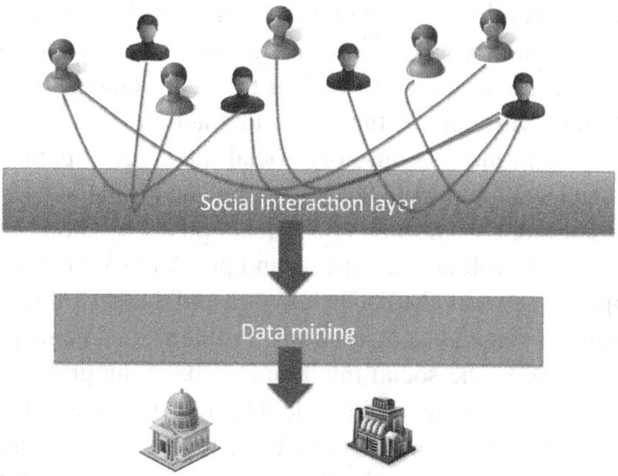

Fig. 1 Social computing and computational social science paradigm

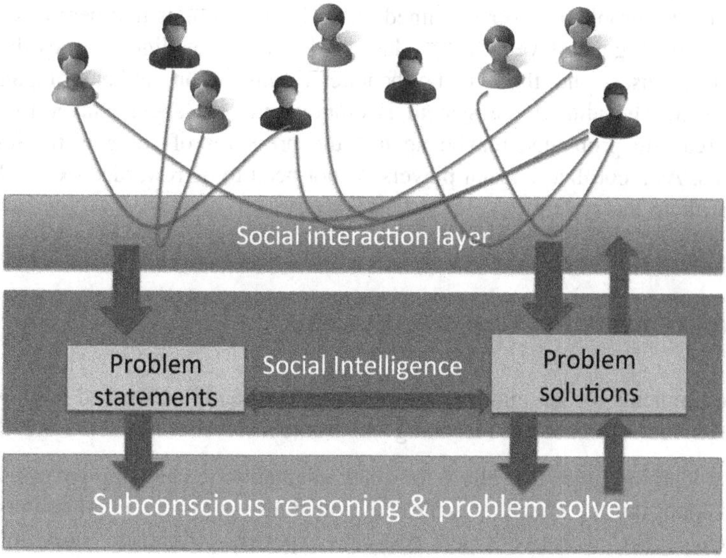

Fig. 2 Subconscious social intelligence paradigm

3 Subconscious Social Intelligence

In the social computing and computational social science paradigms illustrated in Fig. 1, there is an underlying computational layer that performs data mining over observations of the social interactions. The social player is unaware of it and does not directly benefit from it. This lack of benefit motivates research in rewarding mechanisms [7]. The results of these computations are delivered to a third party, either government institutions or industry. The productive model of social computing relies on the technological prowess of this data mining layer that provides the benefit from the investment in the social interaction layer.

On the other hand, in the subconscious social intelligence paradigm illustrated by Fig. 2, the social intelligence layer below the social interaction layer is dedicated to provide solutions to problem statements posed by the social players. To that end, repositories of problem statements and problem solutions are maintained, along with a mapping between them. The upward and downward red arrows model the flow of problem statements and solutions. Problem statements posed by social players flow downward to the social intelligence, matching problem solutions are searched in the repository (horizontal red double-headed arrow), if one is found, it flows upward to the social interaction layer to be retrieved by the interested social player. If there is no solution matching the statement, the statement is percolated further down to the subconscious reasoning and problem solver layer that works to produce a solution that will be pushed upward to the social intelligence layer solutions repository, and the user at the social interaction layer. The subconscious reasoning and problem solver is trained on problem solutions that percolate from the social intelligence layer. The product of the social intelligence goes directly back to the user, and there is no beneficiary institution, either company of government. The aim of the system is empowering the social player to solve his/her real-life problems, maybe against the pressures of some institution, or within its. As a corollary, social players do not need to be rewarded externally to use/contribute the system.

4 Requirements for Learning Systems

The requirements for learning systems meeting the SandS networked intelligence and the general subconscious reasoning and problem solver of Fig. 2 are as follows:

- Quick learning times that allow for quick adaptation to changing environments and supporting the effects of scale that potentially big social communities will introduce. Social network services can experience dramatic rises in user involvement and subsequent computational load. Moreover, changes in problem specification may involve addition/removal of variables with ensuing retraining processes.
- Flexibility to cope with diverse data representations and desired outputs. The desired responses may be categorical and continuous, involving both classification and regression, even in the same problem-solving process.
- Robust performance when dealing with multidimensional heterogenous output. Most machine learning approaches have serious degradation when the desired output is multivariable, and even worse when it is composed of diversely typed variables.

- Minimal uncertainty: In the development of subconscious social intelligence, we want to perform one-shot training with minimal uncertainty about the achieved performance. Machine learning papers often report average or peak results of extensive computational experiments. These results do not provide a performance guarantee for a specific instance of the learning process, nothing prevents it to be catastrophically stupid.
- Robust incremental learning to process incoming batches of user feedback driving the adaptation process. Incremental learning, on the fly adaptation, is not an optional feature in this setting. Social systems and the needs of the social players are continuously evolving. Training systems with a sample of data are meaningless after some period of time.
- Easy implementation/learning of forward and backward mappings, the former to provide solutions and the latter to translate the user feedback into error measures driving learning processes. Social players want to be able to understand why a solution works, which is the chain of reasoning that produces this improved solutions, and to have some control on the responses of the system to required adaptive changes.
- Hybridization of diverse computational paradigms to allow the composition of selection/classification/regression modules to cope with the complex landscape of user problem statement. It is not likely that a single learning paradigm will be able to cope with all kinds of social player requests and needs. Many kinds of intelligence may need to be called upon to provide answers at diverse levels.

5 The SandS Project

The EU-funded SandS project (http://www.sands-project.eu) aims to build an instance of the subconscious social intelligence in the domotic domain. SandS social players are users of household appliances that exchange information about them in the form of "recipes" of use. The software structure under development in the project follows the pattern of Fig. 2. The SandS social network has a repository of household tasks that have been posed by the users and a repository of appliance recipes, which are related by a map between (to and from) tasks and recipes. This map needs not to be one-to-one. User queries interrogate the database of known/solved household tasks. If the queried task is already known, then the corresponding recipe can be send to the user appliance. After recipe execution, the user can express its satisfaction with the results. When the queried task is unknown and unsolved, it is forwarded to the underlying SandS networked intelligence to produce a new recipe by an intelligent system reasoning able to learn and predict new recipes maximizing user satisfaction. The source of recipes filling the repository is, therefore, twofold. On the one hand, engaged user and/or appliance manufacturing companies consciously provide new recipes. On the other hand, the underlying networked intelligence is the subconscious generator of new solutions.

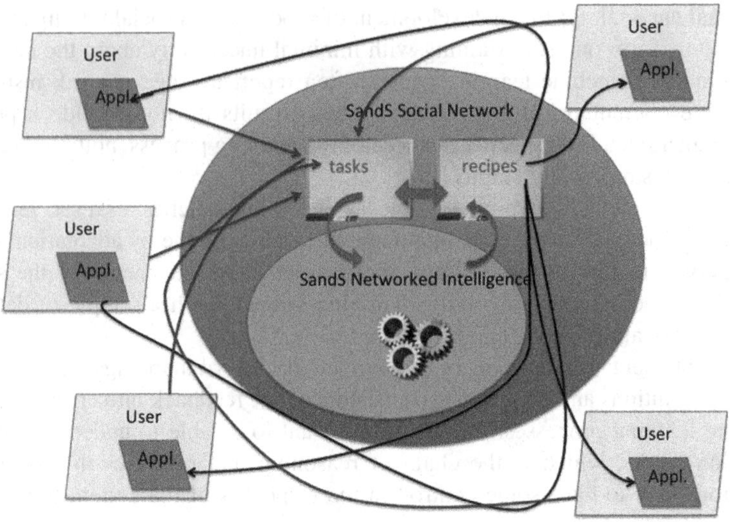

Fig. 3 Social and smart system prototypical architecture

More specifically, Fig. 3 shows an intuitive representation of the architecture and interactions between the system elements. The SandS social network mediates the interaction between populations of users, each owning a set of appliances. The SandS social network has a repository of tasks that have been posed by the eahoukers and a repository of recipes for the use of appliances. These two repositories are related by a map between (to and from) task and recipes. This map needs not to be one-to-one. Blue arrows correspond to the path followed by the eahouker queries, which are used to interrogate the database of known/solved tasks. If the task is already known, then the corresponding recipe can be returned to the eahouker appliance (black arrows). The eahouker can express its satisfaction with the results (blue arrows). When the queried task is unknown and unsolved, the social network will request a solution from the SandS networked intelligence that will consist in a new recipe deduced from the past knowledge stored in the recipe repository. This new solution will be generated by intelligent system reasoning. The eahouker would appreciate some explanation of the sources and how it has been reasoned to be generated; therefore, explicative systems may be of interest for this application.

In the SandS social network, input data should be in the form of household task codifications, while the output may correspond to recipe parameter settings, which may be continuous variables, i.e., water temperature in the washing machine, or categorical, i.e., steps in the washing process. The user feedback may be expressed in simple terms, such a Likert scale of satisfaction, which needs to be translated into an error measure that may drive the recipe learning. Household tasks performed by different appliances need to be solved by specific learned systems, which amounts to perform some partition in the task/recipe space by a selection mechanism driven by the task specification.

6 Conclusions

Subconscious social intelligence is a new way to pose the problem-solving power of social networks, combining conscious social computing built from explicit interactions from social players and subconscious problem solving trained from the experiences percolated from the social interaction down to a subconscious reasoning layer. The consideration of this kind of systems amounts to a radical shift on how social Web services are designed and deployed. It would no longer be the needs and requirements of the large corporations owning huge computational facilities that drive the system computational intelligence. Instead of the social players, the individual users of the system are the ones reaching the benefits of the social interaction for a better personal and social life.

Acknowledgments Work performed under Grant agreement 317947 of the EU, SandS project, the research and funds unit UFI11/07 of the UPV/EHU, and university research group Grant IT874-13 from the Basque Country Government.

References

1. B. Apolloni, M. Fiasche, G. Galliani, C. Zizzo, G. Caridakis, G. Siolas, S. Kollias, M. Grana-Romay, F. Barriento, and S. San-Jose. Social things - the sands instantiation. *In Internet of Things: Smart Objects and Services IoT-SoS 2013*. IEEE PRESS, 2013.
2. D. Lazer et al. Computational social science. *Science,* 323(5915):721–723, 2009.
3. M. Grana, B. Apolloni, M. Fiasche, G. Galliani, C. Zizzo, G. Caridakis, G. Siolas, S. Kollias, F. Barriento, and S. San Jose. Social and smart: towards an instance of subconscious social intelligence. In H. Papadopoulos L. Iliadis and C. Jayne (Eds.), editors, *EANN 2013,* volume part II, pages 302-3011. Springer Berlin Heidelberg, 2013.
4. M Grana and I Rebollo. Instances of subconscious social intelligent computing. In *CASON 2013.* IEEE PRESS, 2013.
5. W. Mao, A. Tuzhilin, and J. Gratch. Social and economic computing. *IEEE Intelligent Systems,* 26(6):19-21, 2011.
6. M. Grana I. Marques, A. Savio, and B. Apolloni. A domestic application of intelligent social computing: the sands project. In *SOCO 2013.* Springer Berlin Heidelberg, 2013.
7. Ognjen Scekic, Hong-Linh Truong, and Schahram Dustdar. Incentives and rewarding in social computing. *Communications of the ACM,* 56(6):72-82, 2013.
8. F.-Y. Wang, K.M. Carley, D. Zeng, and W. Mao. Social computing: From social informatics to social intelligence. *Intelligent Systems, IEEE,* 22(2):79-83, 2007.

Modeling Heavy Tails in Traffic Sources for Network Performance Evaluation

Vaidyanathan Ramaswami, Kaustubh Jain, Rittwik Jana and Vaneet Aggarwal

Abstract Heavy tails in work loads (file sizes, flow lengths, service times, etc.) have significant negative impact on the performance of queues and networks. In the context of the famous Internet file size data of Crovella and some very recent data sets from a wireless mobility network, we examine the new class of LogPH distributions introduced by Ramaswami for modeling heavy-tailed random variables. The fits obtained are validated using separate training and test data sets and also in terms of the ability of the model to predict performance measures accurately as compared with a trace-driven simulation using NS-2 of a bottleneck Internet link running a TCP protocol. The use of the LogPH class is motivated by the fact that these distributions have a power law tail and can approximate any distribution arbitrarily closely not just in the tail but in its entire range. In many practical contexts, although the tail exerts significant effect on performance measures, the bulk of the data is in the head of the distribution. Our results based on a comparison of the LogPH fit with other classical model fits such as Pareto, Weibull, LogNormal, and Log-*t* demonstrate the greater accuracy achievable by the use of LogPH distributions and also confirm the importance of modeling the distribution in its entire range and not just in the tail.

Keywords Network performance · Heavy tailed random variables · LogPH distribution · Markov chain

V. Ramaswami (✉) · R. Jana · V. Aggarwal
Florham Park, New Jersey, USA
e-mail: ram@ramaswami.com

K. Jain
College Park, Maryland, USA

G. S. S. Krishnan et al. (eds.), *Computational Intelligence, Cyber Security and Computational Models*, Advances in Intelligent Systems and Computing 246,
DOI: 10.1007/978-81-322-1680-3_4, © Springer India 2014

1 Introduction

The negative impact of heavy tails in work loads on the performance of systems is well known in the queuing literature. Indeed, many new scheduling strategies came to be invented primarily to avoid these bad effects of very large work loads (even if they be infrequent and from a small set of customers) for systems with schedules such as the First-in-First-Out discipline. Concern about heavy tails nevertheless holds even in the context of modern-day systems such as high-speed and wireless networks. Indeed, the increasing presence of bandwidth-intensive video and streaming audio has heightened the concern particularly in wireless networks as evidenced, for example, by the AT&T experience soon after the introduction of the iPhone.

An early work drawing attention to the presence of heavy tails in Internet file sizes is that of Crovella [4]. We use Crovella's data set and model the distribution as a LogPH distribution and also in terms of classical models such as Pareto, Weibull, LogNormal, and Log-t. The LogPH distribution was proposed by Ramaswami [6] who identified it to have a power law tail and dense (in the weak convergence metric) in the class of all distributions on $[1, \infty)$. A LogPH random variable Y is a random variable that can be written as $Y = e^X$ where X is a phase-type random variable as defined by Neuts [9, 10]; see also [7] for a discussion on phase-type (PH) distributions. The first formal reference to LogPH was made in the paper by Ghosh et al. [6] on modeling traffic to a public Wi-Fi network. A detailed mathematical treatment of the LogPH class of distributions has been given in Ahn et al. [1]. That work of Ahn et al. demonstrates the power of the LogPH class to model heavy-tailed distributions in their entire range in the context of several financial examples. This paper demonstrates its power in the context of network performance modeling.

Needless to say, there are many attempts in the literature (see [1, 5, 8]) in queuing, performance analysis, risk theory, and finance to model heavy-tailed distributions; a key idea is to use some distributions (such as Pareto or Weibull) with a known heavy tail to model the tail and then a mixture to obtain a fit across the entire range. Unfortunately, these attempts have not resulted in a single class of models that can be used dependably in a large number of contexts, and furthermore, many aspects of the fitting methodology appear to be adhoc. Also, it would appear prudent to adopt more stringent standards in assessing the statistical quality of the fits in terms of a test data set that is separate from the training data set used for fitting a model. Also, it is desirable not only to consider the fitted random variable, but also to assess the quality of the fit in terms of its ability to predict performance of systems in which the models are used. Judged in this context, this paper may be found interesting and useful by many.

This paper is organized as follows. In Sect. 2, we provide a quick discussion of various classical models used in the context of heavy tails and also of the LogPH distribution. A brief discussion of a method based on the EM algorithm to fit LogPH distributions is given. In Sect. 3, we fit LogPH distributions to two data sets, the Crovella Internet file size data set [4] and a very recent (2012) data set on file sizes downloaded by mobile phone users in a cellular mobility network. We also provide a comparison of the LogPH fit with various other models such as the Pareto, Weibull, LogNormal, and Log-*t*. In addition to visual comparisons of the empirical with fitted distributions, we also provide some quantitative measures that aid such comparison. We wish to note that in the comparisons related to wireless mobility, since several data sets were available, we have taken a more stringent approach of having a "training set" based on which the model fits were made and a separate "test set" for comparing the models. Unfortunately, we had only one data set in the case of the Crovella data; we did comparisons with bootstrapped samples generated from this data set and found the fit to hold good for them as well. This step was undertaken primarily to make sure that we did not run the risk of overfitting. In Sect. 4, we take the Crovella data set and make a trace-driven simulation of a bottleneck Internet link and compare the performance results (queue lengths, throughput) against simulations run with fitted models using LogPH, Pareto, Weibull, LogNormal, and Log-*t*. Our results show that the LogPH gives more accurate results and thereby increase our confidence in the LogPH model class.

Our results give us great confidence in the ability of the LogPH class to model heavy-tailed distributions in a way to yield more accurate performance predictions in the network context. Much further work is needed on this class of models with regard to various issues including metrics for assessing goodness of fit, comparing different fits as well as certain issues related to the use of heavy traffic distributions with an infinite support. We will discuss some of these open issues. It is our hope that this work and the success of the LogPH class reported in Ahn et al. [1] will draw the attention of researchers and help improve our understanding of this class and our ability to model heavy-tailed phenomena more accurately. With this perspective, we will present not only the results obtained by us, but we shall also dwell on some of the gaps that need to be filled through further research.

2 Background

2.1 Phase-Type Distribution

A Phase-Type (PH) distribution is defined as the distribution of the time until absorption of a Markov chain with an absorbing state. This general class was introduced by M.F. Neuts [9, 10]. To be specific, consider a Markov chain with states 0,1,..., n, initial probability vector $(0, \boldsymbol{\tau})$ and infinitesimal generator Q. The row vector $\boldsymbol{\tau}$ is of size n and satisfies $\boldsymbol{\tau}\mathbf{1} = 1$, where $\mathbf{1}$ is a column vector of 1's. Assuming state 0 is an absorbing state, Q can be denoted as

$$Q = \begin{bmatrix} 0 & \mathbf{0} \\ \boldsymbol{t} & T \end{bmatrix}$$

where $\boldsymbol{t}$ is a column vector of size n and T is a $n \times n$ non-singular matrix satisfying, $T(i, i) < 0$, $T(i, j) > 0$ for $i \neq j$, and $T\mathbf{1} + \boldsymbol{t} = 0$. Thus, the Markov chain is completely characterized by parameters τ and T. The random variable X describing the time until absorption of the Markov chain into state 0 is called a PH random variable, denoted PH (τ, T). The number of non-absorbing states (n) is called the order of the PH random variable. The distribution and density functions of the PH random variable defined above are given by

$$F(x) = 1 - \boldsymbol{\tau}\exp(Tx)\mathbf{1}, \quad \text{for } x \geq 0, \tag{1}$$

$$f(x) = \boldsymbol{\tau}\exp(Tx)\boldsymbol{t}, \quad \text{for } x > 0. \tag{2}$$

PH distributions are known to be dense in the class of all distributions on $[0, \infty)$. That is, they can approximate any distribution arbitrarily closely. Furthermore, they have many interesting closure properties and are highly tractable due to the connection with a Markov chain, which makes conditioning arguments easy. For these reasons, they have attracted much attention in applied probability. A property of the PH distribution is that its tail is asymptotically exponential; more specifically, for the distribution above, $P(X > x) \approx Ke^{-\eta x}$ for large x, where $-\eta < 0$ is the eigenvalue of T closest to zero.

2.2 LogPH Distribution

The LogPH distribution, denoted by LogPH($\boldsymbol{\tau}$, T), is defined as the distribution of the random variable Y that can be written as $Y = \exp(X)$ where X has a PH distribution with ($\boldsymbol{\tau}$, T). The LogPH random variable Y has its distribution function and density function as

$$F_Y(y) = 1 - \tau e^{T\log y}\mathbf{1}, \quad y \geq 1$$

and

$$f_Y(y) = \frac{1}{y}\tau e^{T\log y}t, y \geq 1, \quad t = -T\mathbf{1}.$$

From the exponential decay of the tail of the PH distribution, it easily follows that the LogPH random variable has a power law tail. Specifically, for large y, $P(Y > y) \approx K/y^{\eta}$, where η is as defined earlier. Also, from the fact that PH-type distributions are dense on $[0, \infty)$, it follows by standard continuity theorems governing weak convergence (see Whitt [12]) that LogPH distributions are dense in the set of all distributions defined on $[1, \infty)$. These properties make LogPH an attractive candidate class for modeling heavy-tailed random variables. In this context, we wish to note that the restriction of its range to $[1, \infty)$ is not particularly limiting for two reasons: (a) In many cases, one could rescale the data and fit LogPH to the scaled data set or (b) one can use a similar construction based on the bilateral PH random variables, which generalize the PH distribution to the entire real line; see Ahn and Ramaswami [2].

2.3 Some Classical Heavy-Tailed Distributions

Traditionally, modeling of heavy-tailed random variables has been focused mainly on the tail of the distribution. Some of the commonly used distributions are Pareto, Weibull, and LogNormal. Pareto has the following tail distribution:

$$\Pr[Z > z] = \left(\frac{b}{z}\right)^a, \quad z \geq b \tag{3}$$

where a and b are the shape and the scale parameters, respectively. Various enhancements of the Pareto distribution have also been used to match the mean, to select the cutoff at which the asymptotic power law takes over. A Pareto random variable Y can be realized as exp (X) where X is an exponential random variable; in this sense, one may consider the LogPH class as a natural generalization of the Pareto distribution since the exponential distribution is the most trivial example of a PH distribution.

The Weibull distribution does not have a power law in the tail distribution, but still the tail decays more slowly than the exponential. Denoted by a and b are the shape and scale of the distribution,

$$\Pr[Z > z] = \exp\{-(z/b)^a\}, \quad z > 0, \quad a < 1. \tag{4}$$

LogNormal distribution is modeled as a distribution, which is normal in the log scale. Specifically, Z has a LogNormal distribution if we can write

$$Z = e^X, \quad \text{where } X \sim \mathcal{N}(\mu, \sigma^2). \tag{5}$$

The normal distribution has a fast decaying tail $e^{-x^2/2}$, and this has led some researchers to use the t-distribution in place of the normal and define a Log-t distribution as a model for heavy-tailed distributions. Like the normal, the t-distribution also is symmetric about the origin and that could limit some of its applicability.

While the literature abounds in many applied examples where the above distributions and mixtures involving them have been used successfully for specific situations, there are some basic challenges in their use. These distributions do not form a dense class that provides a guarantee that one may effect a fit from any one of the members to a desired accuracy. Also, often one is forced to make a trade-off between matching the tail and matching the head of the distribution and to come up with ad hoc procedures for fitting a mixture that attempts to give a good model in the entire range of the data. We refer to Ahn et al. for a discussion of the issues in light of a famous data set—the Danish fire insurance data—that has been used as an important test data set in the statistical literature.

2.4 Fitting a LogPH Distribution

A LogPH distribution is fitted to the data by fitting a PH distribution to the logarithms of the data values (to the base e). If only a very small fraction of data values exist that are less than 1, we may discard them, or alternately, we may rescale the data by dividing all the elements by the minimum value so that all logarithms are positive and then fit a LogPH to these log values.

The standard approach to fitting a phase-type distribution is to use the EM algorithm whose details are provided in a paper by Asmussen et al. [3]. The EM algorithm is based on the following observations. Suppose the true distribution is a phase-type distribution of order n. If one knew the number of visits to each of the n transient states in the Markov chain and the amount of times spent in each of them before the Markov chain gets absorbed, these values together would constitute a set of sufficient statistics for the unknown parameters of the Markov chain. Now, the EM algorithm starts with a trial phase-type distribution of order n and iterates on the following two steps: (1) E-step: Consider the number and duration of visits to the transient states as missing values and replace them with their conditional expectation evaluated with respect to the current estimate of the parameters; (2) M-step: Now considering as though we have a complete sample on the sufficient statistics, maximize the likelihood function to obtain an improved estimate of the Markov chain parameters. The general theory of EM guarantees convergence to (a local) maximum of the likelihood function.

3 Data Sets and Fitted Models

This paper deals with two data sets. The first is the well-known World Wide Web file size traces collected by Crovella in 1995 commonly used by researchers to model heavy-tailed data. We will demonstrate that LogPH provides a much better fit for the entire data range as compared with previously used models, particularly in its ability to predict network performance more accurately as evaluated in the context of a bottleneck Internet link. Secondly, we will also use a very recent data set of file sizes from the *mobile Web*. We show once again that LogPH provides a good fit to the entire range of the mobile data set. Our interest in the second data set is due to our current focus on wireless and mobile networks and the fact that the Crovella data are now quite dated and predate some major new bandwidth-intensive applications such as video and streaming audio that have become much more popular. (In fact, there is a strong need to revisit the heavy tail issue with newer data even in the wireline Internet.) A priori, it was not even clear if there would be any need for heavy-tailed models in the wireless context, but as our results show, heavy tails seem to pervade the wireless and mobile networks as well. The mobile Web data set is obtained from a national cellular service provider. The data set is collected over a period of 24 h during 2012 inside a geographic area (spanning hundreds of cell sectors within a large metropolitan area).

In the case of the mobility data, the presence of many different data sets collected over different time intervals gave us the unique opportunity to fit the model with one data set but test it with another one not used in testing. This is obviously a more stringent comparison in tune with practices in the area of machine learning and modern data analysis using separate training and test data sets. For the Crovella data set itself, we used a simulation where one of the inputs was driven by the empirical distribution of the Crovella data; that gave us the opportunity to treat the simulated values as a bootstrapped sample and to assess the quality of the fit against that. Unfortunately, Crovella does not provide a multitude of data sets. We did run our algorithms for fitting on a random subset of the Crovella data using the rest for testing, and the results were entirely satisfactory and as good as in the mobility data sets; more about these later.

3.1 Crovella's WWW Data Set

Crovella's data set [4] contains the file sizes for a large number of HTTP requests. There are around 130,000 data points for file sizes. Figure 1 shows the empirical density of the file sizes on a natural log scale. Figure 2 shows the tail distribution on the log–log scale, clearly indicating the presence of an asymptotic linear decay which for the file sizes entails a power law decay of the form $O(1/x^{\eta})$. A regression fit for the tail portion of the complementary distribution on the log–log scale gave an estimate $\eta = 1.15$; we have shown the regression line for the tail along with the plot of the complementary distribution.

Now, we describe some key steps taken by us in fitting a LogPH to the file size data. Recall that the log of the file size is to be modeled as a PH random variable. A preliminary analysis of the data set showed that an overwhelming majority (98.7 %) of the file sizes were over 200 bytes. We took the set of values above 200, divided them by 200, took the logarithm, and fitted a phase-type distribution to the resulting set of values. This rescaling is equivalent in the log scale to shifting the density (Fig. 1) to the left so as to start from zero (see the density in Fig. 3a). The original distribution of the data was then approximated by the mixture of a point mass at 200 with probability 0.013 and the distribution of a random variable that is 200 times the fitted LogPH random variable, the latter with weight 0.987. The motivation for not expending much effort on the lower tail below 200 (which is possible through the use of a LogPH distribution) was that these observations constituted a small percentage and from the perspective of the performance

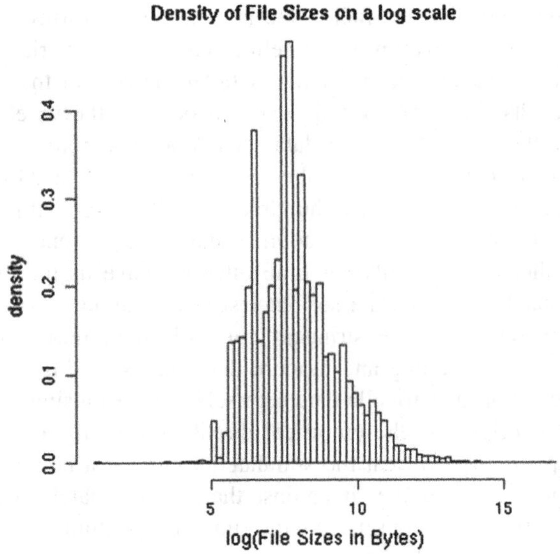

Fig. 1 Crovella data—density

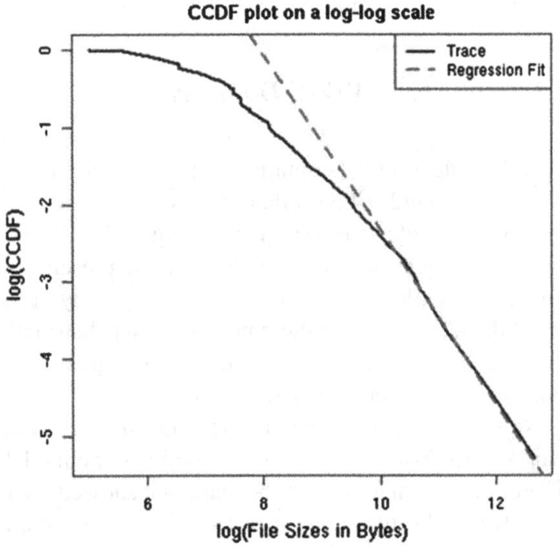

Fig. 2 Crovella data—CCDF and regression fit

comparisons in a queuing model would not make much of a difference. Indeed, we followed a similar procedure for fitting all other distributions we tried. To be mathematically precise, denoted by X the file size, we model $Y = \max(1, X/200)$ as a LogPH random variable and sweep the mass to the left of 200 bytes to a point mass at 200.

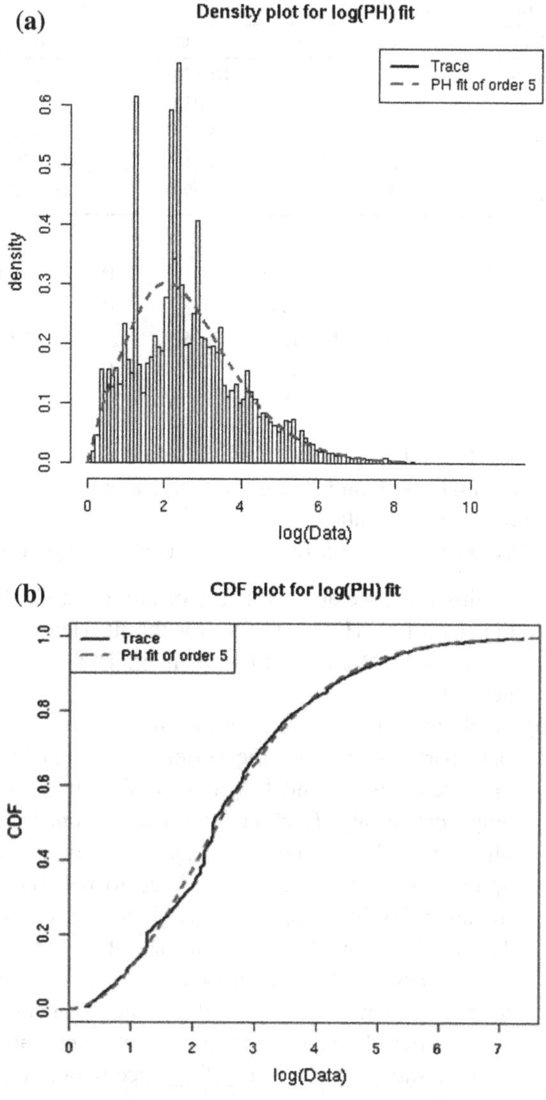

Fig. 3 Fitting PH to log (file size in bytes—200, Crovella data set). **a** Density plot and its LogPH fit and **b** CDF and its LogPH fit

The EM algorithm for fitting a PH distribution was tried with the different number of phases in the range 3–7. We stopped at a value of $n = 5$ at which point the improvement in the maximum-likelihood value appeared negligible. A PH fit of order 5 was found to provide a good match as illustrated in Table 1 by means of log likelihoods. The PH parameter estimates are $\boldsymbol{\tau} = (0, 0, 0, 0, 1)$ and

Table 1 Log-likelihood values (Crovella and mobile Web)

Model	Uncond. Crovella	Cond. Crovella	Uncond. mobile Web	Cond. mobile Web
Weibull	−9.874	−10.176	−10.148	−10.816
Pareto	$-\infty$	−10.034	$-\infty$	−10.809
LogNormal	−9.670	−10.033	−10.801	−10.924
Log-*t*	−9.683	−10.017	−10.059	−10.665
LogPH	−9.554	−9.980	−10.057	−10.682

$$T = \begin{bmatrix} -1.31 & 0 & 0 & 1.31 & 0 \\ 1.31 & -1.31 & 0 & 0 & 0 \\ 0 & 1.79 & -2.11 & 0.32 & 0 \\ 0 & 0 & 0 & -2.11 & 0 \\ 0 & 0 & 2.0 & 0.11 & -2.11 \end{bmatrix}.$$

The fitted PH density function and the cumulative distribution function are shown in Figs. 1 and 2, respectively, along with their corresponding empirical figures based on data.

We wish to make a few comments with respect to this fit:

(a) At this time, we are not aware of any methods that will automatically choose the size of a PH distribution for the fit. Hence, we have had to try out a few values and stop in an ad hoc manner. This is an area where further research is needed.
(b) While the fitted distribution of order 5 appears to do a fairly good job overall (and particularly from the system performance perspectives to be discussed later; see Figs. 13 and 14), it does miss out on a few pronounced peaks in the empirical density. To capture these peaks, one would need a very high order of a phase-type distribution and that may run the risk of overfitting. In many applications, such peaks occur due to peculiarities in the application being modeled. For instance, in the wireless data, we saw a pronounced peak around 1,500 bytes coinciding with the size of an Ethernet packet size. Perhaps some pre-processing of the data to remove these peaks and model them as point distributions may be warranted in some instances. This is also an area for further examination. In the present case, with the probability mass near the peak being small, it did not make much difference in our network performance evaluation,
(c) Note that the tail-decay parameter for the fitted PH distribution is 1.31 (by computing the eigenvalue of T closest to 0) and is reasonably close to the estimate 1.15 based on a crude regression model for the tail. The tail parameter estimate based on regression is highly sensitive to what is considered the tail, and therefore, the value 1.15 should be taken only as a very crude estimate. We need to develop some sound procedures for obtaining reliable tail estimates and to compare them, as well as methods to modify the EM algorithm so that the fitted LogPH distribution will have a given tail decay. Once again, these are topics for further research and beyond the scope of this paper.

3.2 Mobile Web Data Set

The mobile Web data set is obtained from a national cellular service provider. The training and test data sets comprised of the sizes of approximately 11.3 million files that were downloaded by all smart phones over a certain period of 24 h during 2012 inside a metropolitan area with hundreds of cell sectors. Figure 4 shows the empirical histogram and various LogPH model order fits to it. We stopped with a ninth-order fit since we found that the gain in the likelihood function in increasing the order did not justify the risk of overfitting. While a crude regression model provided a tail-decay parameter estimate of 1.14, the resulting value from our ninth-order LogPH fit turned out to be 0.91.

Figure 5a–c shows the various fits to the mobile Web trace. Note that LogPH provides a good fit across the entire distribution. Pareto and Weibull distributions, well known for modeling heavy-tailed distributions, perform poorly at the tails as well as at the head. Using the EMpht code, we obtained a ninth-order fit.

3.3 Comparison of Models

For both data sets under consideration, we also fitted the Pareto, Weibull, Log-Normal, and Log-*t* models to the data. A similar preprocessing of the data in terms of scaling was done for the Crovella data as done for the LogPH fit to make the comparisons meaningful. In this section, we make some comparisons of the fits based on some visual plots and the likelihood ratio values.

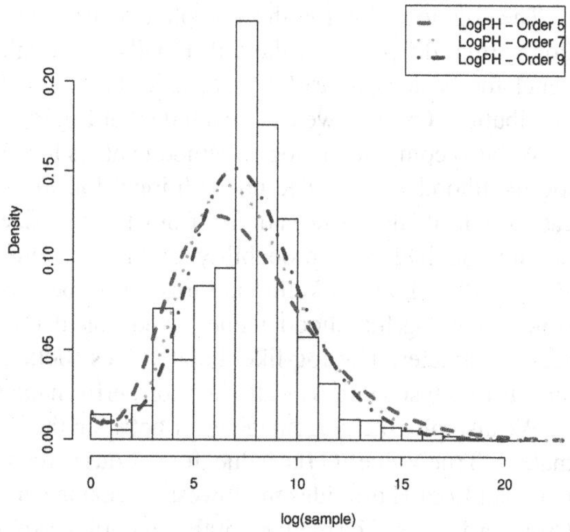

Fig. 4 Density plot and LogPH fit for different model orders of mobile Web data set

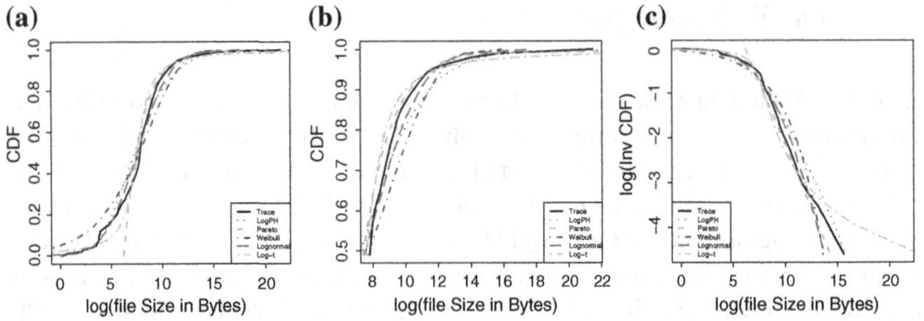

Fig. 5 LogPH fit (9th order) compared to other classical heavy tail distribution fits for mobile Web data set. **a** CDF, **b** CDF tail (magnified), and **c** complementary CDF

1. *The Crovella Data*: The parameters of the fitted classical distributions are as follows. The Weibull distribution (Eq. 4) has shape $a = 0.57$ and scale $b = 5{,}680$. The LogNormal distribution (Eq. 5) has $\mu = 7.87$ and $v = 1.46$. The log-t distribution has $\mu = 7.87$ and $v = 3.66$. Using EM algorithm directly for Pareto distribution gives a poor fit. Hence, for the Pareto distribution (Eq. 3), we use the shape $a = 1.15$ from the power law in the tail (Fig. 2) and the scale parameter $b = 1{,}788$ so that the mean for the Pareto distribution and empirical file size mean are equal (13,710 bytes).

Figure 6 shows the distribution plots for various fits along with the empirical file size distribution from the Crovella's trace. Figure 6a shows CDF over the entire support (with x-axis on the log scale). For the CDF, Log-t and LogPH appear to match the entire distribution well. However, a closer examination of the tail through a log–log plot given in Fig. 6b shows that Log-t does not capture the tail as well as the LogPH.

The log–log plot for the complementary CDF also shows that Weibull and LogNormal do poorly in the tail. Finally, although Pareto appears to give a good match for the decay rate of the tail, nevertheless, the fit does not match the bulk of the distribution. Overall, we can conclude that LogPH does indeed provide the best fit.

We also compute the log-likelihood values for the various fits. Table 1 gives the log-likelihood values. The unconditional log-likelihood values correspond to the case when all the observations (from the file size trace) are used. Since Pareto distribution had zero probability of taking value less than its scale parameter ($b = 1{,}788$), the log likelihood turns out to be $-\infty$. Hence, we also compare the conditional log-likelihood values conditioned on observations above the Pareto scale parameter. The log-likelihood values (both unconditioned and conditioned) appear to attest to an overall superior performance by the LogPH fit.

We also plot the percentage error between the modeled and true file sizes, (Estimate − True Value)/(True value) for various percentiles of the file sizes in Fig. 7. We note that LogPH provides the lowest percentage error compared to other fits for the Crovella data set. The median (50th percentile) error is 5.5 % for LogPH compared to 57.9, 44.5, 26.8, and 26.7 % for Pareto, Weibull, LogNormal, and Log-t, respectively.

Fig. 6 Comparing different fits (Crovella WWW trace). **a** CDF and **b** CCDF

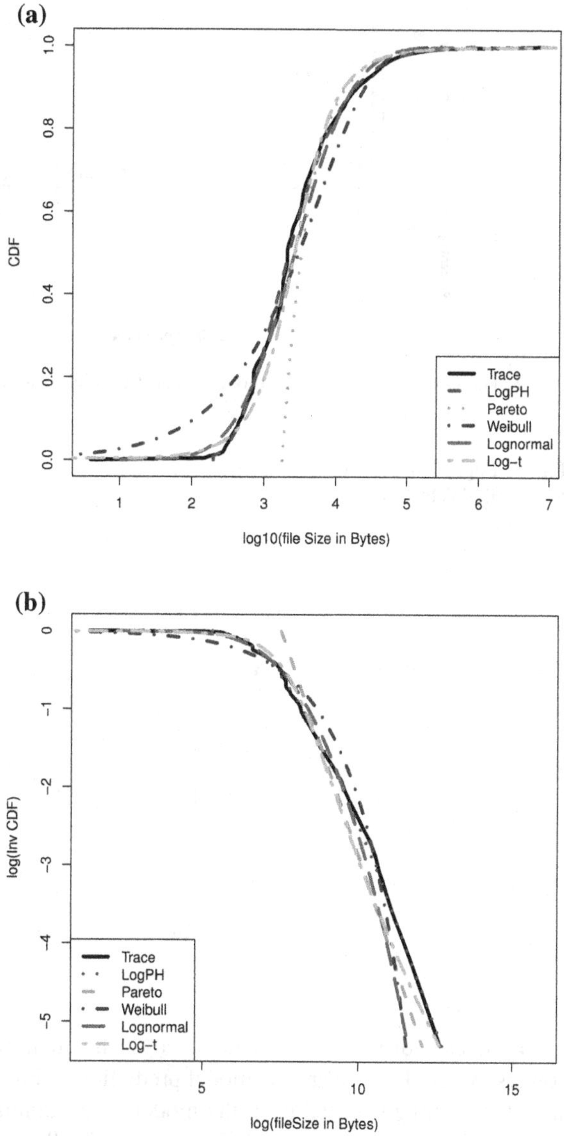

2. *Mobile Web Data*: Figures 8, 9 and 10 show a comparison of the fitted model quantiles with 5, 7, and 9 phases against the corresponding quantiles of the training data set. The improvement that results as we increase the number of phases in the fit as well as the fact that marginal improvement keeps decreasing are both clear from these graphs. As noted earlier, we stopped with an order 9 phase-type fit to the log data.

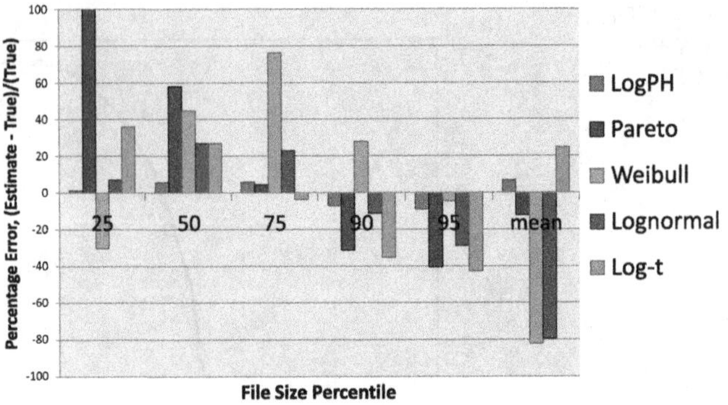

Fig. 7 Percentage error between modeled and true file sizes, varying percentiles

Fig. 8 LogPH model order 5—mobile Web data

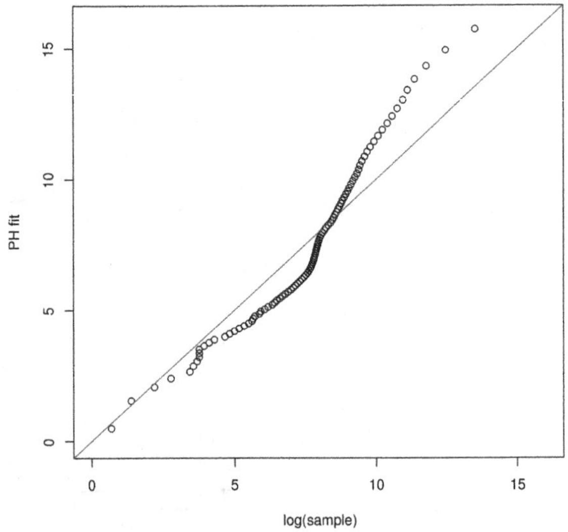

In the case of the mobile data, since we had data sets from many different cell towers, we could compare the model predictions with a test data set that is different than the training set used to fit the model. The training data set and the validation data set were thus from two different sets of cell towers (i.e., spatially separated and different users requesting mobile content). Figure 11, which compares the fitted model quantiles with the quantiles of the test data set, shows an excellent match.

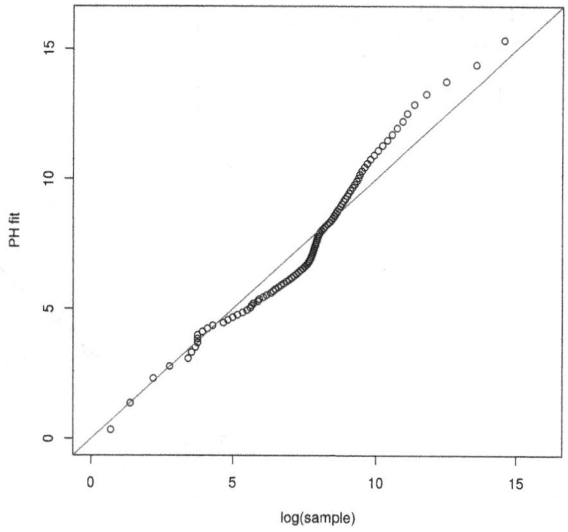

Fig. 9 LogPH model order 7—mobile Web data

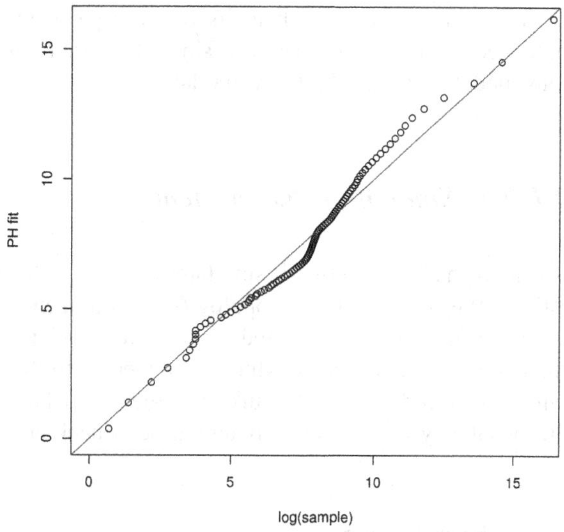

Fig. 10 LogPH model order 9—mobile Web data

4 Impact on Network Performance

Our interest in LogPH as a model class is from its ability to model heavy-tailed data in its entire range. But there is a fundamental question as to if it even matters whether the entire distribution is matched or not. The goal of this section is to examine this issue in the context of the data and the various models fitted to it. We do this with some NS-2 simulations of a bottleneck link implementing the TCP Reno protocol. The simulation is first run with a stream generated by sampling the

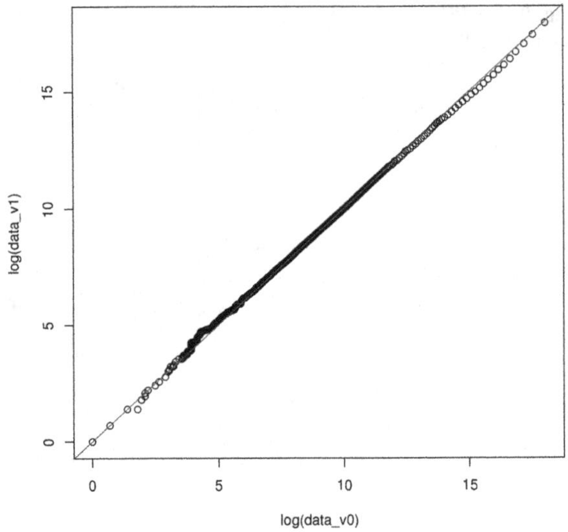

Fig. 11 Test data quantiles (y-axis) versus modeled data quantiles (x-axis), mobile Web data set

Crovella data trace, and then, its results (queue lengths and throughput of the bottleneck link) are compared to simulations run with streams generated by various models fitted to the Crovella data.

4.1 The Simulation Experiment

We used the NS-2 network simulator [11] to perform an extensive set of simulations. We set up a simple topology of a homogeneous set of clients and a server as shown in Fig. 12 connected by one intermediate router. In this configuration, there are N clients each of which is connected to the router using a 10 Mbps link with 1 ms link latency. The link from the server to the router is 1 Mbps link with 30 ms latency and depicts a potential bottleneck condition.

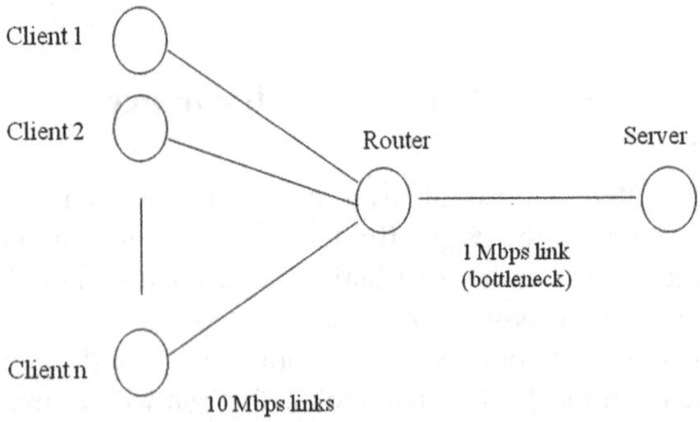

Fig. 12 Network scenario

We use two sets of client–server interaction. In the first set, a client can have at most one pending request at any given time. Thus, in this set, a client generates a request, and upon receiving a download from the server, it will generate the next request after a random amount of idle time. In the second set, a client can have multiple requests pending at the server. Here, each client generates a sequence of requests with a random inter-arrival time distribution without regard to what happened to previous requests. We call the first set as the *finite sources* scenario and the second set as the *infinite sources* scenario.

The connections use the default TCP-Reno protocol. The request size is fixed to a value of 40 bytes, and the interrequest time is set to be an exponential random variable with a mean of 10 s. For each request, the server replies with a file size drawn from an assigned file size distribution. The server think time for processing the request is set to (1 ± 0.1) ms. The response file is divided into packets of size 1,040 bytes and passed to the transport layer, which then passes the packets onto the link from the server to the router, depending upon the congestion on the bottleneck link (from server to router). We vary the number of clients to simulate different load conditions. The buffer sizes for all links are set to a very high value so as to mimic an infinite buffer scenario.

For each configuration—whether finite source or infinite source—we first determine the number of clients corresponding to an assigned load level. Then, for each scenario, we run six different simulations each using a different file size distribution at the server. The six different distributions used are as follows: trace (empirical distribution based on the Crovella data), LogPH, Pareto, Weibull, LogNormal, and Log-*t*, where the latter are the fitted models to the Crovella data. We run each simulation for a clock time of 3,600 s. We measure the queue length for the bottleneck link (from server to router) at the end of the simulation. We also measure the connection throughput (response file size/request completion duration) and take its average over all the connections in the simulation run. Depending on the load, there are about 10,000–25,000 connections per simulation run. We repeat this experiment 1,000 times using different random seeds. Thus, we obtain a distribution for queue length (at $T = 3{,}600$ s) and an average connection throughput. We then compare the results across different choices of the file size distribution.

1. *Results and Analysis*: We run 4 sets of experiments. The first two sets use 30 and 75 clients in the *finite sources* configuration so as to have a link utilization of 40 and 80 % on the bottleneck link, respectively. The third and the fourth sets use 30 and 60 clients in the *infinite sources* configuration to again have a link utilization of 40 and 80 %, respectively. As mentioned in the previous section, we measure the queue length at epoch 3,600 s and average connection throughput over 1,000 runs for file size distributions (at the server) from Crovella's trace or one of the five modeled fits.

Figure 13 shows the results for finite sources with 40 % link utilization. The LogNormal and Weibull fit show a very low queue length. This can be attributed to the lighter tail for LogNormal and Weibull, which results in lower load. Similarly,

Fig. 13 Finite sources with 40 % link utilization. **a** Queue length (Crovella data set) and **b** average connection throughput (Crovella data set)

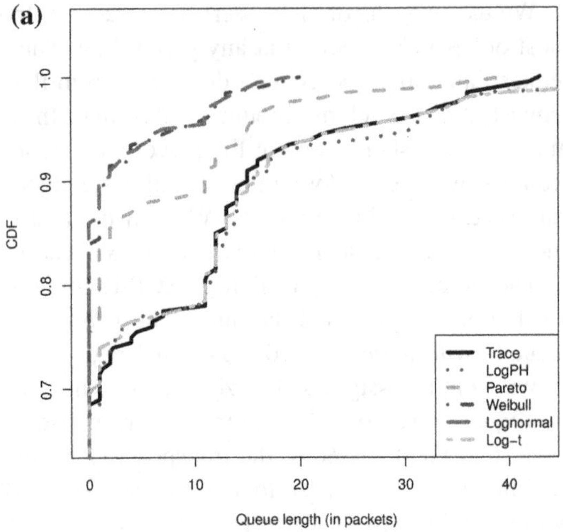

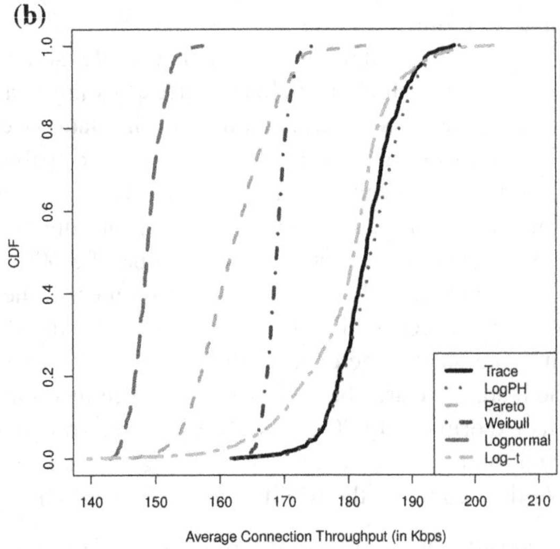

Log-t overestimates the tail and results in higher queue occupancy. Pareto fit seems to predict the queue length slightly better, but still is not as close as the predictions from LogPH. Looking at the average connection throughput, again LogPH predicts the throughput very well. Surprisingly, even though the queue lengths for LogNormal, Weibull, and Pareto runs are low, the corresponding throughputs are also low. This may seem counterintuitive. However, we believe that this may be because of the way TCP behaves. For smaller file sizes, the setup time and the slow-start phase reduce the throughput. For larger file sizes, the TCP connection has enough time to reach the congestion avoidance phase, thereby increasing the throughput.

In Fig. 14, we repeat the experiment with 80 % link utilization. Compared to the previous set, the queue length has increased and the throughput has decreased. This is expected because of the increased load. Even in this experiment, Pareto, LogNormal, and Weibull seem to give poor estimates. Log-*t* gives good estimates in this scenario; however, they are not as accurate as LogPH. Also, unlike previous experiment, the low queue length for LogNormal, Weibull, and Pareto leads to higher average connection throughput, when compared to the trace. This can be attributed to the connection throughput being dominated by the queueing delays rather than the TCP dynamics.

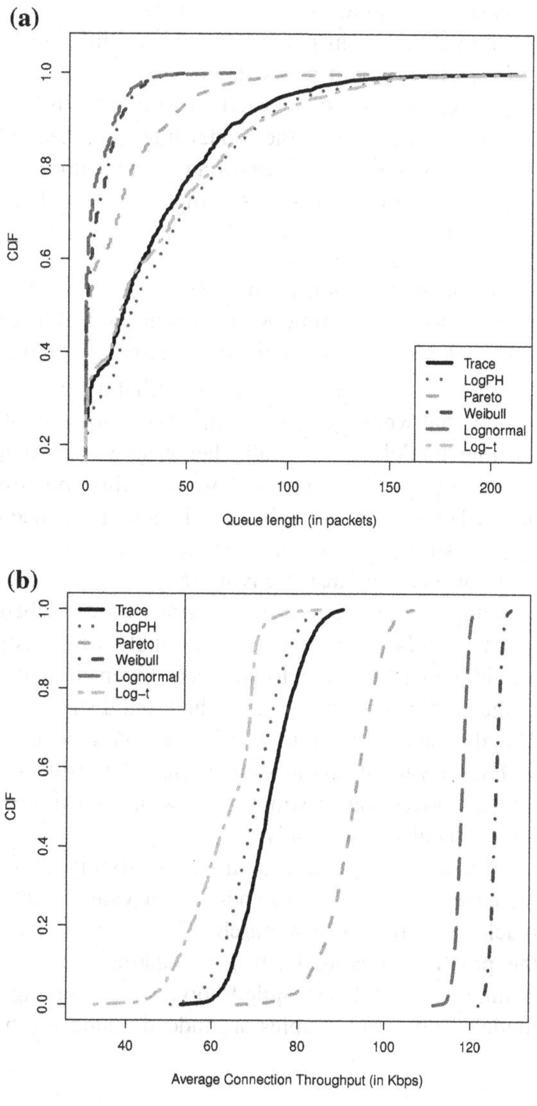

Fig. 14 Finite sources with 80 % link utilization. **a** Queue length (Crovella data set) and **b** average connection throughput (Crovella data set)

Next, we look at similar medium- and high-load scenarios with infinite sources. Figure 15 shows the results for infinite sources with 40 % link utilization. Once again, LogNormal and Weibull give inaccurate estimates. Even though Log-t seems to give very accurate performance estimates in some cases, it does not do so in all the scenarios. This shows that the mismatch due to constrained symmetry (of t-distribution) leads to inaccurate results in some cases. However, LogPH estimates the queue length and throughput very well in all the cases.

2. *A Modeling Issue*: In the context of the discussion of the simulation results, it behooves us to raise an important issue we discovered that raises some interesting challenges for mathematical research. All models fitted to the data sets have an infinite support, while the data sets themselves span only a finite interval. Thus, in our methodology while using the fitted model, there is the inherent problem of extrapolating far beyond available data. This type of a problem has always existed when model fits are made, but it has some serious consequences when the model fitted is a heavy-tailed one as we learned from our experiments. The use of an infinite support model often leads to extremely pessimistic predictions as compared with what would obtain with the sample data trace.

To make the point clearer, consider a FIFO queue and suppose we have a heavy-tailed service time with a power law tail index $\eta = 0.5$; i.e., the tail is of the order $O(1/x^{0.5})$ say. For very large x, now the conditional probability $P(X > 2x|X > x) \approx 1/\sqrt{2} \approx 0.71$. What this result points to is an interesting fact, namely that when we use an infinite support heavy-tailed distribution, the resulting queuing model is potentially being subjected to much larger excursions of the underlying queue length and waiting time processes when such excursions do occur. This affects the tails of various performance measures leading to an overall over-pessimistic assessment. Such effects are negligible when the underlying distributions have fast decaying tails.

Indeed, in the network simulations we described earlier, our use of the heavy-tailed models all resulted in extremely pessimistic results, with none of them providing a good match to the trace-driven simulations. Having guessed the source of the problem, we normalized the fitted distribution in the range (0, 1.2 M) where M is the maximum value seen in the data trace and used the resulting distributions with finite support in our simulations. The results reported above and the attending figures correspond to simulations with such finite support modifications, and the results speak for themselves.

Obviously, our choice of the truncation index is quite arbitrary and we had no recourse to any theory to choose that value appropriately. This is a (difficult but much needed) area of research: What kind of steps should one take to ameliorate the problems arising from extrapolating far from the data through the use of infinite support heavy-tailed models in queuing systems when in reality the modeled random variables are indeed bounded albeit at a high value?

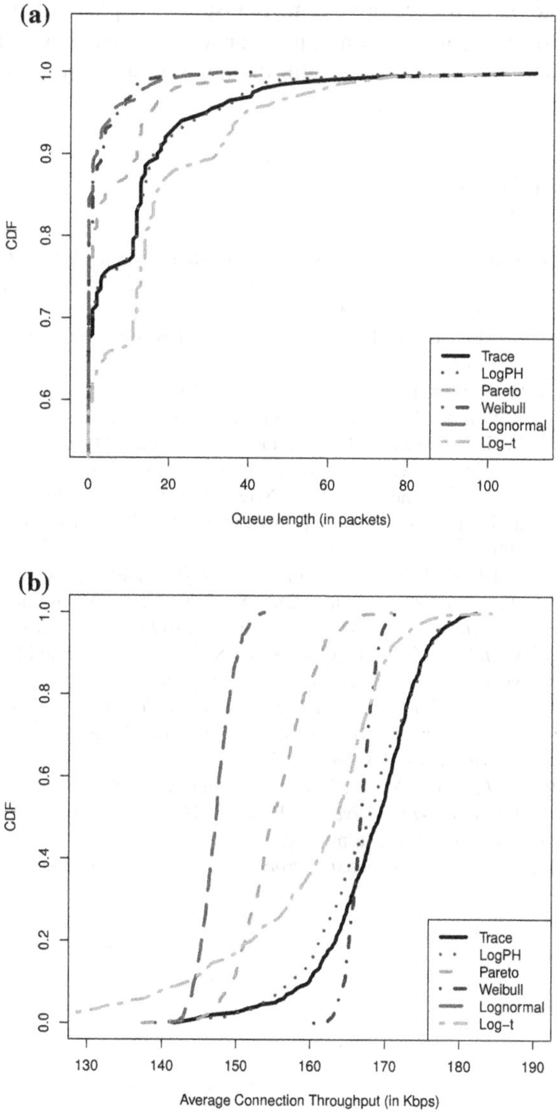

Fig. 15 Infinite sources with 40 % link utilization. **a** Queue length (Crovella data set) and **b** average connection throughput (Crovella data set)

5 Conclusions

We have demonstrated the power of the LogPH class of models to model heavy-tailed network data and provided a set of comparisons with other approaches that provide a compelling argument for its use. More importantly, we have identified an important set of research questions for examination whose resolution will enhance our ability to model heavy-tailed distributions and thereby assess more accurately the performance of systems that suffer the phenomena of heavy-tails. As we report this work, we are continuing our work in this area both at a theoretical level and at an experimental level with the hope of shedding more light in the future.

References

1. S. Ahn, Joseph H. T. Kim, and V. Ramaswami. A New Class of Models for Heavy Tailed Distributions in Finance and Insurance Risk. In *Insurance: Mathematics and Economics*, volume 51, pages 43-52, 2012.
2. S. Ahn and V. Ramaswami. Bilateral phase type distributions. *Stochastic Models*, pages 239-259, 2005.
3. S. Asmussen, O. Nerman, and M. Olsson. Fitting Phase-Type Distributions via the EM Algorithm. *Scandinavian Journal of Statistics*, 1996.
4. M. E. Crovella and A. Bestavros. Self-similarity in world wide web traffic: Evidence and possible causes. *Performance Evaluation Review*, 1996.
5. Chlebus E. and Divgi G. A Novel Probability Distribution for Modeling Internet Traffic and its Parameter Estimation. In *IEEE Global Telecommunications Conference - GLOBECOM*, 2007.
6. A. Ghosh, R. Jana, V. Ramaswami, J. Rowland, and N. K. Shankaranarayanan. Modeling and characterization of large-scale Wi-Fi traffic in public hot-spots. In *INFOCOM*, 2011.
7. G. Latouche and V. Ramaswami. *Introduction to Matrix Analytic Methods in Stochastic Modeling*. ASA-SIAM Series on Statistics and Applied Probability, 1999.
8. Mitzenmacher M. and Tworetzky B. New models and methods for file size distributions. In *Allerton Conference on Communications, Control and Computing*, 2003.
9. M. F. Neuts. Probability distributions of phase type. *Liber Amicorum Prof. Emeritus H. Florin*, pages 173-206, 1975.
10. *M. F. Neuts*. Matrix-Geometric Solutions in Stochastic Models: An Algorithmic Approach. *Johns Hopkins University Press, 1981.*
11. NS-2. http://isi.edu/nsnam/ns/.
12. Ward Whitt. *Stochastic-Process Limits*. Springer, 2002.

The Future of Access Control: Attributes, Automation, and Adaptation

Ravi Sandhu

Abstract Access control has been and will always be one of the center pieces of cyber security. This talk will focus on three necessary characteristics of access control in future systems: attributes, automation, and adaptation. Future access control policies will be built around attributes, which are properties of relevant entities, so they can apply to large numbers of entities while being fine-grained at the same time. This transition to attribute-based access control has been in process for about two decades and is approaching a major inflection point. Automation and adaptation, however, are newer concepts. Automation seeks to break away from requiring human users to configure access control policies, by delegating more of the routine tasks to smart software. Adaptation recognizes that access control must adjust as circumstances change. This talk will speculate on a future built around these three synergistic elements and on the research and technology challenges in making this vision a reality.

Keywords Access control · Attributes · Automation and Adaptation

---

R. Sandhu (✉)
University of Texas, San Antonio, TX, USA
e-mail: ravi.sandhu@utsa.edu; ravi.texan@gmail.com

G. S. S. Krishnan et al. (eds.), *Computational Intelligence, Cyber Security and Computational Models*, Advances in Intelligent Systems and Computing 246,
DOI: 10.1007/978-81-322-1680-3_5, © Springer India 2014

Optimal Control for an $M^X/G/1/N + 1$ Queue with Two Service Modes

Rein D. Nobel and Adriaan A. N. Ridder

Abstract A finite-buffer queueing model is considered with batch Poisson input and controllable service rate. A batch that upon arrival does not fit in the unoccupied places of the buffer is partially rejected. A decision to change the service mode can be made at service completion epochs only, and vacation (switch-over) times are involved in preparing the new mode. During a switch-over time, service is disabled. For the control of this model, three optimization criteria are considered: the average number of jobs in the buffer, the fraction of lost jobs, and the fraction of batches not fully accepted. Using Markov decision theory, the optimal switching policy can be determined for any of these criteria by the value-iteration algorithm. In the calculation of the expected one-step costs and the transition probabilities, an essential role is played by the discrete fast Fourier transform.

Keywords Finite-buffer model · Value iteration · Fast Fourier transform

1 Introduction

In this paper, we consider a control problem in the finite $M^X/G/1/N + 1$ queueing model. Optimal control of queueing models has always been an important feature in the queueing literature and may go back as far as Erlang's study to determine the minimal number of servers c in the $M/G/c/c$ queue such that the loss probability remains below some predefined level. In fact, optimal control of real-world systems is the driving force to analyze most queueing models. A common theme is the

R. D. Nobel (✉) · A. A. N. Ridder
Department of Econometrics, Vrije Universiteit, De Boelelaan 1105,
1081 HV Amsterdam, The Netherlands
e-mail: r.d.nobel@vu.nl

A. A. N. Ridder
e-mail: a.a.n.ridder@vu.nl

G. S. S. Krishnan et al. (eds.), *Computational Intelligence, Cyber Security and Computational Models*, Advances in Intelligent Systems and Computing 246,
DOI: 10.1007/978-81-322-1680-3_6, © Springer India 2014

determination of parameters (such as number of servers, system capacity, and service rate) that optimize a certain performance measure of the system. This paper falls in the same area. We address an operational problem in a system with finite buffer, batch arrivals, and single server, where the service of jobs can be done in one of two modes, regular or high (fast). Switching between modes is possible after service completion and takes a random duration. In other words, the controllable parameter is the choice of service mode after each service completion. We focus on three performance measures: the long-run average number of jobs in the buffer, the long-run fraction of lost jobs, and the long-run fraction of batches with losses.

An application of this queueing model appears in a telecommunication system with a controllable transmission rate: Connections are given less bandwidth (transmission rate) when the transmission line is near full capacity. The question is at which capacity the rate should be slowed down. Because the performance criteria in our queueing model are different than those in the communication problem, we seem to answer the opposite question: how many jobs should be waiting in the buffer before working harder. Although this question is quite natural, we found that there is not much attention in the literature for finite-buffer queueing models with controllable service. Though, infinite-buffer systems with controllable service have been studied extensively, see, e.g., Dshalalow [1] for a recent overview. Nishimura and Jiang developed in [2] an algorithm for the steady-state probabilities of an infinite-buffer, single-arrival $M/G/1$ queue with two service modes and switch-over times for any so-called two-level hysteretic switching rule, i.e., changes in the service mode take place only if the number of jobs in the system rises above a fixed high level or falls below a fixed low level. In [3], Nobel and Tijms extended this model to compound Poisson input, and they used Markov decision theory to calculate the optimal one-level hysteretic rule (i.e., once a change from the slower to the faster service mode is made, this mode is continued until the system is empty) that minimizes the long-run average number of jobs in the system. By exploiting well-known results for the standard $M^X/G/1$ queue, a tailor-made policy-iteration algorithm was developed by an embedding technique to cope with the problem of the infinite state space. Also, in Nobel [4], these results for the $M^X/G/1$ queue formed the basis for a regenerative analysis, so that for every two-level hysteretic switching rule, the average number of jobs in the system could be calculated. In both [4] and [3], the discrete fast Fourier transform (FFT) played an essential role in the calculation of the probability distributions of the number of individual arrivals during a service time and a switching time via their generating functions. Because the application of the FFT requires that these generating functions are given explicitly, some minor restriction with respect to the generality of the service-time and switching-time distributions was required in these papers.

As said, in this paper we discuss the finite-buffer equivalent. The essential differences in the infinite-buffer model are twofold. First, the state space is now finite, suggesting a simplification compared to the infinite model, but, second, the queueing results analogous to those of the standard $M^X/G/1$ model, which were so crucial in the analysis of both [3] and [4], are not available for the standard $M^X/G/1/N+1$ model. As a consequence, we cannot simply adapt the policy-iteration algorithm of [3] or the regenerative approach of [4] to the finite-buffer model. Besides that, now also other performance measures become relevant for practical applications, e.g., the loss probability of a job.

The traditional approach for analyzing a controlled queueing model is to formulate a stochastic dynamic program (Markov decision model) and then prove that the optimal policy has certain structural properties, see Koole [5] and Stidham and Weber [6] for recent surveys. For an overview of this approach, we refer to the monographs of Kitaev and Rykov [7] or Sennot [8], and the review paper by Cavazos-Cadena and Sennot [9]. We remark that most results obtained by this approach are valid only in queueing models with exponential services. The contribution of this paper is that we are able to determine the optimal switching strategy (not necessarily hysteretic!) for the performance criteria mentioned above and for a wide range of service and switching distributions. Our approach combines the three usual features: modeling, analyzing, and calculating.

Modeling: We reformulate the control problem as a semi-Markov decision model. The value-iteration algorithm will be applied for determining optimal policies.

Analyzing: We use general results from Nobel [10] to find expressions of the one-step costs in the semi-Markov decision model. These expressions are formulated in terms of coefficients of different generating functions.

Calculating: We apply the discrete fast Fourier transform for inverting these generating functions.

The paper is organized as follows. In Sect. 2, the queueing model is described, and in Sect. 3, the semi-Markov decision model is formulated with the associated value-iteration algorithm. In Sects. 4 and 5, we explain how to calculate the one-step costs, which are required in the decision model. In Sect. 6, some numerical results are given.

2 Description of the Queueing Model

We consider a finite-buffer $M^X/G/1/N + 1$ queueing model with two service modes. Batches of jobs arrive at a single-server station according to a Poisson process. The batch sizes are identically distributed, independent random variables. The single server can handle one job at a time and serves the jobs in the order of their arrival (within a batch in random order). The service times are independent random variables, but they are typically not identical. Before servicing a job, the server chooses between a regular speed service mode (R) and a high-speed service mode (H). Only at service completion epochs, the server can switch from one mode to another. Every time the server decides to change mode, a vacation (or switch-over) time is required to prepare for the new mode. These switch-over times are independent random variables and again one of two types: switching from regular to high, or vice versa. In accordance with practical applications, the model assumption is made that the system has always to switch from the high-speed mode to the regular speed mode when the system becomes empty and the last service was done in high-speed mode. The service is disabled during a switch-over time, and the next job to be served stays in the buffer until the switch-over time expires and the service can start. The buffer is a waiting room for N places. Batches that upon arrival do not find sufficient places in the buffer are partially rejected.

This completes the description of the model. The purpose of a controller of the system is to specify the decisions when to switch to the other service mode, in such a way that an optimization criterion is met. The three optimization criteria which we propose are as follows:

- Minimize the long-run average number of jobs in the buffer,
- Minimize the long-run fraction of jobs that are lost, and
- Minimize the long-run fraction of batches that are not fully accepted.

Clearly, we consider in all three cases the so-called expected (long-run) average cost optimality criterion. Because we deal with finite state and action spaces—which we will specify in the sequel—an optimal policy is stationary deterministic [11]. An optimal (stationary deterministic) policy can be found by the value-iteration algorithm of Markov decision theory [11, 12]. Before we give this algorithm in the context of our control problem, we specify the notation.

Notation

λ rate of Poisson arrivals of batches;
β_k Prjbatch contains k jobs} ($k = 1, 2,\ldots$);
$\beta^{(1)}$ mean batch size;
A generic random variable representing the interarrival time of batches;
S_R generic random variable representing the service time of a job under the regular speed service mode;
S_H similarly under the high-speed mode;
V_R generic random variable representing the switch-over time for changing from regular speed to high-speed service mode;
V_H similarly vice versa;
$\alpha_j^{(L)}$ Pr{j jobs arrive during a service time S_L} ($L = R, H$);
$\phi_j^{(L)}$ Pr{j jobs arrive during switch-over time V_L} ($L = R, H$).

3 The Semi-Markov Decision Model

The control problem of finding the best policy is formulated in a semi-Markov decision model. The elements are decision epochs, state space, action space, transition probabilities, and one-step costs.

- The decision epochs occur at the end of the service times and at the end of the switching times. The latter epochs are actually no decision epochs. They are introduced merely to simplify the transition probabilities and the one-step expected costs (see below).
- State space:

$$S = \{(i, K, L) \mid i = 0, \ldots, N; K = S,\ V; L = R, H\}$$

where state $(i,\ S,\ L)$ describes the situation that a service in L-mode has been completed leaving behind i jobs in the buffer and state (i, V, L) denotes that a switching time V_L has expired with i jobs in the buffer.

- Action spaces $A(i,\ K,\ L)$ of feasible actions in state $(i,\ K,\ L)$:

$$\begin{array}{lll} A(0,\ S,\ L) = \{R\}, & A(i, S, L) = \{R, H\} & (i > 1,\ L = R, H), \\ A(i,\ V,\ R) = \{H\}, & A(i, V, H) = \{R\} & (i > 0). \end{array}$$

- One-step transition probabilities $p_{st}(a)$, denoting the probability that the system jumps to state t at the next decision epoch when given that the state is s at the current decision epoch and that action a is chosen. For instance, suppose that a service in regular mode has been completed leaving behind $i \geq 1$ jobs in the buffer and that the action is to switch to high-speed mode. Then, the probability of a full buffer at the next decision epoch is

$$\begin{aligned} p_{(i,S,R)(N,V,H)}(H) &= \Pr\{\text{at least } N-i \text{ jobs arrive during switch-over time} V_R\} \\ &= \sum_{k > N-i} \phi_k^{(R)}. \end{aligned}$$

In this way, all transition probabilities are expressed by the densities (β_k), $\alpha_k^{(L)}$ and $\phi_k^{(L)}$. We summarize here the transition probabilities under action $a = R$ only.

$$\begin{aligned} p_{(0,S,R)(j,S,R)}(R) &= \textstyle\sum_{k=1}^{j+1} \beta_k \alpha_{j+1-k}^{(R)} \qquad (j \leq N-1), \\ p_{(0,S,R)(N,S,R)}(R) &= \textstyle\sum_{k>N} \beta_k + \sum_{k=1}^{N} \beta_k \quad \sum_{l>N-k} \alpha_l^{(r)}, \\ p_{(i,S,R)(j,S,R)}(R) &= \alpha_{j+1-i}^{(R)} \qquad (i \geq 1,\ i-1 \leq j \leq N-1), \\ p_{(i,S,H)(N,S,R)}(R) &= 1 - \textstyle\sum_{k=0}^{N-i} \alpha_k^{(R)} \qquad (i \geq 1), \\ p_{(i,S,H)(j,V,H)}(R) &= \phi_{j-i}^{(H)} \qquad (i \geq 0, i \leq j \leq N-1), \\ p_{(i,S,H)(N,V,H)}(R) &= \textstyle\sum_{k \geq N-i} \phi_k^{(H)} \qquad (i \geq 0), \\ p_{(0,V,H)(j,S,R)}(R) &= \textstyle\sum_{k=1}^{j+1} \beta_k \alpha_{j+1-k}^{(R)} \qquad (j \leq N-1), \\ p_{(0,V,H)(N,S,R)}(R) &= \textstyle\sum_{k>N} \beta_k + \sum_{k=1}^{N} \beta_k \quad \sum_{l>N-k} \alpha_l^{(R)}, \\ p_{(i,V,H)(j,S,R)}(R) &= \alpha_{j+1-i}^{(R)} \qquad (i \geq 1,\ i-1 \leq j \leq N-1), \\ p_{(i,V,H)(N,S,R)}(R) &= \textstyle\sum_{k>N-1} \alpha_k^{(R)} \qquad (i \geq 1). \end{aligned}$$

All other one-step transition probabilities under action R are zero. The set of transition probabilities under action H is constructed similarly. Note that when service is enabled, the system can hold up to $N + 1$ jobs, while only up to N jobs (the buffer size) can be present when service is disabled.

- Expected times between decision epochs: $\tau_s(a)$ denotes the expected time until the next decision epoch given that in state s action a is chosen. Again we give them here for $a = R$ only:

$$\begin{aligned}
\tau_{(0,S,R)}(R) &= E[A + S_R] = \tfrac{1}{\lambda} + E[S_R], \\
\tau_{(i,S,R)}(R) &= E[S_R] \qquad (i \geq 1), \\
\tau_{(i,S,H)}(R) &= E[V_H] \qquad (i \geq 0), \\
\tau_{(0,V,H)}(R) &= E[A + S_R] = \tfrac{1}{\lambda} + E[S_R], \\
\tau_{(i,V,H)}(R) &= E[S_R] \qquad (i \geq 1).
\end{aligned}$$

- Expected one-step costs: $c_s(a)$ are the total expected costs incurred until the next decision epoch if in state s action a is taken. Because we consider three different optimization criteria, we deal with three different one-step cost specifications. They will be discussed in the following section.
- Expected average cost criterion: Let D be a stationary deterministic policy. The process that describes the consecutive states of the control system (under D) is a Markov chain. The transition probabilities of this chain are $p_{st}(D(s))$ given above ($D(s)$ is the action in state s). Clearly, the chain has only one closed (sub)set of states. Hence, the expected average cost of policy D is

$$g(D) := \frac{\sum_{s \in S} c_s(D(s))\pi(s|D)}{\sum_{s \in S} \tau_s(D(s))\pi(s|D)},$$

where $\pi(s|D)$ is the stationary distribution in state s under policy D, cf. [12, Sect. 3.5]. The goal is to find an optimal policy D^*, i.e., a stationary policy for which

$$g^* := g(D^*) \leq g(D)$$

for all stationary policies D.

3.1 The Value-Iteration Algorithm

Once all the elements of the Markov decision model are known, we can use the value-iteration algorithm to calculate the optimal stationary deterministic policy D^*. We give the formulation of the algorithm in general terms (see also [12, Sect. 3.5]).

First choose a positive number τ with $\tau \leq \min_{s,a} \tau_s(a)$ and a tolerance number ϵ, e.g., $\epsilon = 10^{-6}$.

INIT For all $s \in S$, choose non-negative numbers $W_0(s)$ with

$$W_0(s) \leq \min_a\{c_s(a)/\tau_s(a)\}.$$

Let $n := 1$.

LOOP For all $s \in S$, calculate

$$W_n(s) = \min_{a \in A(s)} \left[\frac{c_s(a)}{\tau_s(a)} + \frac{\tau}{\tau_s(a)} \sum_{t \in S} p_{st}(a) W_{n-1}(t) + \left\{ 1 - \frac{\tau}{\tau_s(a)} \right\} W_{n-1}(s) \right],$$

and let $D_n(s) \in A(s)$ be the action that minimizes the right-hand side.

EVAL Compute the bounds,

$$m_n = \min_{s \in S}\{W_n(s) - W_{n-1} - (s)\}, \quad M_n = \max_{s \in S}\{Wn(s) - W_{n-1}(s)\}.$$

TEST $M_n - m_n \leq \epsilon m_n$ then STOP with the resulting policy D_n, else $n := n + 1$ and go to **LOOP.**

This algorithm returns after, say, n iterations a stationary deterministic policy D_n. Let $g_n := (m_n + M_n)/2$. Then, $g_n \to g^*$ if $n \to \infty$ and $|g_n - g^*| \leq \in g^*$. In other words, g_n is an approximation of the minimum expected average costs.

4 The Expected One-Step Costs

In this section, we specify the expected one-step costs $c_s(a)$ for the three optimization criteria.

Consider an arbitrary feasible state-action pair (s, a) of the Markov decision model. Immediately after the action, there is a unique buffer content q. For instance, suppose $(s, a) = ((i, S, R), H)$ and $i \geq 1$. Then, $q = i$ because the server is disabled due to switching. On the other hand, if $(s, a) = ((i, S, H), H)$ and $i \geq 1$, then $q = i - 1$ because the server takes a job out of the buffer. Also, the time it takes until the next decision epoch depends on the state-action pair (s,a). We denote an arbitrary interdecision time by X. To illustrate, consider the same examples: $(s, a) = ((i, S, R), H)$ yields $X = V_R$ (switching and preparing for the other service mode); $(s, a) = ((i, S, H), H)$ yields $X = S_H$ (service duration). We summarize the buffer content q and interdecision time X for all feasible state-action pairs (s, a) in Table 1. Notice that we introduced already in Sect. 3 the different expectations of X as $\tau_s(a)$. The calculation of the expected one-step costs, though, requires the complete distributions of interdecision time X.

The one-step costs are costs incurred during the interdecision time X after the buffer started with q jobs. Therefore, we may denote the expected one-step costs by

$$c_s(a) = w(X, q),$$

where X and q are given in Table 1. Because the buffer content does not change during an interarrival time A, we can express $W(A + S_L, 0)$ in terms of $w(S_L, q)$:

Table 1 Buffer content and interdecision time, given the state-action pair

s	a	X	q	
(0, S,R)	R	$A + S_R$	0	
(i, S, R)	R	S_R	$i - 1$	$(i \geq 1)$
(i, S, R)	H	V_R	i	$(i \geq 1)$
(i, S, H)	R	V_H	$i - 1$	$(i \geq 1)$
(i, S, H)	H	S_H	i	$(i \geq 0)$
(0, V, R)	H	$A + S_H$	0	
(i, V, R)	H	S_H	$i - 1$	$(i \geq 1)$
(0, V, H)	R	$A + S_R$	0	
(i, V, H)	R	S_R	$i - 1$	$(i \geq 1)$

$$w(A + S_R, 0) = \sum_{k=1}^{N} \beta_k w(S_R, k-1) + \sum_{k>N} \beta_k w(S_R, N),$$

$$w(A + S_H, 0) = \sum_{k=1}^{N} \beta_k w(S_H, k-1) + \sum_{k>N} \beta_k w(S_H, N).$$

In the following section, we shall determine the functions $w(X, q)$ in detail. We call these functions the expected interdecision cost functions.

5 The Expected Interdecision Cost Functions

As said, it remains to calculate the functions $w(X, q)$ in the three optimization criteria for $X = S_R, S_H, V_R, V_H$, and $q = 0, 1, \ldots, N$. This problem is simplified by applying some general results from Chap. 6 in [10]. There, similar functions in the context of a lost-sales production/inventory model are studied. For that purpose, we have to introduce the following two sets of generating functions.

5.1 Generating Functions Associated with the Model

$$\beta(z) := \sum_{k=1}^{\infty} \beta_k z^k$$

$$B(z) := \lambda(1 - \beta(z)),$$

$$A_L(z) := \sum_{j=0}^{\infty} \alpha_j^{(L)} z^j \text{ the generating function of } (\alpha_j^{(L)})_j (L = H, R),$$

$$\Phi_L(z) := \sum_{j=0}^{\infty} \phi_j^{(L)} z^j \text{ similarly for } (\phi_j^{(L)})_j.$$

5.2 Generating Functions Associated with X

Let X be any of the two service times S_R, S_H or any of the two switch-over times V_R, V_H. It has cumulative distribution function $G(x) := \Pr(X \le x)$ and Laplace–Stieltjes transform $\widehat{G}(s) := \int_0^\infty \exp(-sx)G(dx)$.

$$\Psi(z) = \sum_{k=0}^{\infty} \psi_k z^k := \widehat{G}(B(z)),$$

$$\Gamma(j,z) = \sum_{k=0}^{\infty} \gamma_k(j) z^k := j\widehat{G}'(B(z)) - \frac{1}{2} zB'(z)\widehat{G}''(B(z)), \quad j = 1, 2, \ldots,$$

$$\Delta(z) = \sum_{k=0}^{\infty} \delta_k z^k := \frac{z}{(1-z)B(z)} \left\{1 - \Psi(z) + B(z)\widehat{G}'(\mathrm{B}(z))\right\},$$

$$\Theta(z) = \sum_{k=0}^{\infty} \theta_k z^k := \frac{-zB'(z)}{(1-z)B^2(z)} \left\{1 - \Psi(z) + B(z)\widehat{G}'(B(z)) - \frac{1}{2}B^2(z)\widehat{G}''(B(z))\right\},$$

$$\Xi(z) = \sum_{k=0}^{\infty} \xi_{k^{z^k}} := \frac{1}{1-z}\left\{E[X] + \frac{\Psi(z) - 1}{B(z)}\right\}.$$

Notice that if $X = S_R$, then $\Psi(z) = A_R(z)$: the generating function of the number of arrivals during service time S_R. Similarly, if $X = V_R$, then $\Psi(z) = \Phi_R(z)$: the generating function of the number of arrivals during switch-over time V_R. (Also, the high-speed equivalents are valid.)

To calculate the expected interdecision cost functions *w(X, q)*, we consider the three optimization criteria separately.

- Average number of jobs in the buffer.

Let *Q*(*t*) be the number of jobs in the buffer at time *t*, given that at time 0, a decision is taken. Hence,

$$w(X,q) = E\left[\int_{(0,X)} Q(t)dt | Q(0+) = q\right], \quad q = 0, 1, \ldots, N$$

Notice that the buffer content *Q*(*t*) on the time interval (0, *X*) varies due to the arriving of batches only. The following theorem gives the exact expression.

Theorem 1 *Let the random variable X be either a service time or a switching time. Then, w(X, N) = NE [X], and for q = 0, 1,…,N − 1:*

$$w(X,q) = NE[X] + \sum_{k=0}^{N-q-1} \gamma_k(N-q) - (N-q)\delta_{N-q} + \theta_{N-q-1},$$

where $\gamma_k(N-q)$, δ_{N-q}, θ_{N-q-1} are the coefficients in the terms of the generating functions associated with X.

Proof The proof follows directly from Theorem 6.4.2 of [10]. We only need to equate the number of empty places in the buffer with the inventory on hand in the lost-sales production/inventory model discussed in [10, Chap. 6]. □

• Loss probability of jobs.

This is the problem how to find the switching policy that minimizes the loss probability, i.e., the long-run fraction of jobs that is rejected. The rejection rate is the long-run average number of jobs that is rejected per unit time. Hence, the loss probability is simply the rejection rate divided by the average number of individual arrivals per unit time, which is $\lambda\beta^{(1)}$. Therefore, it suffices to find the switching policy that minimizes the rejection rate. For that problem, the expected interdecision cost function is

$w(x, q) :=$ the expected number of rejected jobs during random time X, given that at the start of this interval q jobs are in the buffer.

This is easy:

$$w(X, q) = \sum_{k>N-q} (k - N + q)\psi_k, \quad q = 0, 1, \ldots, N,$$

where the probabilities ψ_k are defined earlier in this section. As a side remark, the infinite sum can be easily rewritten as a finite sum, because $\sum_{k>0} k\psi_k = \lambda\beta^{(1)}E[X]$.

• Fraction of not fully accepted batches.

The optimal switching policy that minimizes this fraction will be calculated by solving the equivalent problem of finding the switching policy that minimizes the long-run average number of batches that is not fully accepted per unit time. So,

$w(x, q) :=$ the expected number of batches not fully accepted during random time X, given that at the start of this interval q jobs are in the buffer.

$w(X, q) =$ the expected number of batches not fully accepted during random time X, given that at the start of this interval, q jobs are in the buffer. From Theorem 6.6.1 in [10], we borrow the following result.

Theorem 2 *Let the random variable X be either a service time or a switching time. Then,*

$$w(X, q) = \sum_{k>N-q} \psi_k + \lambda\xi_{N-q}, \quad q = 0, 1, \ldots, N,$$

where the numbers ψ_k and ξ_k are the coefficients in the terms of the generating functions associated with X.

6 Numerical Results

We have chosen the following $M^X/G/1/N + 1$ model with two service modes. The parameters of arrivals, batches, and service and switching times are

$$\lambda = 0.15,\ \beta^{(1)} = 4,\ E[S_R] = 1.7,\ E[S_H] = 1.5\ E[V_R] = 1,\ E[V_H] = 1$$

We consider both a constant and a geometrically distributed batch size. The buffer size is varied from $N = 10$ to 150. The regular service time S_R is taken to be Erlang-2, while the high-speed service time S_H is given a squared coefficient of variation $C^2_{S_H} = 4$ with a Coxian-2 distribution. The switch-over times are both taken exponentially distributed.

These distributions give us analytic expressions of the generating function $\beta(z)$ of the batch size and of the Laplace–Stieltjes transform $\widehat{G}(s)$ of any of the service and switching times. Hence, all generating functions mentioned in Sect. 5 can be given explicitly. Then, the numerical values of the coefficients of these functions (i.e., $\alpha_k^{(L)}$, $\phi_k^{(L)}, \psi_k, \lambda_k(j), \delta_k, \theta_k, \xi_k$) can be calculated efficiently by the discrete fast Fourier transform. And so, all ingredients of the semi-Markov decision model are ready for use.

The results are presented in Table 2 (constant batch sizes) and Table 3 (geometric batch sizes). We have considered the criteria "average number of jobs present" and "fraction of lost jobs." The stopping ε in the value-iteration algorithm was set to 10^{-6}. Then, the number of iterations varied from 216 (buffer size $N = 10$) to 11,234 (buffer size $N = 150$) for the loss probability criterion with the constant batch sizes and from 112 (buffer size $N = 10$) to 10,243 (buffer size $N = 150$) for the average buffer content criterion with the constant batch sizes. The number of iterations in case of geometric batch sizes is slightly less (e.g., $N = 150$ requires 9,106 and 8,473 iterations). In all cases, we found that the optimal switching rules are of the hysteretic type (m, M), where $m < M$. That is, when the service mode is regular and the buffer content becomes M or more, the

Table 2 Constant batches. Optimal loss probabilities π^*_{loss} and optimal buffer contents Q^*

N	π^*_{loss}	(m, M)	Q	Q^*	(m, M)	π_{loss}
10	0.188936	(2, 11)	4.3246	3.9832	(0, 7)	0.200883
20	0.100803	(3, 17)	8.1254	7.7478	(0, 9)	0.105156
30	0.0634553	(4, 19)	11.5640	11.1519	(1, 11)	0.0651914
40	0.0430466	(4, 20)	14.5821	14.2138	(1, 12)	0.0438801
50	0.0303526	(4, 20)	17.2363	16.9395	(1, 13)	0.0307808
60	0.0219347	(4, 20)	19.5914	19.3384	(2, 14)	0.0221218
75	0.0138796	(4, 20)	22.5853	22.3717	(2, 14)	0.0139907
100	0.0067750	(4, 20)	26.3133	26.1332	(2, 15)	0.00681112
125	0.00340616	(4, 20)	28.7955	28.6334	(2, 15)	0.00342380
150	0.00173715	(4, 20)	30.3896	30.2368	(2, 16)	0.00174311
∞				33.6333	(2, 16)	

Table 3 Geometric batches. Optimal loss probabilities π^*_{loss} and optimal buffer contents Q^*

N	π^*_{loss}	(m, M)	Q	Q^*	(m, M)	π_{loss}
10	0.253907	(1, 11)	3.7093	3.5385	(0, 8)	0.261979
20	0.153002	(3, 17)	7.6869	7.5004	(0, 10)	0.155402
30	0.101877	(3, 19)	11.4571	11.2581	(1, 11)	0.103287
40	0.0728404	(3, 20)	14.9550	14.7725	(1, 13)	0.0734649
50	0.0542501	(3, 20)	18.1796	18.0388	(1, 14)	0.0545850
60	0.0414826	(3, 20)	21.1666	21.0536	(1, 14)	0.0417198
75	0.0286625	(3, 20)	25.1999	25.1109	(2, 15)	0.0287550
100	0.0163363	(3, 20)	30.7864	30.7195	(2, 16)	0.0163693
125	0.0096780	(3, 20)	35.1103	35.0551	(2, 16)	0.00969651
150	0.0058582	(3, 20)	38.3814	38.3334	(2, 17)	0.00586518
∞				47.6220	(2, 17)	

server switches to high speed. On the other hand, when the service mode is high and the buffer content becomes m or less, the server switches to regular mode. ($M = N + 1$ means that always the regular mode is applied.) We conjecture that there are examples in which the optimal policy is not of the hysteretic type. We have not yet found such an example.

We see from the figures in the tables that the optimal switching rules for the two optimization criteria are quite different. To see the deteriorating effect of minimizing the loss probability on the average buffer contents and vice versa, we give, apart from the optimal values, also the results of the non-optimized criterion. As expected, the difference between these values fades away for large buffer sizes, when the loss probability becomes small.

Furthermore, we compare the results with those for the infinite buffer. With the method of [4], we determined the policy that minimizes the number of jobs in the system for the infinite model with the same parameter values. This differs by at most one job with the average number in the buffer. We see that the optimal policies are the same for the infinite model and the finite model with large buffer size, but the average number of jobs in the system (in the infinite model) is quite different, especially for the geometric case.

Finally, we remark that while the criterion values for geometric batches and constant batches are completely different, the optimal switching rules are very similar. This insensitivity of the optimal policy for the batch size distribution has already been observed in [3].

7 Conclusions

In this paper, we considered a natural optimization problem in the context of a queueing system with finite buffer and two service modes: "at which levels of buffer content should the server switch service mode?" We formulated the optimization problem as a semi-Markov decision model with fictitious decision epochs and solved it by the value-iteration algorithm. In all cases we evaluated numerically, we found that the optimal policy is of the hysteretic type. We doubt whether always the optimal policy is hysteretic, but we cannot provide counter-examples. Furthermore, we saw that one has to be careful in approximating the performance of the finite-buffer system by the corresponding infinite-buffer system, even for large buffers (with small loss rate). The average number of jobs in the system can be far off, as we noticed in the case of geometric batch sizes. Further study and analysis may give more insights of this—to us—surprising phenomenon.

References

1. J.H. Dshalalow (1997). Queueing systems with state dependent parameters. In *Frontiers in Queueing,* edited by J.H. Dshalalow. CRC Press, New York.
2. S. Nishimura and Y. Jiang (1994). An *M/G/1* vacation model with two service modes. *Probability in the Engineering and Informational Sciences,* 9, 355–374.
3. R.D. Nobel and H.C. Tijms (1999). Optimal control for a $M^X/G/1$ queue with two service modes. In *European Journal of Operational Research*, 113, 610–619.
4. R.D. Nobel (1998). A regenerative approach for an $M^X/G/1$ queue with two service modes. *Automatic Control and Computer Science,* 1, 3–14 (in Russian).
5. G.M. Koole (1998). Structural results for the control of queueing systems using event-based dynamic programming. In *Queueing Systems and its Applications*, 30, 332–339.
6. S. Stidham and R.R. Weber (1993). A survey of Markov decision models for control of networks of queues. *Queueing Systems and its Applications*, 13, 291–314.
7. M.Y. Kitaev and V.V. Rykov (1995). *Controlled queueing systems.* CRC Press, New York.
8. L.I. Sennot (1998). *Stochastic dynamic programming and the control of queueing systems.* Wiley, New York.
9. R. Cavazos-Cadena and L.I. Sennot (1992). Comparing recent assumptions for the existence of average cost optimal stationary policies. *Operations Research Letters,* 11, 33–37.
10. R.D. Nobel and M. van der Heeden (2000). A lost-sales production/inventory model with two discrete production modes. *Stochastic Models*, 16, 453–478
11. M.L. Puterman (1994). *Markov decision processes*. Wiley, New York.
12. H.C. Tijms (1994). *Stochastic models: an algorithmic approach,* Wiley, New York.

Part II
Computational Intelligence

A Novel Approach to Gene Selection of Leukemia Dataset Using Different Clustering Methods

P. Prasath, K. Perumal, K. Thangavel and R. Manavalan

Abstract Gene datasets from microarray comprise large number of genes. Clustering is a widely used approach for grouping similar kind of genes. The main objective of this paper is to identify the optimal subset of genes from the leukemia dataset in order to classify the leukemia cancer. Different clustering approaches such as *K*-means (KM) clustering, fuzzy *C*-means (FCM) clustering, and modified *K*-means (MKM) clustering have been adopted in this research. The clusters obtained from these methods are further clustered using *K*-means sample-wise (by omitting class values), and the results are compared with ground truth value to evaluate the performance of the different clustering methods. The highly correlated genes are selected from the cluster that produces more accurate classification results. It is observed that the FCM (gene-wise clustering) with *K*-means (sample-wise clustering) produces better accuracy, and the resultant genes have been identified.

Keywords FCM · MKM · *K*-means · Leukemia · Microarray

P. Prasath (✉) · K. Perumal
Department of Biotechnology, Periyar University, Salem 636 011, India
e-mail: prasathbiotech@rediffmail.com

K. Perumal
e-mail: perumaldr@gmail.com

K. Thangavel
Department of Computer Science, Periyar University, Salem 636 011, India
e-mail: drktvelu@yahoo.co.in

R. Manavalan
Department of Computer Science, K. S. Rangasamy College of Arts and Science, Thiruchengode, India
e-mail: manavalan_r@rediffmail.com

G. S. S. Krishnan et al. (eds.), *Computational Intelligence, Cyber Security and Computational Models*, Advances in Intelligent Systems and Computing 246,
DOI: 10.1007/978-81-322-1680-3_7,

1 Introduction

Gene expression data can be obtained by high-throughput technologies such as microarray and oligonucleotide chips under various experimental conditions, at different developmental stages. This technique promises to allow for the detection of networks of correlated genes, which are characteristic of phenomena such as diseases. However, classification of samples according to phenotypes or other criteria is not necessarily precise; therefore, it is desirable to use unsupervised methods to classify samples according to gene expression similarity and to detect networks of correlated genes that discriminate those sample classes [1].

The gene expression data are usually organized in a matrix of *n* rows and *m* columns, which is known as a gene expression profile. Due to the large amount of gene expression data available on various cancerous samples, it is important to construct classifiers that have high predictive accuracy in classifying cancerous samples based on their gene expression profiles [2]. Microarrays contain precisely positioned DNA probes that are designed to specifically monitor the expression of genes in parallel. Data mining often utilizes mathematic techniques that are traditionally used to identify patterns in complex data. Here, the unsupervised benchmark *K*-means clustering method is adopted from [7] for comparative analysis.

The rest of the paper is organized as follows: Sect. 2 describes the FCM clustering method, Sect. 3 deals with MKM clustering, Sect. 4 proposes novel proposed approach to gene selection, Sect. 5 provides experimental environment, Sect. 6 includes experimental results, and Sect. 7 concludes this paper with direction for further research.

2 Fuzzy *C*-Means Clustering

In FCM clustering method, an object can simultaneously be a member of multiple clusters. The objective function, which is minimized iteratively, is a weighted within-group sum of distances. The weight is computed by multiplying the squared distances with membership values. After computing the membership values for all calibration objects, the cluster centers are described by prototypes, which are fuzzy weighted means. This fuzzy clustering method allows intermediate logical assignments whereby genes are placed into multiple groups by assigning a membership value for each group that is compared between 0 (not in group) and 1 (completely in group). The use of membership values has the advantage of allowing a gene or sample to belong to multiple clusters, which may better reflect the underlying biology [4].

3 Modified *K*-Means Clustering Algorithm

This algorithm calculates the cluster centers that are quite close to the desired cluster centers. It first divides the dataset into K subsets according to some rule associated with data space patterns and then chooses cluster centers for each subset [5].

4 Proposed Approach

In this paper, the gene datasets have been clustered using K-means, modified K-means, and fuzzy C-means by setting $K = 5$, $K = 10$, and $K = 15$. The clusters, which are obtained using the above methods, are further clustered using K-means clustering by taking $K = 2$. Hence, it is a novel approach to gene selection using clustering methods.

5 Experimental Environment

The description of leukemia cancer dataset is as follows [6]: This has 7,129 genes with 34 samples and consists of 2 classes: acute lymphoblastic leukemia (ALL) and acute myeloid leukemia (AML). They are various types of cancer, and each of them has different characteristics. Each patient is represented as one row. First column is the patient number in the dataset, columns 2 to 34 denote the gene expression values corresponding to each patient, and 7,130th column indicates the type of cancer (ALL, AML) that each patient is classified. In order to ease the algebraic manipulations of data, the dataset can also be represented as a real two-dimensional matrix S of size $7{,}129 \times 34$; the entry s_{ij} of S measures the expression of the jth gene of the ith patient. Each patient is determined by a sequence of 34 real numbers, each measuring the relative expression of the corresponding gene [3].

6 Computational Results

The K value is arbitrarily fixed as 5, 10, and 15, FCM clustering is performed, and the results are provided in Tables 1, 2, and 3, respectively. The best results are indicated in bold letters. Similarly, the K value is arbitrarily fixed as 5, 10, and 15, MKM clustering is performed, and the results of the clusters are provided in Tables 4, 5, and 6, respectively. The best results are indicated in bold letters.

Table 1 Experimental results for $K = 5$

Run(s)	K	FCM clustering		
		Sensitivity	Specificity	Accuracy
1	5	0.94	0.76	0.85
2	5	0.76	0.69	0.74
3	5	0.76	0.69	0.74
4	5	0.90	0.92	0.91
5	5	0.94	0.76	0.85
6	5	0.59	0.42	0.53
7	5	0.90	0.92	0.91
8	5	0.90	0.92	0.91
9	5	0.76	0.69	0.74
10	5	0.90	0.92	0.91

Table 2 Experimental results for $K = 10$

Run(s)	K	FCM clustering		
		Sensitivity	Specificity	Accuracy
1	10	0.84	0.73	0.79
2	**10**	**0.95**	**0.87**	**0.91**
3	10	0.62	0.46	0.56
4	10	0.86	0.85	0.85
5	10	0.95	0.93	0.94
6	10	0.94	0.72	0.82
7	10	0.86	0.85	0.85
8	10	0.94	0.72	0.82
9	10	0.86	0.85	0.85
10	10	0.86	0.85	0.85

Table 3 Experimental results for $K = 15$

Run(s)	K	FCM clustering		
		Sensitivity	Specificity	Accuracy
1	15	0.90	0.92	0.91
2	15	0.93	0.65	0.76
3	15	0.75	0.56	0.65
4	**15**	**0.81**	**0.77**	**0.79**
5	15	0.95	0.87	0.91
6	15	0.95	0.93	0.94
7	**15**	**0.95**	**0.93**	**0.94**
8	**15**	**0.83**	**1.00**	**0.88**
9	15	0.73	0.53	0.62
10	**15**	**0.86**	**0.85**	**0.85**

Table 4 Experimental results for $K = 5$

Cluster(s)	*K*	MKM clustering		
		Sensitivity	Specificity	Accuracy
1	5	0.36	0.30	0.32
2		*0.87*	*1.00*	*0.91*
3		0.62	0.60	0.62
4		0.66	0.80	0.68
5		0.61	0.45	0.56

Table 5 Experimental results for $K = 10$

Cluster(s)	*K*	MKM clustering		
		Sensitivity	Specificity	Accuracy
1	10	0.36	0.30	0.32
2		0.79	0.67	0.74
3		*0.91*	*1.00*	*0.94*
4		0.55	0.20	0.50
5		0.56	0.33	0.50
6		0.58	0.38	0.53
7		0.65	0.63	0.65
8		0.57	0.33	0.53
9		0.50	0.33	0.41
10		0.67	0.54	0.62

Table 6 Experimental results for $K = 15$

Cluster(s)	*K*	MKM clustering		
		Sensitivity	Specificity	Accuracy
1	15	0.36	0.30	0.32
2		0.79	0.67	0.74
3		**0.83**	**1.00**	**0.88**
4		0.77	0.52	0.62
5		0.55	0.20	0.50
6		0.67	0.71	0.68
7		0.52	0.27	0.44
8		0.60	0.44	0.56
9		0.72	0.78	0.74
10		0.68	0.83	0.71
11		0.62	0.60	0.62
12		0.59	0.40	0.56
13		0.67	0.54	0.62
14		0.57	0.38	0.50
15		0.50	0.33	0.41

Table 7 Relative performance measure of experimental analysis

Run	*K* value	Sensitivity	Specificity	Accuracy	Number of gene selected
1	5	0.94	0.76	0.85	75
4	5	0.90	0.92	0.91	203
5	5	0.94	0.76	0.85	42
7	5	0.90	0.92	0.91	75
8	5	0.90	0.92	0.91	75
10	5	0.90	0.92	0.91	42
2	10	0.94	0.72	0.82	23
5	10	0.95	0.93	0.94	219
6	10	0.94	0.72	0.82	37
8	10	0.94	0.72	0.82	37
10	10	0.86	0.85	0.85	34
1	15	0.90	0.92	0.91	104
5	15	0.95	0.87	0.91	20
6	15	0.95	0.93	0.94	19
7	15	0.95	0.93	0.94	189
8	15	0.83	1.00	0.88	29

Table 8 Significant genes selected

FCM algorithm	42, 43, 291, 1032, 1370, 1930, 2159, 2290, 2727, 2797, 2801, 3314, 4624, 5105, 5308, 5716, 6168, 6184, 6209

The accuracy of 91 % is achieved when $\boldsymbol{K} = \mathbf{5}$; 203 genes are selected in Run 4, and the same is achieved for selecting 75, 75, and 42 genes in Runs 7, 8, and 10, respectively, in FCM clustering. The accuracy of 94 % is achieved when $\boldsymbol{K} = \mathbf{10}$; 219 genes are selected in Run 5, and the accuracy of 85 % is achieved for 34 genes in Run 10. Also, an accuracy of 82 % is achieved for 23, 37, and 37 genes in Runs 2, 6, and 8, respectively, in FCM clustering. The accuracy of 91 % is achieved when $\boldsymbol{K} = \mathbf{15}$; 104 genes are selected in Run 1, and 20 genes are selected in Run 5 in FCM clustering. The accuracy of 94 % is achieved when $K = 15$; 19 genes are selected in Run 6, and 189 genes are selected in Run 7. The accuracy of 88 % is achieved for the same value of K; 29 genes are selected in Run 10. The results with best accuracy obtained for $K = 5$, $K = 10$, and $K = 15$ using FCM clusters are given in Table 7.

The results of the significant genes selected by FCM algorithm are given in Table 8.

7 Conclusion

In this paper, clustering-based gene selection methods have been proposed and analyzed. It is a novel approach, since genes were selected through sequence processing of clustering approaches. The FCM and MKM clustering algorithms have been applied for different values of *K*. Again KM clustering algorithm has been performed for all the clusters produced by FCM and MKM methods. The highly correlated genes were selected from the clusters of the high accurate classification results. Out of 7,129 genes, 19 genes were selected by the proposed novel gene selection method, and it is enough to consider only such 19 genes to predict the leukemia cancer. It was observed that FCM clustering method outperformed.

The further research direction is to identify a single gene for diagnosing the leukemia cancer.

Acknowledgments The third author gratefully acknowledges the UGC, New Delhi, for partial financial assistance under UGC-SAP (DRS) Grant No. F3-50/2011.

References

1. Stanislav Busygin, Gerrit Jacobsen, and Ewald Kramer. Double conjugated clustering applied to leukemia microarray data. In Proceedings of the 2nd SIAM International Conference on Data Mining, Workshop on Clustering High Dimensional Data, 2002.
2. Aik Choon Tan and David Gilbert, Ensemble machine learning on gene expression data for cancer classification: Applied Bioinformatics 2003:2 (3 Suppl) S75–S83.
3. Cherie H. Dunphy (2006) Gene Expression Profiling Data in Lymphoma and Leukemia: Review of the Literature and Extrapolation of Pertinent Clinical Applications. Archives of Pathology & Laboratory Medicine: April 2006, Vol. 130, No. 4, pp. 483–520.
4. Yoo CK, Vanrolleghem PA. Interpreting patterns and analysis of acute leukemia gene expression data by multivariate statistical analysis. In: Barbosa Povoa A, Matos H, editors. Computer-Aided Chemical Engineering. Elsevier Science; 2004. pp. 1165–70.
5. Wei Li, Modified K-means clustering algorithm, Congress on Image & Signal Processing, IEEE, 2008, pp. 618–621.
6. T.R. Golub et al. Molecular classification of cancer: class discovery and class prediction by gene expression monitoring, Science, 1999, Vol. 286, pp. 531–537.
7. Palanisamy, P.; Perumal; Thangavel, K.; Manavalan, R., "A novel approach to select significant genes of leukemia cancer data using K-Means clustering," Pattern Recognition, Informatics and Medical Engineering (PRIME), 2013 International Conference on, pp. 104, 108, 21–22 Feb. 2013.

A Study on Enhancing Network Stability in VANET with Energy Conservation of Roadside Infrastructures Using LPMA Agent

T. Karthikeyan and N. Sudha Bhuvaneswari

Abstract Designing an intelligent transportation system (ITS) for vehicular ad hoc networks (VANETs) with seamless connectivity and appropriate energy conservation of roadside units (RSU) that takes part in communication between vehicular nodes is a challenging task because of the high mobility or dynamic nature of these networks. The nodes in the VANET whether stationary or moving are limited in terms of storage capacity, reliability, and energy. The main concern about these networks is whether the existing protocols meet the demands of feasibility of this high mobility VANETs and make them less dynamic or stable. In this paper, we discuss a clustering-based architecture designed with mobile agents and a combination of MH-LEACH and power-efficient gathering in sensor information systems (PEGASIS) protocols for building stability in VANET for seamless communication along with energy conservation of other RSU to improve quality of service.

Keywords MH-LEACH · PEGASIS · LPMA · MALP · EEC

1 Introduction

There is a great impact of vehicular networks and communication system in the very near future with the ultimate goals of improvising safety, efficiency, and ease of travel. This demand is because of the time-critical application of vehicular

T. Karthikeyan
Department of Computer Science, PSG College of Arts and Science, Coimbatore, India
e-mail: t.karthikeyan.gasc@gmail.com

N. Sudha Bhuvaneswari (✉)
School of IT and Science, Dr. G.R. Damodaran College of Science, Coimbatore, India
e-mail: sudhanarayan03@yahoo.com

G. S. S. Krishnan et al. (eds.), *Computational Intelligence, Cyber Security and Computational Models*, Advances in Intelligent Systems and Computing 246,
DOI: 10.1007/978-81-322-1680-3_8,

networks that necessitates direct communication between nodes than relaying on a centralized architecture [3].

Vehicles in a VANET have many constraints with which they move around communicating with each other through wireless links. This network does not depend on any special hardware setup, and the entire network stability is managed by each node in the network [6]. Moreover, to build an optimized communication system within the network and to maintain the scalability and stability of this network researches show that hierarchical clustering is the best approach to offer efficient solution [4]. Clustering is necessary to achieve basic network performance. The grouping process of nodes in a dense population is known as clustering [1]. This clustering process should be dynamic and should be periodically updated to reflect changes in its environment and the time taken to form the cluster should be minimal [2].

It is a fact that in VANET reconfiguration is often required in cluster formation. Therefore, a good VANET clustering algorithm should seek to regulate rather than eliminate cluster changes. The algorithm we apply should provide an optimal and stable solution; otherwise, the overhead associated with reconfiguration and information exchange will result in high computation cost [6].

This paper proposes hybrid architecture for building stability in VANET for seamless communication along with energy conservation in VANET using mobile agent technology with MH-LEACH and power-efficient gathering in sensor information systems (PEGASIS) called the MALP approach.

2 Problem Formulation for MALP Approach

In VANET, the nodes have limited processing power, communication bandwidth, and storage space that demands efficient resource utilization. Clustering offers the necessary features like stability and energy-saving attributes for roadside units (RSU) and devices fed by limited energy sources such as sensors, smart phones, and traffic signs. Clustering is done using the protocols MH-LEACH and PEGASIS. MH-LEACH is the protocol used for finding the optimal number of clusters to save energy. MH-LEACH stands for multihop–low energy adaptive clustering hierarchy [8]. This protocol uses multihop approach for optimal path selection by providing energy efficiency [5]. The PEGASIS routing algorithm supports MH-LEACH by forming a chain of headers using greedy algorithm, so that the movement of messages from cluster to base station (BS) is optimized by chaining the clusters and exchanging the message to BS from the nearest cluster head (CH).

The MALP approach proposed in this paper mainly focus on cluster formation and message transfer using efficient energy conservation (EEC) clustering algorithm.

2.1 Cluster Formation

In clustering of VANET, moving nodes are divided into different groups and stand together in one single cluster based on certain rules [4]. The size of each cluster represents the number of vehicles in a street or highway within a communication range (CR) and velocity to form a cluster. The vehicles that form a cluster are identified using distance formula and estimating their velocity. The nodes are said to be connected when they fall within a CR [7], and this is estimated using distance formula between two nodes ($x1$, $y1$) and ($x2$, $y2$).

$$D = \sqrt{(x1 - x2)^2} + \sqrt{(y1 - y2)^2} \tag{1}$$

The above formula calculates the distance of two moving vehicular nodes, and it is compared with the predetermined CR to identify cluster elements [2].

```
//Estimate the Communication Range(CR)
   //Estimate the Velocity Limit(VL)
   While (Vehicle Node Status==Active)
   {//Calculate relative distance between Nodes using equation (1);
   Find the relative velocity among nodes;
       If (D<=CR && velocity==VL)
         {//Create cluster;
        Bandwidth();
        NoOfAvailableNodes();          }
        service_announcement() }
```

2.2 Cluster Head Selection

In VANET, a vehicle qualified to become a CH depends on consumption of resources and computational time. Therefore, it is of great importance that the cluster size should be neither too large nor too small. The first option to choose a head is selection of RSU since like other nodes frequent change of CHs are not required [7]. If RSU is not present, then it is necessary to choose the CH using MH-LEACH energy tolerance method. In EEC algorithm, the LPMA agent will perform CH selection based on MH-LEACH energy computations. LPMA agent acquires the initial energy and residual energy of cluster node's (CN) and node with the highest energy E is computed:

$$E = E_r \div E_i \times \mathrm{CH_p} \tag{2}$$

$$E^1 = \sum E_r \div \sum E_i \times \mathrm{CH_p} \tag{3}$$

$$\mathrm{CH_p} = \mathrm{NH_{net}} \div \mathrm{NN_{net}} \tag{4}$$

where E is the energy and E^1 is the energy of current CH, E_r is the residual energy, E_i is the initial energy, CH_p is the proportion of number of CH nodes (NH_{net}) to the number of all nodes in the network (NN_{net}). Here, CH_p is used as constant to calculate the tolerance limit of the header, and the tolerance limit is assumed in this work to be 5 %, which is the default limit used by MH-LEACH.

```
//Compute the cluster head tolerance limit CHp=NHnet/NNnet
  //calculate the energy of the new node  E=Er/Ei *CHp
  //calculate the energy of the existing node E1=∑ Er/∑ Ei *CHp
  While (Vehicle Node Status==Active)
  {    If (E > E1)
          {ClusterHead(CH)=E;
           ClusterNode(CN)= E1;         } }
```

2.3 Cluster Head Chaining

After identification of the CHs, the heads are chained together for exchange of messages. The algorithm used for this chaining process is greedy algorithm of PEGASIS [9]. Here, PEGASIS is used in coordination with MH-LEACH to chain all the headers and help in energy-efficient data transfer.

Here, all the member nodes in each cluster transmit its data to the header, all the headers in turn send the data to the leader node along the chain, and it is work of the leader node to handover the message to the BS. The leader node of the chain is dynamically selected based on the order of the residual energy to avoid one node getting its energy fully exhausted.

```
//LastClusterHead sends token to NextClusterHead
While (Vehicle Node Status==Active)
{ BS receives token from ClusterHead;
  BS broadcast msg "Chain Completion";
//Cluster head collects the entire message from cluster nodes
    ClusterHead=msg(ClusterNodes)
//All the Cluster Heads transmit the message to the cluster leader
    ClusterLeader=∑ClusterHead(Data);
//Cluster Leader collects the message and send it to the base station
    BaseStation=∑ClusterLeader(Data);}
```

2.4 Information Retrieval Using LPMA Agent

The LPMA mobile agent works along with MH-LEACH and PEGASIS for efficient information retrieval in clusters. The LPMA agent migrates in the network for getting the information from the appropriate node via the cluster chain formed using PEGASIS algorithm. If the node searches the information in all the clusters

C1 to Cn retrieve information from the CH, if CH is not found, then find the CH and retrieve information after service announcement.

```
//Search information from Node[j] for handoff
If node[j] € Cluster Ci
{For all existing Clusters C1 to Cn
If (ClusterHead not found)
{//find the cluster head and give the service announcement
ClusterHead();
ServiceAnnouncement();}
Else
{//retrieve information from all nodes with LPMA agent
While (ClusterHead[i] is active)
For all ClusterHead[i]
RetriveInfo=InfoNode[j] [ClusterHead[i]]
ServiceAnnouncement();}}
```

2.5 *Mirroring Effects*

A problem in information retrieval is retrieval of same message from two different clusters. The message has been received by both the clusters from the same node due to the dynamic topology of the network will cause a mirroring effect, and this problem can be overcome by retrieving the latest information by finding the cluster in which the node currently resides.

```
While retrieving information
{If (more than one cluster has the info [Node[j]])
For all existing cluster C1 to Cn
Find ();
If (Ci have info [Node[j]]) //Retrieve information from the Current node
{ Find (Ci)
{For all existing clusters C1 to Cn
If (Ci contains Node[j] )
Retrieve information ();}}}
```

3 Performance Evaluations

The performance of the MALP approach is compared with two other known clustering algorithms: position-based clustering and lowest ID algorithm. The simulation is performed based on the constraints transmission delay, delivery ratio, and energy consumption. From the study, it is inferred that out of the existing approaches, MALP offers better performance (Figs. 1, 2, 3).

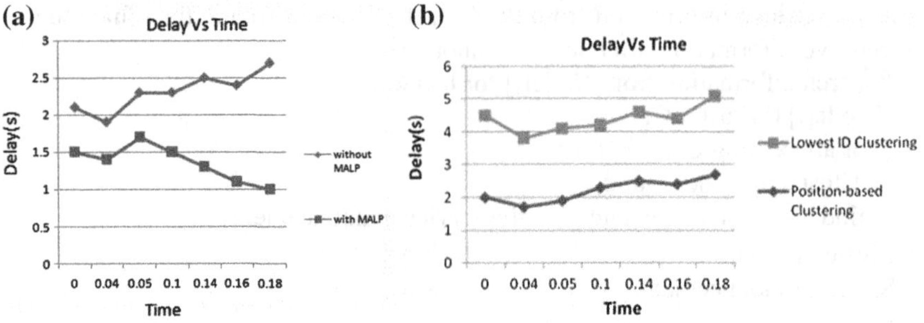

Fig. 1 **a** Delay versus time using MALP approach. **b** Delay versus time using other approaches

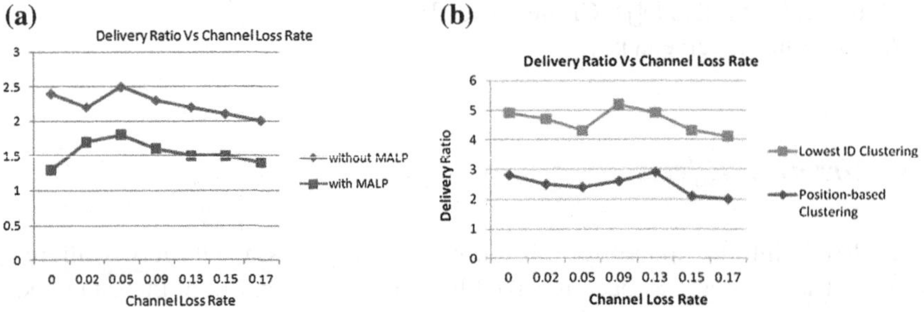

Fig. 2 **a** Delivery ratio versus channel loss rate using MALP approach. **b** Delivery ratio versus channel loss rate using other approaches

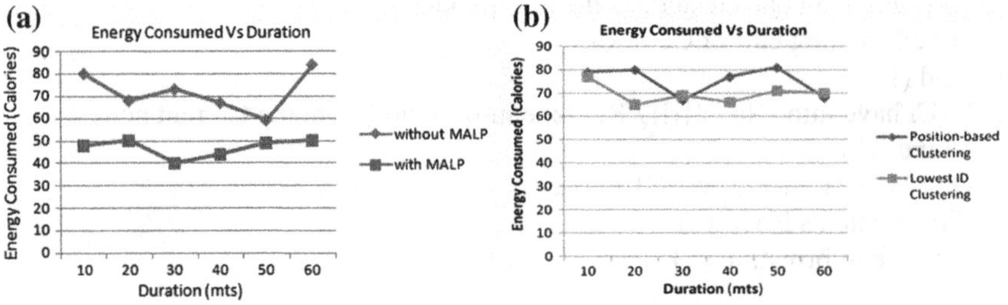

Fig. 3 **a** Energy consumed versus duration using MALP approach. **b** Energy consumed versus duration using other approaches

4 Conclusion

This paper proposes clustering approach to build a somewhat stable vehicular ad hoc network for establishing seamless communication between the nodes along with LPMA agent. It is a hybrid approach using agent, MH-LEACH and PEGASIS protocol that is also capable of building an energy-efficient architecture for roadside sensor units that are energy constrained in VANET. In future, this work can be extended to included data aggregation and data dissemination for effective message passing between clusters and BS.

References

1. Marco Gramaglia, Ignacio Soto, Carlos J. Bernardos, Maria Calderon, "Overhearing-Assisted Optimization of Address Autoconfiguration in Position-Aware VANETs", IEEE Transactions on Vehicular Technology, Vol. 60, No. 7, (2011).
2. Peng Fan, James G. Haran, John Dillenburg, Peter C. Nelson, "Cluster-Based Framework in Vehicular Ad-Hoc Networks", ADHOC-NOW, Springer Verlag" (2005) pp. 32–42.
3. Mehrnaz Mottahedi1, Sam Jabbehdari1, Sepideh Adabi1, "IBCAV: Intelligent Based Clustering Algorithm in VANET", IJCSI International Journal of Comp.Sc Issues, Vol. 10, Issue 1, No 2 (2013).
4. Anshu Garg1,Sanjeev Bansal,"A Handoff Mechanism in VANET Using Mobile Agent", International Journal of Comp.Science & Management Studies, Special Issue of Vol. 12 (2012).
5. Johansson T, Carr-Motyckova, L.,"Bandwidth-constrained Clustering in Networks" (2004).
6. Vinay Kumar, Raghuvansi, S.Tiwari, "LEACH and its derivatives in Wireless Sensor Networks(WSNs):A Survey", Proceedings of the International Conference on Communication and Computational Intelligence (2010).
7. Mo Xiaoyan, "Study and design on cluster routing protocols of wireless sensor networks", Dissertation, Hang Zhou, Zhe Jiang University (2006).
8. Prashant Sangulagi, Mallikarjun Sarsamba, Mallikarjun Talwar, Vijay Katgi, "Recognition and Elimination of Malicious Nodes in Vehicular Ad hoc Networks (VANET's)", Indian Journal of Computer Science and Engineering, Vol. 4 No.1 (2013).
9. Young Han Lee, Kyoung Oh Lee, Hyun Jun Lee, Aries Kusdaryono," CBERP:Cluster Based Energy Efficient Routing Protocol for Wireless Sensor Network", Proceedings of the 12th International Conference on VLSI and Signal Processing (2010).

An Intuitionistic Fuzzy Approach to Fuzzy Clustering of Numerical Dataset

N. Karthikeyani Visalakshi, S. Parvathavarthini and K. Thangavel

Abstract Fuzzy c-means (FCM) clustering is one of the most widely used fuzzy clustering algorithms. However, the main disadvantage of this algorithm is its sensitivity to noise and outliers. Intuitionistic fuzzy set is a suitable tool to cope with imperfectly defined facts and data, as well as with imprecise knowledge. So far, there exists a little investigation on FCM algorithm for clustering intuitionistic fuzzy data. This paper focuses mainly on two aspects. Firstly, it proposes an intuitionistic fuzzy representation (IFR) scheme for numerical dataset and applies the modified FCM clustering for clustering intuitionistic fuzzy (IF) data and comparing results with that of crisp and fuzzy data. Secondly, in clustering of IF data, different IF similarity measures are studied and a comparative analysis is carried out on the results. The experiments are conducted for numerical datasets of UCI machine learning data repository.

Keywords Clustering · Fuzzy c-means · Intuitionistic fuzzy data · Intuitionistic fuzzy similarity measure

N. Karthikeyani Visalakshi (✉) · S. Parvathavarthini
Kongu Engineering College, Perundurai, Erode, Tamil Nadu, India
e-mail: karthichitru@yahoo.co.in

S. Parvathavarthini
e-mail: varthinis@gmail.com

K. Thangavel
Periyar University, Salem, Tamil Nadu, India
e-mail: drktvelu@yahoo.com

G. S. S. Krishnan et al. (eds.), *Computational Intelligence, Cyber Security and Computational Models*, Advances in Intelligent Systems and Computing 246,
DOI: 10.1007/978-81-322-1680-3_9, © Springer India 2014

1 Introduction

Clustering algorithms seek to organize a set of objects into clusters such that objects within a given cluster have a high degree of similarity, whereas objects belonging to different clusters have a high degree of dissimilarity. Clusters can be hard or fuzzy in nature based on whether each data object has to be assigned exclusively to one cluster or allowing each object to be assigned to every cluster with an associated membership value.

The Fuzzy C-Means (FCM) algorithm is sensitive to the presence of noise and outliers in data [1]. To enhance robustness of FCM, different researchers proposed different methodologies [1–3]. Intuitionistic fuzzy sets (IFSs) [4] are generalized fuzzy sets, which use the hesitancy originating from imprecise information. Pelekis et al. [5] introduced an Intuitionistic Fuzzy Representation (IFR) scheme for color images and an Intuitionistic Fuzzy (IF) similarity measure through which a new variant of FCM algorithm is derived. But this cannot be directly used for clustering numerical datasets. Hence, robust fuzzy clustering is proposed in this paper to make FCM algorithm as noise insensitive, by dealing with IF data. Real data are converted into IFR, before clustering, in order to achieve the benefit of IFSs in fuzzy clustering. A comparative study is made on fuzzy clustering of crisp, fuzzy, and IF data, and the performance of IF clustering is measured using four different IF similarity measures.

The rest of this paper is organized as follows: Sect. 2 provides discussions on IFS and IF similarity measures. Section 3 reviews the related works. The proposed method of clustering numerical dataset is described in Sect. 4. Section 5 summarizes the experimental analysis performed with benchmark datasets. Section 6 concludes the paper.

2 Background

2.1 Intuitionistic Fuzzy Sets

Fuzzy sets are designed to manipulate data and information possessing non-statistical uncertainties. Since Zadeh [6] introduced the concept of fuzzy sets, various notions of high-order fuzzy sets have been proposed. Among them, IFSs, introduced by Atanassov [4], can present the degrees of membership and non-membership with a degree of hesitancy.

Definition 2.1 An IFS A is an object of the form:

$$A = \{\langle x, \mu_A(x), v_A(x)\rangle | x \in E\} \tag{1}$$

where $\mu_A : E \rightarrow [0,1]$ and $v_A : E \rightarrow [0,1]$ define the degree of membership and non-membership, respectively, of the element $x \in E$ to the set $A \subset E$. For every element $x \in E$, it holds that $0 \leq \mu_A(x) + \nu_A(x) \leq 1$. If A represents a fuzzy set, for every $x \in E$, if $\nu_A(x) = 1 - \mu_A(x)$ and

$$\pi_A(x) = 1 - \mu_A(x) - \nu_A(x) \tag{2}$$

represents the degree of hesitancy of the element $x \in E$ to the set $A \subset E$.

2.2 Intuitionistic Fuzzy Similarity Measures

Similarity measure determines the degree of similarity between two objects. Many of them are proposed by different researchers [5, 7, 8] and are applied in a wide range of applications. In the following, four IFS similarity measures used in this work for comparative analysis are reviewed.

Pelekis [5] proposed a similarity measure S_1between the IFSs A and B as

$$S_1(A,B) = \frac{S'(\mu_A(x_i), \mu_B(x_i)) + S'(\nu_A(x_i), \nu_B(x_i))}{2} \tag{3}$$

$$\text{where } S'(A', B') = \begin{cases} \frac{\sum_{i=1}^{n} \min(A'(x_i), B'(x_i))}{\sum_{i=1}^{n} \max(A'(x_i), B'(x_i))}, & A' \cup B' \neq \Phi \\ 1, & A' \cup B' = \Phi \end{cases} \tag{4}$$

where Φ is a fuzzy set for which the membership function is zero for all elements. This measure uses the aggregation of the minimum and maximum membership values in combination with those of the non-membership values.

Hung and Yang [8] extended some similarity measures of FS to IFSs,

$$S_2(A,B) = \frac{\sum_{i=1}^{n} (\min(\mu_A(x_i), \mu_B(x_i)) + \min(\nu_A(x_i), \nu_B(x_i)))}{\sum_{i=1}^{n} (\max(\mu_A(x_i), \mu_B(x_i)) + \max(\nu_A(x_i), \nu_B(x_i)))} \tag{5}$$

It focuses on the ratio of the aggregation of minimum of membership and non-membership values to the aggregation of maximum of membership and non-membership values. They also proposed a new similarity measure S_3 as in Eq. (6), which adopts exponential operation to the Hamming distance between IFSs A and B.

$$S_3(A,B) = 1 - \frac{1 - \exp(-\frac{1}{2}\sum_{i=1}^{n} |\mu_A(x_i) - \mu_B(x_i)| + |\nu_A(x_i) - \nu_B(x_i)|}{1 - \exp(-n)} \tag{6}$$

The similarity measure S_4 as in Eq. (7) considers the hesitancy values of IFSs in computing similarity between IFSs A and B, based on normalized Hamming distance [7].

$$S_4(A,B) = \frac{1}{2n}\sum_{i=1}^{n} |\mu_A(x_i) - \mu_B(x_i)| + |\nu_A(x_i) - \nu_B(x_i)| + |\pi_A(x_i) - \pi_B(x_i)| \tag{7}$$

3 Related Works

There are different variants of FCM clustering in the literature.

D'Urso and Giordani [9] proposed a FCM clustering model for LR-type fuzzy data, based on a weighted dissimilarity measure for comparing fuzzy data objects, using center distance and spread distance. Leski [1] introduced a new ε-insensitive Fuzzy C-Means (εFCM) clustering algorithm in order to make FCM as noise insensitive. In [10], the fuzzy clustering based on IF relation is discussed. The clustering algorithm uses similarity-relation matrix, obtained by n-step procedure based on max-t and min-s compositions.

Bannerji et al. [2] proposed robust fuzzy clustering methodologies to deal with noise and outliers, by means of mega-cluster concept and robust error estimator. In [5, 11], Pelekis et al. clustered IF representation of images and proposed a clustering approach based on the FCM using a novel similarity metric defined over IFSs, which is more noise tolerant and efficient as compared with the conventional FCM clustering of both crisp and fuzzy image representations.

4 Intuitionistic Fuzzy Approach to Fuzzy Clustering

The proposed methodology for the fuzzy clustering using IFSs involves two stages, viz., intuitionistic fuzzification to convert the real scalar values into IF values and using modified FCM algorithm based on IF similarity measure to cluster IF data. Additionally, four IF similarity measures are also used for comparative analysis.

4.1 Intuitionistic Fuzzification

Following [12], a new procedure for intuitionistic fuzzification of numerical dataset is derived where the crisp dataset is first transferred to fuzzy domain and sequentially into the IF domain, where the clustering is performed.

Let X be the dataset of N objects, and each object contains d features. The proposed IF data clustering requires that each data element x_{ij} belongs to an IFS X' by a degree $\mu_i(x_j)$ and does not belong to X' by a degree $v_i(x_j)$, where i and j represent objects and features of the dataset, respectively.

A membership function $\overline{\mu_i}(x_j)$ for intermediate fuzzy representation is defined by

$$\overline{\mu_i}(x_j) = \frac{x_{\mathrm{ij}} - \min(x_j)}{\max(x_j) - \min(x_j)} \quad \text{where} \quad i = 1, 2, \ldots, N \,\text{and}\, j = 1, 2, \ldots, d \qquad (8)$$

The intuitionistic fuzzification based on the family of parametric membership and non-membership function, used for clustering, is defined, respectively, by

$$\mu_i(x_j;\ \lambda) = 1 - (1 - \overline{\mu_i}(x_j))^{\lambda} \tag{9}$$

and

$$\nu_i(x_j;\ \lambda) = (1 - \overline{\mu_i}(x_j))^{\lambda(\lambda+1)} \quad \text{where} \quad \lambda \in [0, 1] \tag{10}$$

The intuitionistic fuzzification converts crisp dataset $X(x_{\mathrm{ij}})$ into IF dataset $X'(x_{\mathrm{ij}}, \mu_i(x_j), \nu_i(x_j))$.

4.2 Fuzzy Clustering of IF Data

In this stage, Pelekis's modified FCM [5] is applied to cluster IF data. Instead of euclidean distance in conventional FCM, the modified FCM applies IF similarity measure. The modified FCM algorithm is as follows:

Step 1. Determine initial centroids by selecting c random IF objects.
Step 2. Calculate the membership matrix U_{ij}, using

$$\underset{\substack{1 \le i \le c \\ 1 \le j \le N}}{\forall} \quad U_{\mathrm{ij}} = \begin{cases} \dfrac{\left(S_1(x_j - C_i)\right)^{\frac{1}{1-m}}}{\sum\limits_{l=1}^{c}\left(S_1(x_j - C_l)\right)^{\frac{1}{1-m}}}, & I_j = \phi \\ \begin{cases} 0, & i \notin I_j \\ \sum\limits_{i \in I_j} U_{\mathrm{ij}} = 1, & i \in I_j,\ I_j \neq \phi \end{cases} \end{cases} \tag{11}$$

$$\text{where} \quad \underset{\forall 1 \le j \le N}{I_j} = \left\{ i \middle| 1 \le i \le c;\ S_1(x_j, C_i) = 0 \right\}$$

Step 3. Update the centroids' matrix C_i using

$$\underset{1 \le i \le c}{\forall} \quad C_i = \frac{\sum\limits_{j=1}^{n} (U_{\mathrm{ij}})^m x_j}{\sum\limits_{j=1}^{n} (U_{\mathrm{ij}})^m} \tag{12}$$

Step 4. Compute membership and non-membership degrees of C_i

Step 5. Repeat step 2 to step 4 until converges.

Initially, c number of centroids are randomly selected from the IF objects, which contain both membership and non-membership values. Next, the membership degree of each object to each cluster U_{ij} is computed using IF similarity measure as in Eq. (11). The centroids are then updated using cluster membership matrix, and corresponding membership and non-membership degrees of centroids C_i are also computed. Repeat the above two steps until convergence.

5 Experimental Analysis

This work explores the role of intuitionistic fuzzification of numerical data and IF similarity measures in the process of FCM clustering. The experimental analysis is carried out with five benchmark datasets in two aspects. First, the results of FCM clustering on crisp, fuzzy, and IF data are compared, and the fuzzification is done using Eq. (8). The λ value is set as 0.95, for the computation of membership and non-membership values using Eqs. (9) and (10). Next, the performance of four different IF similarity measures is evaluated.

Experiments are conducted using the breast cancer, dermatology, image segmentation, satellite image, and wine datasets available in the UCI machine learning data repository [13]. The conventional FCM algorithm is used to cluster crisp and fuzzy data, and the modified FCM using IF similarity measure is used to cluster IF data. Experiments are run 50 times on each dataset, and average values are taken for evaluation.

5.1 Cluster Evaluation Criteria

Here, the performance of fuzzy clustering algorithm is measured in terms of two external validity measures, [12, 14] the Rand index, F-measure and two fuzzy internal validity measures, fuzzy DB (FDB) index, and Xie–Beni (XB) index. The maximum value indicates good performance for Rand index and F-measure. The minimum value indicates the better performance for both FDB and XB indices.

5.2 Comparative Analysis on Crisp, Fuzzy, and IF Data

Two sets of experiments are conducted to evaluate the performance of IF representation.

5.2.1 Hard Cluster Evaluation

The first experiment compares the efficiency using hard cluster validity measures. Table 1 depicts the performance of FCM clustering on crisp, fuzzy, and IF data. It is observed that the number of iterations required for FCM clustering is highly

Table 1 Comparative analysis based on Rand index, F-measure, and number of iterations

S. No.	Dataset	Number of iterations			Rand index			F-measure		
		Crisp data	Fuzzy data	IF data	Crisp data	Fuzzy data	IF data	Crisp data	Fuzzy data	IF data
1	Breast cancer	43	17	**5**	0.521	**0.928**	0.907	0.615	0.653	**0.664**
2	Dermatology	46	12	**6**	0.701	0.672	**0.910**	0.307	0.642	**0.823**
3	Image segmentation	100	97	**14**	0.818	0.635	**0.881**	0.488	0.459	**0.691**
4	Satellite image	100	101	**34**	0.848	0.594	**0.853**	0.673	0.549	**0.713**
5	Wine	55	22	**5**	0.717	0.683	**0.911**	0.674	0.749	**0.934**

reduced, when IF data are used in clustering for all datasets. With Rand index, the performance of clustering IF data dominates that of clustering crisp data, for all datasets, and is outstanding for dermatology and wine datasets. With F-measure, the performance improvement of IF data is higher for all datasets. The F-measure is highly appreciable for wine and dermatology datasets with IF data.

5.2.2 Soft Cluster Evaluation

The second experiment compares the efficiency using fuzzy cluster validity measures. Table 2 depicts the performance of modified FCM clustering on crisp, fuzzy, and IF data in terms of FDB index and XB index. It is proved that performance of clustering IF data is better than other two approaches, for all datasets.

From the analysis, it is observed that the representation of IF data before clustering is more suitable for all numerical datasets. However, satellite image and image segmentation datasets yield better results for fuzzy data representations.

Table 2 Comparative analysis based on FDB index and XB index

S. No.	Dataset	FDB index			XB index		
		Crisp data	Fuzzy data	IF data	Crisp data	Fuzzy data	IF data
1	Breast cancer	0.525	0.468	**0.389**	0.276	0.489	**0.074**
2	Dermatology	1.256	1.115	**1.002**	0.349	0.298	**0.265**
3	Image segmentation	0.898	0.813	**0.761**	1.125	0.265	**0.722**
4	Satellite image	1.562	1.455	**1.232**	0.704	0.381	**0.671**
5	Wine	0.501	0.478	**0.312**	0.126	0.117	**0.101**

Table 3 Comparative analysis of IF similarity measures

S. No.	Dataset	Validity measure	S_1	S_2	S_3	S_4
1	Breast cancer	Rand index	0.907	0.626	0.626	0.630
		F-measure	0.664	0.763	0.763	0.762
2	Dermatology	Rand index	0.910	0.817	0.821	0.201
		F-measure	0.823	0.693	0.712	0.469
3	Image segmentation	Rand index	0.881	0.875	0.872	0.879
		F-measure	0.691	0.678	0.669	0.686
4	Satellite image	Rand index	0.854	0.828	0.831	0.185
		F-measure	0.713	0.665	0.669	0.381
5	Wine	Rand index	0.911	0.872	0.872	0.342
		F-measure	0.934	0.900	0.900	0.570

5.3 Comparative Analysis on Intuitionistic Fuzzy Similarity Measures

This experiment compares the quality of clusters obtained by the modified FCM with IF similarity measure S_1, S_2, S_3, and S_4. Table 3 shows the effect of four similarity measures based on Rand index and F-measure. From the results of the Table 3, it is identified that the similarity measure S_1 is more suitable than the other three measures for all datasets. The impacts of all four similarity measures are almost same, for image segmentation datasets.

6 Conclusion

In this paper, a novel procedure for intuitionistic fuzzification of numerical dataset is proposed and the IF data are applied to the modified FCM clustering algorithm to obtain fuzzy clusters. Experiments are conducted to study the impact of using IF data representation and IF similarity measures in FCM clustering. It can be concluded that the conversion of crisp data into IF data before clustering leads to obtain better quality clusters. It is observed that the IF similarity measure S_1 may be suitable for achieving competent fuzzy clusters. In future, applying optimization algorithm for tuning of parameter λ will help in producing superior quality clusters. Proposed algorithm may be enhanced to produce IF partitions.

References

1. J. Leski, Towards a robust fuzzy clustering. Fuzzy Sets and Systems. 137(2) (2003) 215-233.
2. Banerjee A, Dave R.N, The fuzzy mega-cluster: Robustifying FCM by Scaling down memberships. In: Lecture Notes in Artificial Intelligence, Springer (2005).

3. Bohdan S. Butkiewicz, Robust fuzzy clustering with fuzzy data. In: Advances in web intelligence, Springer, Berlin, 2005.
4. KT. Atanassov, Intuitionistic fuzzy sets: past, present and future. In: Proceedings of the 3rd Conference of the European Society for Fuzzy Logic and Technology, 2003, pp. 12-19.
5. Nikos Pelekis, Dimitrios K. Iakovidis, Evangelos E. Kotsifakos, Ioannis Kopanakis, Fuzzy clustering of intuitionistic fuzzy data. International Journal of Business Intelligence and Data Mining 3(1) (2008) 45-65.
6. L.A. Zadeh, Fuzzy sets. Information and Control 8(3) (1965) 338-353.
7. Szmidt E, Kacprzyk J, A measure of similarity for intuitionistic fuzzy sets. In: Proceedings of 3rd conference of the European Society for fuzzy logic and technology, 2003, pp. 206-209.
8. Wen-Liang Hung, Miin-Shen Yang, Similarity measures of intuitionistic fuzzy sets based on Hausdorff distance. Pattern Recognition Letters 25(14) (2004) 1603-1611.
9. Pierpaolo D'Urso, Paolo Giordani A weighted fuzzy c-means clustering model for fuzzy data. Computational Statistics & Data Analysis 50(6) (2006) 1496-1523.
10. Wen-Liang Hung, Jinn-Shing Lee, Cheng-Der Fuh, Fuzzy Clustering Based On Intuitionistic Fuzzy Relations. International Journal of Uncertainty, Fuzziness and Knowledge-Based Systems 12(4) (2004) 513-530.
11. Dimitrios K. Iakovidis, Nikos Pelekis, Evangelos E. Kotsifakos, Ioannis Kopanakis, Intuitionistic fuzzy clustering with applications in computer vision. In: Advanced concepts for intelligent vision system. Springer, Berlin, 2008.
12. Ioannis K. Vlachos, George D. Sergiadis, The Role of Entropy in Intuitionistic Fuzzy Contrast Enhancement. Foundations of fuzzy logic and soft computing, Springer, Berlin, 2007.
13. Asuncion A, Newman DJ, UCI Repository of Machine Learning Databases. Irvine, University of California, http://www.ics.uci.eedu/~mlearn/, 2007.
14. Halkidi M, Batistakis Y, Vazirgiannis M, Cluster validity methods: part I. ACM SIGMOD Record 31(2) (2002) 19-27.

ELM-Based Ensemble Classifier for Gas Sensor Array Drift Dataset

D. Arul Pon Daniel, K. Thangavel, R. Manavalan and R. Subash Chandra Boss

Abstract Much work has been done on classification for the past fifteen years to develop adapted techniques and robust algorithms. The problem of data correction in the presence of simultaneous sources of drift, other than sensor drift, should also be investigated, since it is often the case in practical situations. ELM is a competitive machine learning technique, which has been applied in different domains for classification. In this paper, ELM with different activation functions has been implemented for gas sensor array drift dataset. The experimental results show that the ELM with bipolar function classifies the drift dataset with an average accuracy of 96 % than the other function. The proposed method is compared with SVM.

Keywords ELM · Ensembles · Gas sensor array drift dataset · Bipolar

1 Introduction

The past decade has seen a significant increase in the application of multi-sensor arrays to gas classification and quantification. The idea to combine an array of sensors with a pattern recognition algorithm to improve the selectivity of the single

D. Arul Pon Daniel (✉) · K. Thangavel · R. Subash Chandra Boss
Department of Computer Science, Periyar University, Salem 636011, India
e-mail: apdaniel86@yahoo.com

K. Thangavel
e-mail: drktvelu@yahoo.com

R. Subash Chandra Boss
e-mail: rmsubash_18@yahoo.co.in

R. Manavalan
Department of Computer Application, KSR Arts and Science College, Trichengodu, India
e-mail: manavalan_r@rediffmail.com

G. S. S. Krishnan et al. (eds.), *Computational Intelligence, Cyber Security and Computational Models*, Advances in Intelligent Systems and Computing 246,
DOI: 10.1007/978-81-322-1680-3_10,

gas sensor has been widely accepted and being used by researchers in this field. In fact, an array of different gas sensors is used to generate a unique signature for each gas [1]. A single sensor in the array should not be highly specific in its response but should respond to a broad range of compounds, so that different patterns are expected to be related to different odors [2]. Different methods have been suggested recently to compensate for sensor drift in experiments for gas identification [3]. Chemical sensor arrays combined with read-out electronics and a properly trained pattern recognition stage are considered to be the candidate instrument to detect and recognize odors as gas mixtures and volatiles [4].

After learning the features of the class, the SVM recognizes unknown samples as a member of a specific class. SVMs have been shown to perform especially well in multiple areas of biological analyses, especially functional class prediction from microarray sensors produced data [5].

It is not surprising to see that it may take several minutes, several hours, and several days to train neural networks in most of the applications. Unlike traditional popular implementations, for single-hidden-layer feedforward neural networks (SLFNs) with additive neurons, which is a new learning algorithm called extreme learning machine (ELM) [6].

This paper has been organized into five sections. Section 2 presents the short note about the dataset used. Sect. 3 describes the approach of (ELM). In Sect. 4, experimental results of various activation function and discussion are presented. In Sect. 5, conclusions and further research scope are presented.

2 Dataset

The drift dataset contains 13,910 measurements from 16 chemical sensors utilized in simulations for drift compensation in a discrimination task of six gases at various levels of concentrations. The resulting dataset comprises of six distinct pure gaseous substances, namely ammonia, acetaldehyde, acetone, ethylene, ethanol, and toluene, each dosed at a wide variety of concentration values ranging from 5 to 1,000 ppmv [7]. This dataset is available in http://archive.ics.uci.edu/ml/datasets/Gas+Sensor+Array+Drift+Dataset

3 Extreme Learning Machine

Recently, a new learning algorithm for SLFN named the ELM has been proposed by Huang et al. [6, 8]. The SLFN with randomly chosen input weights and hidden bias can approximate any continuous function to any desirable accuracy. ELM is a single-hidden-layer neural network with good generalization capabilities and extreme learning capacity. The generalization performance of ELM for classification problem depends on three parameters. Number of hidden nodes, the input weights, and the bias values are needed to be optimally chosen [8]. For hidden layer, many activation functions such as sigmoidal, sine, Gaussian, and hard limiting function can be used, and the output neurons have a linear function as an activation function [9–11].

The ELM has several interesting and significant features different from traditional popular learning algorithms.

For the dataset which contain N distinguish objects (x_i, t_i) where $x_i = [x_{i1}, x_{i2}, x_{i3}, \ldots, x_{in}]^T \in R^n$ and $t_i = [t_{i1}, t_{i2}, t_{i3}, \ldots, t_{im}]^T \in R^m$, the relationship between the actual output of SLFN, with an infinite differentiable activation function $g(x)$, and the target outputs t_i is given by

$$\sum_{i=1}^{\tilde{N}} \beta_i g(w_i x_i + b_i) = t_j, \quad j = 1, \ldots, N. \tag{1}$$

Here, $\tilde{N}$ is the number of hidden nodes, $w_i = [w_{i1}, w_{i2}, w_{i3}, \ldots, w_{in}]^T$ and $\beta_i = [\beta_{i1}, \beta_{i2}, \beta_{i3}, \ldots, \beta_{im}]^T$ are the weight vector connecting inputs to the ith hidden neuron and the ith hidden neuron to output neurons, respectively, and b_i is the bias of the ith hidden neuron. Equation (1) can be rewritten compactly as $H\beta = T$

where $H = \begin{bmatrix} g(\mathrm{w}_1.\mathrm{x}_1 + \mathrm{b}_1) & \ldots & g(\mathrm{w}_{\tilde{N}}.\mathrm{x}_1 + \mathrm{b}_{\tilde{N}}) \\ . & \ldots & . \\ . & \ldots & . \\ . & \ldots & . \\ g(\mathrm{w}_i.\mathrm{x}_N + b_1) & \ldots & g(\mathrm{w}_{\tilde{N}}.\mathrm{x}_N + \mathrm{b}_{\tilde{N}}) \end{bmatrix}_{N \times \tilde{N}}$ is called the hidden layer output matrix of the neural network [6], and $\beta_i = [\beta_i, \ldots, \beta_{\tilde{N}}]^T_{\tilde{N} \times m}$, $T = [t_i, \ldots, t_N]^T_{N \times m}$.

Traditionally, training of SLFN has typically applied the back-propagation learning algorithm to adjust the set of weights (w_i, β_i) and biases (b_i). It is common and problem dependent.

ELM is introduced to resolve the issues in back-propagation neural network. Initially, parameters for the hidden node may be randomly specified. The output weights can then be analytically determined. Also, it is shown that the upper bound of the required number of hidden nodes is the number of distinct training objects (i.e., $\tilde{N} \leq N$). Thus, given (pre-specified) N, associated with parameters (w_i, b_i), the hidden nodes can be randomly generated. Determining the output weights β is as simple as finding the least-square solutions to the given linear system.

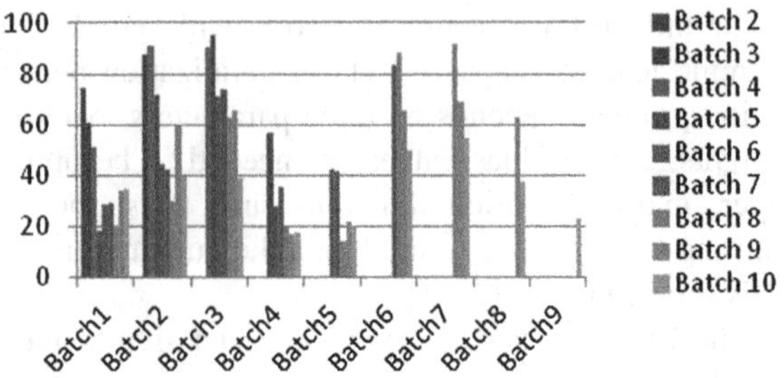

Fig. 1 Classification accuracy of the SVM classifiers with RBF kernel function

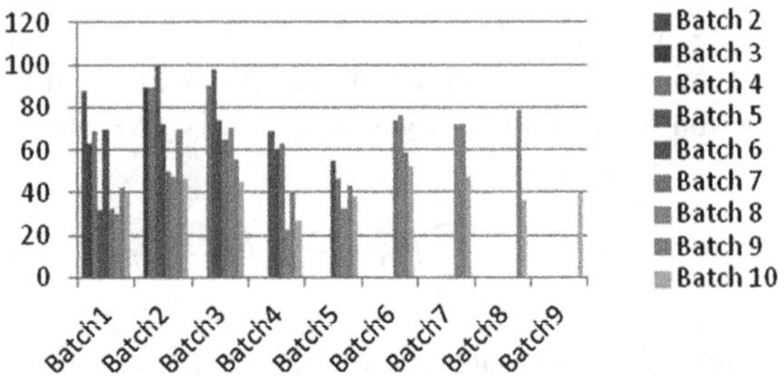

Fig. 2 Classification accuracy of the ELM classifiers with unipolar function

4 Experimental Results

In this experiment, the features in the training datasets are scaled appropriately to lie between −1 and +1. The kernel bandwidth parameter, the SVM parameter, and ELM parameter were chosen using 10-fold cross-validation by performing a grid search in the range $[2^{-10}, 2^{-9},\ldots, 2^{4}, 2^{5}]$ and $[2^{-5}, 2^{-4},\ldots, 2^{9}, 2^{10}]$, respectively. The performance of an SVM trained on batch 1 and tested on batches 2–10 respectively. Note that this curve is estimated with the same SVM model used in Fig. 1 but tested on data from batches instead of months. Similar behaviors were found when we trained several SVMs on batches 2–5 and tested them on successive batches. These results are again shown in Fig. 2. The complete set of results, i.e., the accuracy of classifiers trained on batches 1–9 and tested on successive batches, is given in Table 1. The individual plots correspond to the performance of classifier trained with batch 1 and tested on batches at subsequent time points after applying the component correction method for every one of the six reference gases (Figs. 3, 4 and 5).

In this section, the gas sensor array drift data are classified by the activation function unipolar, bipolar, and radial basis kernel, and they are classified into six classes. Before classifying, features are normalized between −1 and +1.

Table 1 Classification accuracy (in %) on batches 2–10 by SVM and ELM with activation function (unipolar, bipolar, RBF)

Batch ID	Function	Batch 2	Batch 3	Batch 4	Batch 5	Batch 6	Batch 7	Batch 8	Batch 9	Batch 10
Batch 1	SVM RBF	74.36	61.03	50.93	18.27	28.26	28.81	20.07	34.26	34.48
	ELM Unipolar	87.86	62.79	68.94	31.47	69.65	32.07	29.59	42.12	39.72
	ELM Bipolar	85.28	63.36	76.39	44.16	69.91	33.76	31.63	44.04	37.27
	ELM RBF	87.78	69.92	63.97	33.50	65.95	37.83	40.47	42.55	40.72
Batch 2	SVM RBF		87.83	90.68	72.08	44.52	42.46	29.93	59.57	39.69
	ELM Unipolar		89.28	89.44	98.98	72.08	49.51	47.27	69.36	46.13
	ELM Bipolar		84.23	77.64	80.71	73.52	46.00	61.90	60.42	40.61
	ELM RBF		73.70	73.91	89.84	70.43	45.64	59.86	60.85	43.72
Batch 3	SVM RBF			90.06	94.92	70.96	73.73	62.59	65.74	38.89
	ELM Unipolar			90.06	97.46	73.95	64.76	70.74	55.74	45.13
	ELM Bipolar			85.71	98.98	75.26	65.70	70.06	53.61	45.50
	ELM RBF			85.71	97.46	71.60	64.57	66.32	55.74	45.13
Batch 4	SVM RBF				56.35	27.52	35.40	19.73	17.02	17.56
	ELM Unipolar				69.03	60.04	63.35	22.44	40.85	26.41
	ELM Bipolar				98.47	63.91	62.77	32.99	49.36	44.30
	ELM RBF				92.89	63.3	68.53	26.87	42.55	42.86
Batch 5	SVM RBF					42.52	41.32	13.95	21.49	20.11
	ELM Unipolar					54.34	46.38	32.65	42.97	38.61
	ELM Bipolar					58.56	54.30	70.40	50.42	39.05
	ELM RBF					54.47	50.34	54.42	48.08	32.33
Batch 6	SVM RBF						83.53	88.44	65.74	49.97
	ELM Unipolar						73.42	75.85	58.51	52.13
	ELM Bipolar						72.82	77.55	82.12	53.52
	ELM RBF						70.13	74.83	62.76	51.47

(continued)

Table 1 (continued)

Batch ID	Function	Batch 2	Batch 3	Batch 4	Batch 5	Batch 6	Batch 7	Batch 8	Batch 9	Batch 10
Batch 7	SVM RBF							91.84	69.15	54.28
	ELM Unipolar							72.10	71.70	47.36
	ELM Bipolar							78.57	74.89	60.75
	ELM RBF							72.10	71.70	49.47
Batch 8	SVM RBF								62.98	37.69
	ELM Unipolar								78.93	36.86
	ELM Bipolar								83.83	42.00
	ELM RBF								81.27	40.08
Batch 9	SVM RBF									22.64
	ELM Unipolar									39.86
	ELM Bipolar									53.55
	ELM RBF									43.97

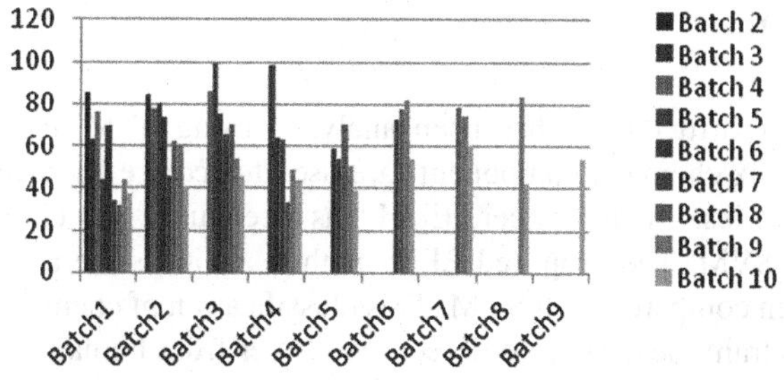

Fig. 3 Classification accuracy of the ELM classifiers with bipolar function

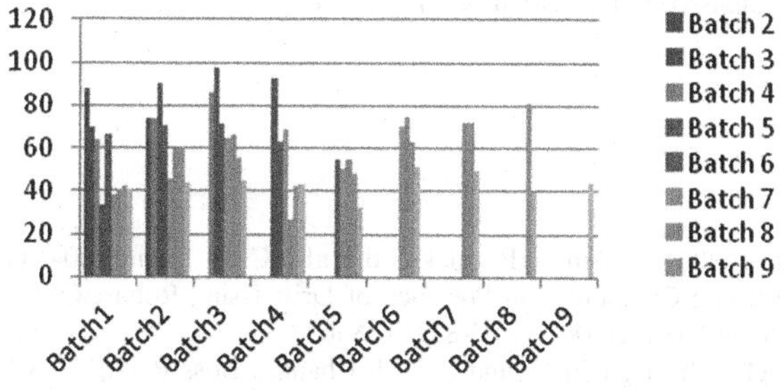

Fig. 4 Classification accuracy of the ELM classifiers with RBF kernel function

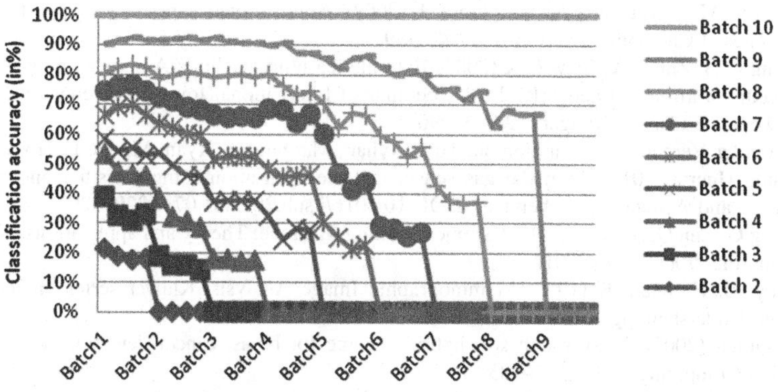

Fig. 5 Classification accuracy (in %) on batches 2–10 by SVM and ELM with activation function (unipolar, bipolar, RBF)

5 Conclusion

Gas sensor array drift dataset has been analyzed using SVM and the proposed ELM methods. Six chemical components are used to acquire the drift dataset with different time series. In this paper, ELM has been used for classification and compared with SVM. The proposed ELM method achieves the average accuracy of 92.23 % when compared with SVM. This classification of chemical components may be used to train the system to defect the cancer from human exhaled breathe in future.

Acknowledgments The first and fourth author immensely acknowledges the partial financial assistance under University Research Fellowship, Periyar University, Salem.

The second author immensely acknowledges the UGC, New Delhi, for partial financial assistance under UGC-SAP (DRS) Grant No. F.3-50/2011.

References

1. Sofiane Brahim-Belhouari, Amine Bermak and Philip C. H. Chan (2004) Gas Identification with Microelectronic Gas Sensor in Presence of Drift Using Robust GMM. IEEE ICASSP 2004, 0-7803-8484-9/04/$20.00, pp. V-833–V-836.
2. Arul Pon Daniel D, Thangavel K, and Subash Chandra Boss R (2012) A Review of Early De-tection of Cancers using Breath Analysis. Proc. IEEE Conf. Pattern Recognition, Infor-matics and Mobile Engineering (PRIME 2012), IEEE Press, DOI: 10.1109/ICPRIME.2013.6208385: 433–438.
3. John-Erik Haugen, Oliver Tomic, Knut Kvaal (1999) A calibration method for handling the temporal drift of solid state gas-sensors. Analytica Chimica Acta, pp. 23–39.
4. Persaud K, Dodd G (1982) Analysis of discrimination mechanisms in the mammalian olfactory system using a model nose. Nature 299 (5881):352–355.
5. Belusov AI, Verkazov SA, von Frese J (2002) Applicational aspects of support vector machines. J Chemometric, 16(8–12):482–489.
6. Huang G B, Zhu Q Y, Siew C K (2004) Extreme learning machine: A new learning scheme of feedforward neural networks. In Proceedings of International Joint Conference on Neural Networks, Budapest, Hungary, 2:985–990.
7. Alexander Vergara, Shankar Vembu, Tuba Ayhan, Margaret A. Ryan, Margie L. Homer and Ramón Huerta (2012) Chemical gas sensor drift compensation using classifier ensembles. Sensors and Actuators B: Chemical, DOI: 10.1016/j.snb.2012.01.074: 320–329.
8. Huang G, Zhu Q, Siew C (2006) Exreme Learning Machine: Theory and applications. Neurocomputing. 70(1-3):489–501.
9. Brady M, Highnam R (1999) Mammographic Image Analysis. Kluwer series on medical image Understanding.
10. Hassanien (2007) Fuzzy rough sets hybrid scheme for breast cancer detection. Image and Vision Computing, 25(2):172–183.
11. Roffilli M (2006) Advanced machine learning techniques for digital mammography. Technical Report UBLCS, University of Bologna. Italy.

Hippocampus Atrophy Detection Using Hybrid Semantic Categorization

K. Selva Bhuvaneswari and P. Geetha

Abstract Medical image analysis plays a vital role in the diagnosis and prognosis of brain-related diseases. MR images are often preferred for brain anatomy analysis for their high resolution. In this work, the components of the brain are analyzed to identify and locate the region of interest (hippocampus). The internal structures of the brain are segmented via the combination of wavelet and watershed approach. The segmented regions are categorized through semantic categorization. The region of interest is identified and cropped, and periodical volume analysis is performed to identify the atrophy. The atrophy detection of the proposed system is found to be more effective than the identification done by the traditional system of radiologist. Performance measures such as sensitivity, specificity, and accuracy are used to evaluate the system.

Keywords MRI · Image analysis · Wavelet · Watershed · Hippocampus detection · Atrophy

1 Introduction

Medical imaging is the method or process used to create the images of human body for clinical purposes or medical science. Medical imaging is a state-of-the-art technology used to see 2-D and 3-D images of the living body. Magnetic resonance imaging (MRI) is a highly developed medical imaging technique that provides more information about the human soft tissue anatomy. The advantages when compared with other imaging techniques enable it to provide three-dimensional

K. Selva Bhuvaneswari (✉) · P. Geetha
Department of CSE, College of Engineering Guindy, Anna University,
Chennai 600025, India
e-mail: bhuvana_ksb@yahoo.co.in

P. Geetha
e-mail: geetha_akp@cs.annauniv.edu

G. S. S. Krishnan et al. (eds.), *Computational Intelligence, Cyber Security and Computational Models*, Advances in Intelligent Systems and Computing 246,
DOI: 10.1007/978-81-322-1680-3_11, © Springer India 2014

data that highlight the high contrast between soft tissues. However, the amount of data are enough for manual analysis/interpretation, and this has been one of the highest obstacles in the effective use of MRI.

Our work concentrates hippocampus atrophy detection in brain MRI. In interpretation of MRI scan, we learn about three different types of images, T1, T2, and photon density images that are the output of different magnetization percentages of input MRI. The high-fat-content tissues look bright in T1-weighted images, and high-water-content tissues look dark. The reverse is true for T2-weighted images. Normally all the diseases are characterized by amount of water content in tissues, T2-weighted images are generally used for pathological investigations and T1-weighted images are dealing with details of anatomy and pathology, if contrast enhancement is applied. In proton-weighted images, bright areas indicate high-proton-density tissues, namely cerebrospinal fluid, and dark areas indicate cortical bone denoted by low-proton density. Proton-density images give details of anatomy and certain pathological information. Hence, a combined analysis and segmentation approach is required for differentiating the three input images (T1, T2, and p) for locating the region of interest and segmentation. Our approach is highly useful because it deals with hyper-intense and hypo-intense areas.

All MRI analysis such as breast cancer, lung cancer, and bone marrow cancer have pathological information or test for ruling the prognosis and diagnosis. But brain MRI is the only case where the basic therapeutic choice starts with MRI analysis and pathological information need to be extracted from MRI alone without any gene tests. Segmenting of MR images for identification of tumors and their class need to be done in a systematic way. Semantics plays a vital role here because all the analysis are done solely by the radiologist. In informatics point of view, the analysis done is epistemic rather than ontological. Hence, we propose semantic categorization of segmented regions for better pathological information.

2 Overview

MR image segmentation is the integral approach for better prognosis and diagnosis of brain disorders. Texture plays a vital role in segmenting the region of interest in the input image.

Wavelet transforms are often preferred for image analysis for their better decomposition capability, feature extraction, and effective segmentation of boundary parameters. Watershed segmentation is used to find the image ridgelines and the primitive regions with their boundaries. Upon direct application of the watershed algorithm on an input image often results in over segmentation [1]. This is because of the local irregularities of the gradient and noise.

The solution to this problem is to normalize the number of regions by incorporating a preprocessing stage with wavelet decomposition [2]. Additionally, semantic categorization is performed to incorporate knowledge into the segmentation procedure for effective image analysis.

3 Methodology

In this work, we propose a system to automatically extract the region of interest from the T1 slice image and perform volume analysis for atrophy identification.

The stages involved in the proposed work are as follows:

- Segmentation (wavelet and watershed)
- Semantic categorization
- Volume analysis for atrophy detection and segmentation.

3.1 Two-Dimensional Discrete Wavelet Transform

Generally, digital images are in the form of 2-D signals, and so a two-dimensional wavelet transform is required for processing. To separate the input image into separate horizontal, vertical, and diagonal details, the 2-D DWT performs analysis on image across its rows and columns [1]. Initially, the rows of an $N \times N$ image are filtered using high- and low-pass filters. The next stage 1-D convolution of the columns of the filtered image with the filters is performed.

The branches in the tree have shown in the diagram yields an $(N/2) \times (N/2)$ sub-image. At each level, four different sub-bands *HH, HL, LH,* and *LL* are obtained [2]. The *LL sub-band* is again filtered to get the next level image; Fig. 1 summarizes the transform for a two-level decomposition (Fig. 2).

The transformed image along with its sub-images is later reconstructed by low-pass approximation using upsampling and convolution. The process of upsampling is done by inserting of a zero row after each present row or a zero column after each present column (Fig. 3).

One of the most fundamental segmentation techniques is edge detection. There are many methods for edge detection. Convolution of the approximation image is done by Sobel mask.

3.2 Watershed Segmentation

Watershed algorithm is employed for the segmentation of region of interest from the input brain MRI. Approach based on immersion technique is adopted to calculate the watershed lines for the segmentation [3]. The principle of this algorithm is described below with the following diagram (Fig. 4):

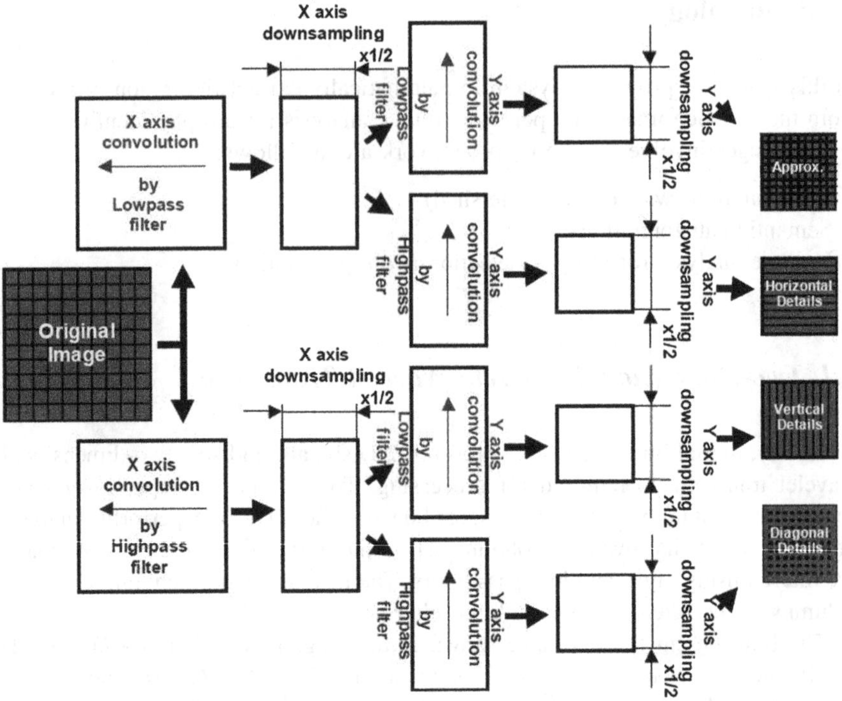

Fig. 1 Bands of 2-D wavelet transform

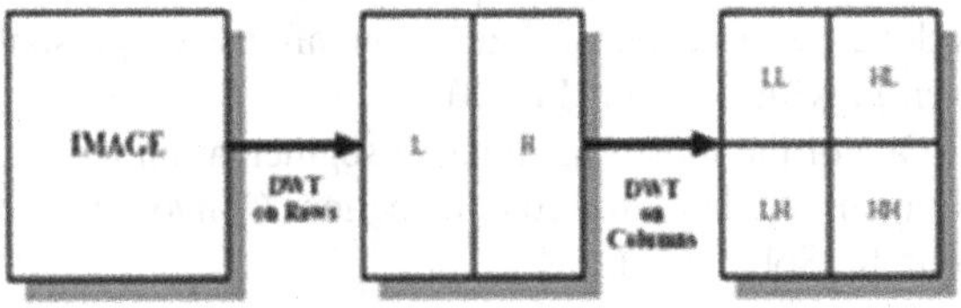

Fig. 2 Block diagram of DWT. Original image followed by output image after the 1-D applied on row input and output image after the second 1-D applied on row input

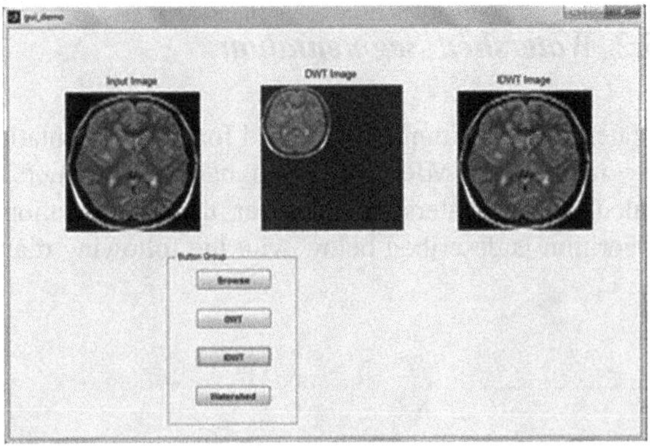

Fig. 3 LL, LH, HL, HH sets of wavelet transform

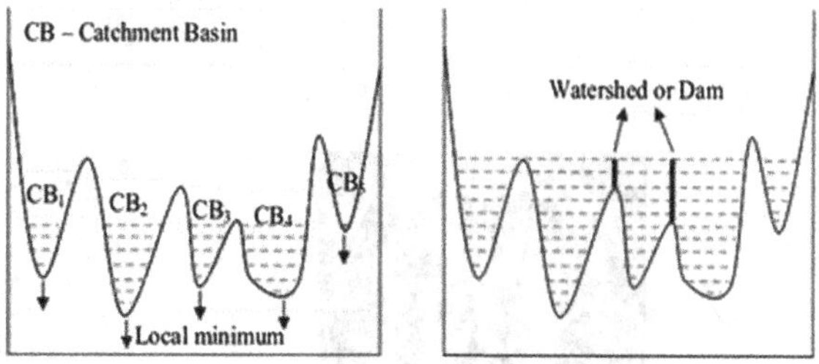

Fig. 4 Gray-level profile of image data, watershed segmentation—local minima define catchment basins; local maxima define the watershed lines

Algorithm:

1. The image $f(x, y)$ is visualized as a topographic surface, with both valleys and mountains.
2. Assuming that there is a hole in each minima and the surface is immersed into a lake.
3. The water will flood the surface by entering through the holes at the minima.
4. The location of merge of flooded water is identified, and a dam is build so as to avoid the meeting of water coming from two different minima.
5. To end with, the surface will visible with the dams and their walls are called the watershed lines.

Watershed segmentation of an brain MR image that contains image regions of different shapes is shown in Fig. 5.

3.3 *Semantic Categorization*

According to our model, each semantic category C_i *such as tumor*, *hippocampus*, *thalamus*, *cortex*, is 'uniquely' described by the set of features S^{C_i}. Ideally, these features can be used to separate C_i from other categories in the set. To test if the region x belongs to the category C_i, we will therefore use only the features that are important for this category. This is done according to the following scheme [4]. The semantics of each category C_i is captured by the following four sets:

The set of features, S^{C_i}, that are representative for the category, and their values, V^{C_i}

$$S^{C_i} = \left\{ \left\{ RF_j^{C_i} \right\}_{j=1,\ldots,M_i}, \left\{ FO_k^{C_i} \right\}_{k=1,\ldots,N_i} \right\} \tag{1}$$

$$V^{C_i} = \left\{ V_j^{C_i} \right\}_{j=1,\ldots,M_i+N_i} \tag{2}$$

where RF and FO are the required and frequently occurring features for the category C_i. For the region x, the values involved in testing if x belongs to C_i are then

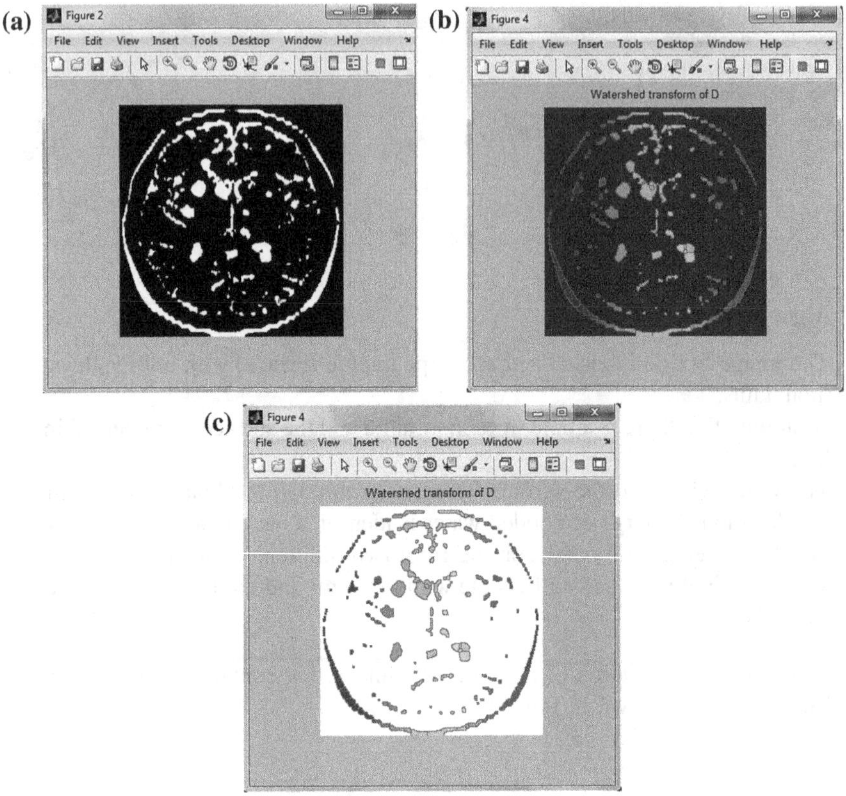

Fig. 5 **a**, **b**, **c** Different stages of output of watershed transform

$$S^{C_i}(x) = \left\{ S_j^{C_i}(x) \right\}_{j=1,\ldots,M_i+N_i} = \left\{ \left\{ RF_j^{C_i}(x) \right\}_{j=1,\ldots,M_i}, \left\{ FO_k^{C_i}(x) \right\}_{k=1,\ldots,N_i} \right\} \quad (3)$$

The set of operators, or functions, O^{C_i}, describing how $S^{C_i}(x)$ will be compared to V^{C_i}.

$$O_{C_i} = \left\{ O_j^{C_i}\left(S^{C_i}(x)_j^{C_i}, V_j^{C_i}, P_j^{C_i} \right) / O_j^{C_i} \in [0, 1] \right\}_{j=1,\ldots,M_i+N_i} \quad (4)$$

where $P_{C_i} = \left\{ P_j^{C_i} \right\}_{j=1,\ldots,M_i+N_i}$ is the set of parameters involved in the comparison. The comparison is performed according to the following formula:

$$\begin{aligned} \mathrm{sim}(x, C_i) = \frac{1}{N_i} &\left[\prod_{j=1}^{M_i} O_j^{C_i}\left(RF_j^{C_i}(x), V_j^{C_i}, P_j^{C_i} \right) \right] \\ &\times \left[\sum_{j=M_i+1}^{j=N_i+M_i} O_j^{C_i}\left(FO_{j-M_i}^{C_i}(x), V_j^{C_i}, P_j^{C_i} \right) \right] \end{aligned} \quad (5)$$

$$\mathrm{sim}(x, C_i) > 0 \Rightarrow x \in C_i$$

Note that, according to the final equation to classify a region into a semantic category, at least one frequently occurring feature and all the required features must be present to classify a region to its semantic category [4].

3.4 Hippocampus: Region Area and Volume Analysis

The desired region of interest, namely the right and left hippocampus, is cropped using the regional maxima image obtained from the watershed algorithm. This region of interest is used to calculate the volume for atrophy detection.

Area of a region is the total number of the pixels present in the region. The product of the number of pixels and the dimension of the pixel n will give the area. The dimension of the pixel is 0.035 × 0.034 cm^2; the pixel area is 0.001190 cm^2. To calculate the number of pixels in the cropped image the function Bw area is used [5].

Volume is then calculated by multiplying the area by the thickness of the image slice. The thickness of the one slice of the image is 0.50 cm. So the volume of the desired area of interest (left and right hippocampus) present in the image slice is given as:

$$V = A \times 0.50\,\mathrm{cm}^3. \tag{6}$$

where 'V' is the volume of the region of interest present in one slice, and 'A' is the area.

3.5 Performance Evaluation

The experimentation of the proposed algorithm has been implemented for the test images obtained from OASIS dataset. The results of the processed brain MR images using the hybrid approach with semantics are analyzed for hippocampus atrophy detection. The volume of the desired region of interest of the brain MR image of a patient in two consecutive visits is listed in Table 1.

Atrophy is the reduction in volume of the desired region of interest. For analysis, three benign and two normal case studies were processed.

Performance evaluation with respect to volume analysis in two consecutive visits is done between the hybrid hippocampal segmentation and the radiologist opinion. Our proposed method shows exact atrophy measurement through automated approach (Fig. 6).

Table 1 Left and right hippocampal volume

Input image	Volume of R and L hippocampus in cm^3	Visit-1	Visit-2
I1	Right hippo	3.42	3.42
	Left hippo	3.29	3.29
I2	Right hippo	4.18	4.18
	Left hippo	4.02	4.02
I3	Right hippo	3.91	3.68
	Left hippo	3.74	3.74
I4	Right hippo	3.38	3.20
	Left hippo	3.59	3.50
I5	Right hippo	4.09	3.50
	Left hippo	3.44	3.44

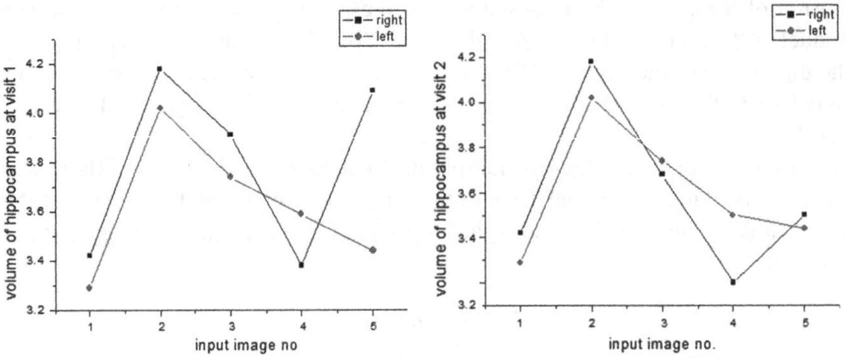

Fig. 6 Atrophy analysis at visit-1 and visit-2

3.6 Statistical Analysis

The performance of the hybrid segmentation with semantic categorization is analyzed by various statistical measures. The statistical performance measures that are obtained for the identification of cortical bone (cortex), cerebrospinal fluid (CSF), atrophy, edema, and tumor using the proposed method are shown in Table 2.

3.7 Comparative Analysis

To analyze the performance of atrophy detection, the proposed method is compared with the traditional watershed segmentation and hybrid watershed (wavelet and watershed) segmentation. Test images are used to validate the result in identifying the atrophy and tumor in consecutive visits. The segmentation performance is shown in Table 3.

Table 2 Statistical performance measures of tissues of brain MR images

Category	Measures		
	Sensitivity	Specificity	Accuracy
Cortex	74.2	99.1	94.45
Hippocampus	72.2	99.125	99.5
Atrophy	68	97.01	94.9
Thalamus	69	97.83	97.65
Tumor	48	95.6	94.1

Table 3 Performance comparison of watershed and hybrid watershed with the proposed method, in identifying (i) atrophy (ii) tumor

Identification	Methods		
	Proposed method	Watershed	Hybrid (wavelet and watershed)
Atrophy	94.9	90.1	92.5
Tumor	94.1	80.6	91

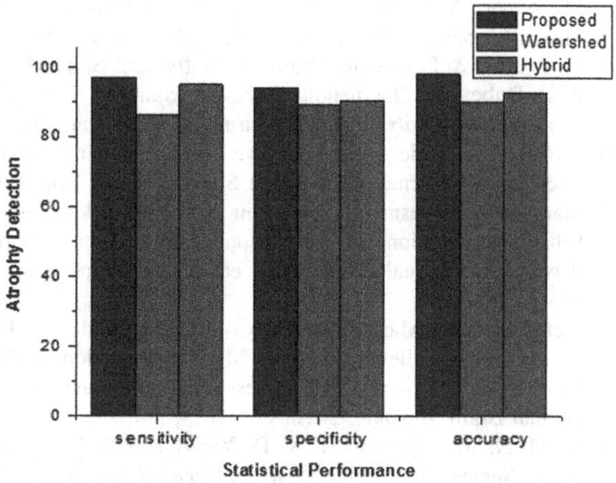

Fig. 7 Graph result for comparing accuracy of identifying atrophy of proposed method with watershed and hybrid watershed

Fig. 7 shows the result of a graph while comparing the accuracy sensitivity and specificity for identifying atrophy with the three methods. From Fig. 7, it is observed that the statistical property of the proposed method is higher than the existing methods such as watershed and hybrid watershed.

4 Conclusion

Atrophy is a decrement in the size of the cell in the human brain. This decrease has to be effectively estimated since it leads to major diseases such as Alzheimer's disease, dementia, disorientation, and language disorders. This work uses semantic segmentation to detect the region of interest using effective categorization and segmentation. This schema detects atrophy early and provides improved results with the help of hybrid approach of transforms. The future work would be incorporating the domain concepts of brain from the reference ontology. Further storing the output in a database and pattern matching can be done, so that degree of atrophy will be identified in improved manner.

References

1. Rafael C. Gonzalez, Richard E. Woods and Steven L. Eddins, *Digital Image Processing using MATLAB* Second edition, Pearson Education, 2011.
2. Jun Zhang, Jiulun Fan, *Medical Image Segmentation Based on Wavelet Transformation and Watershed Algorithm.* In IEEE Proceedings of the International Conference on Information Acquisition, pp 484–489, August 20–23, 2006, Weihai, Shandong, China.
3. Malik Sikandar Hayat Khiyal, Aihab Khan, and AmnaBibi, *Modified Watershed Algorithm for Segmentation of 2D Images.* In Issues in Informing Science and Information Technology, Volume 6, 2009.
4. Mojsilovic, A.; Gomes, J. "Semantic based categorization, browsing and retrieval in medical image databases", Image Processing. 2002. Proceedings. 2002 International Conference on Computing & Processing (Hardware/Software); Signal Processing & Analysis.
5. R. B. Dubey, M. Hanmandlu, S. K. Gupta and S. K. Gupta, *Region growing for MRI brain tumor volume analysis,* Indian Journal of Science and Technology, Vol.2 No. 9 (Sep 2009).
6. Rowayda A. Sadek, "An Improved MRI Segmentation for Atrophy Assessment", IJCSI International Journal of Computer Science Issues, Vol. 9, Issue 3, No 2, May 2012.
7. Maryam Hajiesmaeili, Bashir Bagherinakhjavanlo, JamshidDehmeshki, Tim Ellis,"Segmentation of the Hippocampus for Detection of Alzheimer's Disease", Advances in Visual Computing Lecture Notes in Computer Science Volume 7431, 2012, pp 42–50.
8. Stefan M. Gold, Mary-Frances O'Connor, Raja Gill, Kyle C. Kern, Yonggang Shi, Roland G. Henry, Daniel Pelletier, David C. Mohr[8], "Detection of altered hippocampal morphology in multiple sclerosis-associated depression using automated surface mesh modeling", Journal of Human Brain Mapping, 2012.
9. G. M. N. R. Gajanayake1, R. D. Yapa1 and B. Hewawithana2, *Comparison of Standard Image Segmentation Methods for Segmentation of Brain Tumors from 2D MR Images.*
10. GowriAllampalli-Nagaraj, Isabelle Bichindaritz, "Automatic semantic indexing of medical images using a web ontology language for case-based image retrieval", Engineering Applications of Artificial Intelligence, February 2009, Volume 22 Issue 1.
11. Yifei Zhang, Shuang Wu, Ge Yu, Daling Wang, *A Hybrid Image Segmentation Approach Using Watershed Transform and FCM.* In IEEE Fourth International Conference on FSKD, 2007.
12. Jiang. Liu, Tze-Yun LeongP, Kin Ban CheeP, Boon Pin TanP, *A Set-based Hybrid Approach (SHA) for MRI Segmentation,* ICARCV., pp. 1–6, 2006.

Hyper-Quadtree-Based K-Means Algorithm for Software Fault Prediction

Rakhi Sasidharan and Padmamala Sriram

Abstract Software faults are recoverable errors in a program that occur due to the programming errors. Software fault prediction is subject to problems like non-availability of fault data which makes the application of supervised technique difficult. In such cases, unsupervised techniques are helpful. In this paper, a hyper-quadtree-based K-means algorithm has been applied for predicting the faults in the program module. This paper contains two parts. First, the hyper-quadtree is applied on the software fault prediction dataset for the initialization of the K-means clustering algorithm. An input parameter Δ governs the initial number of clusters and cluster centers. Second, the cluster centers and the number of cluster centers obtained from the initialization algorithm are used as the input for the K-means clustering algorithm for predicting the faults in the software modules. The overall error rate of this prediction approach is compared with the other existing algorithms.

Keywords Hyper-quadtree · K-means clustering · Software fault prediction

1 Introduction

Software fault prediction plays an important role in software quality assessment. It identifies the software subsystems (modules, components, classes or files) which are likely to contain faults. Faults lead to software failures and reduce software

R. Sasidharan (✉) · P. Sriram
Department of Computer Science and Engineering, Amrita University, Kollam, Kerala, India
e-mail: rakhisasidharan@gmail.com

P. Sriram
e-mail: rpadmamla_sriram@gmail.com

G. S. S. Krishnan et al. (eds.), *Computational Intelligence, Cyber Security and Computational Models*, Advances in Intelligent Systems and Computing 246,
DOI: 10.1007/978-81-322-1680-3_12,

quality. Software fault predictions help to improve the quality of the software and reduce overall cost of the software. Some researchers prefer using the term *software quality estimation* for the software fault prediction modeling studies [2]. The aim of building this kind of model is to predict the fault labels (fault prone or not fault prone) of the modules for the next release of the software. A typical software fault prediction model includes two steps. First, a fault model is built using previous software metrics and fault data belonging to each software module. After this training phase, fault labels of program modules can be estimated using this model [3]. From machine learning perspective, it is a supervised learning approach because the modeling phase uses class labels represented as known *fault data* and *false data*. However, there are cases when previous fault data are not available. For example, a software company might start to work on a new domain or might plan building fault predictors for the first time in their development cycle. In such a situation, supervised learning approach cannot be used because of the absence of class labels. In these situations, unsupervised learning approach can be applied.

Unsupervised techniques like clustering can be used for the fault prediction in software modules. The K-means clustering algorithm is used in the previous works for the prediction of the faults in the software modules. K-means clustering is a non-hierarchical clustering procedure that tries to group the set of points until a desired criterion is reached [4]. The partitioning of dataset is such that the sum of intracluster distance is reduced to an optimum value [5]. K-means is a simple and widely used clustering algorithm. However, it suffers from some drawbacks. First, the user has to initialize the number of clusters which is very difficult to identify in most of the situation [1]. Second, it requires selection of suitable initial cluster centers which is again subject to errors. Since the structure depends only on the initial cluster centers, it results in inefficient clustering. Third, the K-means is very sensitive to noise [1]. Forth, it does not scale to large number of data points.

To solve the above-mentioned problem, here, we propose a software fault prediction model using hyper-quadtree-based K-means algorithm. The objective of this model is as follows: First, hyper-quadtree is applied for finding initial cluster centers and the number of clusters. An input parameter Δ governs the initial number of clusters, and by varying Δ, users can generate desired cluster centers and the initial number of clusters. Second, the results obtained from the hyper-quadtree are given as the input of the K-means algorithm for predicting the faults in the software modules. The proposed system is implemented and tested on the three real data sets [12] used for the software fault prediction. We compare the new approach with the existing works using the result of evaluation parameters.

The remaining part of the paper is organized as follows: Sect. 2 presents the related work on this topic. Section 3 presents an overview of the hyper-quadtree. Section 4 presents the system architecture and algorithms. Section 5 presents the analysis of the results, while the Sect. 6 presents the conclusion and the future work.

2 Related Work

There are a few fault prediction studies that do not use prior fault data for modeling. Zhong [6] applied clustering technique and expert-based approach for software fault prediction. They applied K-means and neural gas technique on different real data set; then, an expert explored the representative modules of the cluster and several statistical data in order to label each cluster as fault prone or not fault prone. And, based on their experiment, neural gas-based prediction approach performed slightly worser than K-means clustering-based approach in terms of overall error rate on large data set. But, their approach depends on the availability and capability of the expert.

Bishnu and Bhattacherjee [7] applied a K-d tree and K-means clustering algorithm for the software fault prediction. They did experiments on the various data sets and found that the error rate is high compared to naïve K-means and does not scale to large number of data points. In addition to this, it is necessary to initialize the cluster center and number of clusters by the user. It gives inefficient clustering result in most of the cases. Manalopoulos et al. [8] proposed a K-means algorithm based on the R-Tree. It well scales to large number of data points and low error rate compared with the naïve K-means algorithm. But here, also the user has to initialize the number of clusters and the cluster centroids.

Seliya and Khoshgoftaar [10] proposed a semi-supervised clustering approach for software quality analysis with limited fault proneness data. Most recently, Catal [9] proposed a metric threshold and clustering-based approach for software fault prediction. The result of their study demonstrates the effectiveness of metric threshold and shows that a standalone application is easier than the clustering and metric-based approach because selection of number of clusters is performed heuristically in this clustering-based approach. In this work, the false-negative rates (FNR) for the clustering approach are less than that for the metric-based approach, while the false-positive rates (FPR) are better for the metric-based approach. The overall error rate for both the approaches remains the same.

In our present study, we have presented comparative result performed on the same data as [9], Bhattacherjee and Bishnu applied unsupervised learning approach for fault prediction in software modules.

3 Hyper-Quadtree

The *PR quadtree* or, simply, *quadtree* is a hierarchical spatial subdivision that begins with an axis-parallel bounding square that tightly encloses a set of points in the plane [11]. This bounding square, which is associated with the root of the tree, is subdivided into four congruent sub-squares by two axis-parallel lines that cross at the square's center. Each sub-square is associated with a child node of the root. In general, every non-leaf node has four children and an associated bounding

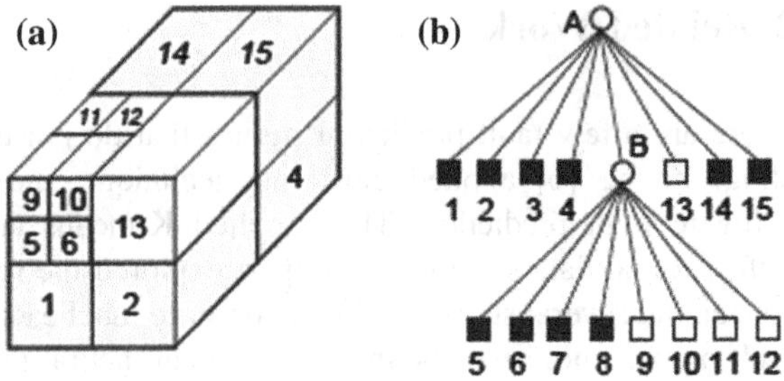

Fig. 1 **a** Points in three-dimensional space; **b** hyper-quadtree representation for $n = 3$

square. This top-down approach continues until reaching squares that each square contains only one point inside it. The three-dimensional analog of the quadtree, constructed over a set of data points on the space, is called an octtree, an eight-way branching tree whose nodes are associated with axis-parallel boxes. The d-dimensional analog is known variously as a *multidimensional quadtree* or a *hyper-quadtree*. Thus, the definition of hyper-quadtree for a set of O data points inside a n-dimensional space μ is as follows: Let $\mu = [d1\mu{:}d1\mu'] \times [d2\mu{:}d2\mu'] \times [d3\mu{:}d3\mu'] \times \cdots \times [dn\mu{:}dn\mu']$. If the number of data points in any bucket is less than threshold, then the Quadtree consists of a single leaf where the set O and the hypercube μ are stored. At each stage, every bucket gets subdivided into 2^n sub-buckets (Fig. 1).

4 System Architecture and Algorithm

The system architecture comprises of four parts. The first part is the construction of the hyper-quadtree. Second part discusses how we can calculate the initial centroids of the clusters and the number of clusters. For that, the hyper-quadtree obtained from the first part is used. The third part is the clustering. That is, how the software module dataset can be clustered. For clustering, we use the K-means clustering algorithm from machine learning. The forth part consists of the measure of faults in the software modules. The overall system architecture is shown in Fig. 2.

4.1 Construction of Hyper-Quadtree

Before explaining the method for the construction of the quadtree, first we explain some parameters and notations used. They are as follows:

- MIN: user-defined threshold for minimum number of data points in a sub-bucket.

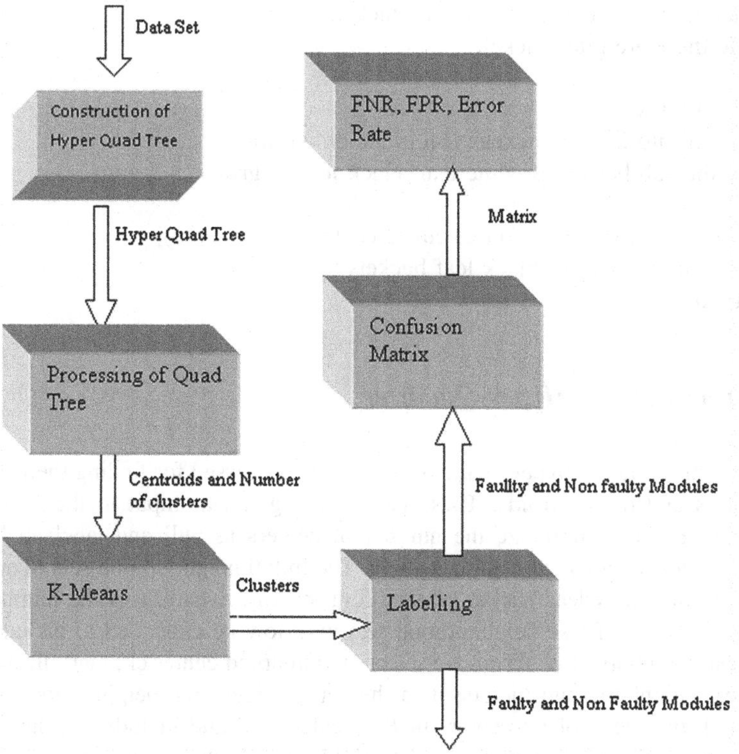

Fig. 2 Overall system architecture

- MAX: user-defined threshold for maximum number of data points in a sub-bucket.
- Δ: user-specified distance for finding nearest neighbors.
- White leaf bucket: A sub-bucket having less than MIN percent of data points of the parent bucket.
- Black leaf bucket: A sub-bucket having more than MAX percent of data points of the parent bucket.
- Gray bucket: a sub-bucket which is neither white nor black.
- Rk: neighborhood set of center ck of a black leaf bucket.
- C: set of cluster centers used for initializing K-means algorithm.

For the hyper-quadtree construction, first, we divide an initial data space into 2^n buckets. These buckets may be black or white. Each bucket is again divided into 2^n buckets. This process continues until all buckets become gray. Algorithm 1 gives the pseudocode for the construction of the hyper-quadtree.

Algorithm 1: Hyper_Quad_Tree (Min %, Max %, D)
Input: Data Set, A Max percentage, A Min Percentage
Output: Hyper Quad Tree.

1. Initialize the data space D as gray bucket.
2. **while** there are gray bucket;
   do
3. select a bucket
4. divide it into 2^n sub buckets; // n is the dimension
5. label the sub bucket as white leaf, black leaf or gray leaf
6. **for** every black leaf bucket calculate center (ci(1 $\leq$ i$\leq$m)) // m is the number of black leaf buckets
7. **end while**.

4.2 Processing of Hyper-Quadtree

After constructing the hyper-quadtree, the tree is processed for finding the number of clusters and the centroids. These centers are given as input to the K-means algorithm. For this, initialize the number of centers as null and label all black leaves as bucket centers and unmark them. The initial neighborhood sets R_i are set to include the black leaf bucket centers ci (1 < *i*<*m*), where *m* is the number of black leaf bucket. These neighborhood sets will now be expanded to include the Δ-nearest neighbors of ci. Then, we select an unmarked center ck, mark it, find its Δ-nearest neighbors, and include it in the set R_i. After one neighborhood set is exhausted, the mean of all centers in R_i is calculated and included in the set of cluster centers *C*. This is done for all the neighborhood sets with unmarked centers.

Note that some neighborhood sets may not get expanded at all if their initial centers get included into other neighborhood sets and eventually get marked. In other words, we group the centers of the black leaf buckets such that each group contains centers of adjacent black leaf buckets. Then, we calculate the mean of each group. The means are used as initial cluster centers for the K-means algorithm. At the end of all iterations, the algorithm returns the set *C* and the number of centers as the number of clusters. Algorithm 2 gives the pseudocode for the calculation of cluster centroid and number of clusters.

Algorithm 2: Proc_Hyper_Quad_Tree (Hyper quad tree, Δ)
Input: Hyper quad tree obtained from the procedure Hyper-Quad_Tree
Output: Cluster Centers C and number of clusters | C |

1. Set C = Ø
2. Label all centers ci (1 < i<m), as unmarked;
3. **for** i = 1 to m do Ri = ci;
4. **for** each neighborhood Ri(1 < i<m), **do**
5. **if** there exist an unmarked center in Ri **then**,
6. **while** there is an unmarked center ck in Ri, **do**
7. select ck and label it as marked;

8. find Δ-nearest unmarked neighbors of ck and include them in Ri.
9. **for** all ck €Ri calculate mean mi and call it cluster Center.
10. C = C U {mi};
11. **end while;**
12. **end for;**
13. **return** C and |C|

Line 1 initializes the set of cluster centers to null set, and line 2 labels all black leaf bucket centers (obtained from the Algorithm 1) as unmarked. The initial neighborhood sets R_i are set to include the black leaf bucket centers c_i $(1 < i <$ m), (line 3). These neighborhood sets will now be expanded to include the Δ-nearest neighbors of c_i. This is done in lines 5–8, where we select an unmarked center c_k. After one neighborhood set is exhausted, in lines 9 and 10, the mean of all centers in <*i* is calculated and included in the set of cluster centers *C*. This is done for all the neighborhood sets with unmarked centers. Note that some neighborhood sets may not get expanded at all if their initial centers (as initialized in line 4) get included into other neighborhood sets and eventually get marked. In other words, we group the centers of the black leaf buckets such that each group contains centers of adjacent black leaf buckets. Then, we calculate the mean of each group. The means are used as initial cluster centers for the K-means algorithm. At the end of all iterations, the algorithm returns the set C and the number of centers. The output of the Proc_Hyper_Qud_Tree algorithm presented in Algorithm 2 is the set of centers and the number of clusters. We use such centers as the initial cluster centers for the original K-means algorithm.

Complexity of the algorithm: The complexity of generating the black leaf bucket is $(b + 1)dc$ where b is the depth of the tree, d is the number of data points, and c is the constant. Lines 4–12 generate the neighborhood set of the m black leaf buckets, and this takes $O(m^2)$ time. Hence, complexity of our algorithm is $O((b + 1)d + m^2)$. Assuming $m \ll d$, the complexity can be assumed to be $O((b + 1)d)$.

Criteria for selecting the parameter Δ: For selecting Δ, we consider the lmin and lmax as the minimum and maximum levels at which the black leaf buckets are created. Let p be the side length of the initial bucket and n is the dimension then $\mathrm{dia}_{\mathrm{max}} = \text{square root(np)}/2^{l}_{\mathrm{min}}$ and $\mathrm{dia}_{\mathrm{min}} = \text{squre root(np)}/2^{l}_{\mathrm{max}}$. As a guiding rule it is suggested that Δ be selected between diamin and diamax.

4.3 Labeling

The K-means clustering algorithm returns the clusters that are equal to the number of clusters obtained from the hyper-quadtree. Each cluster has a centroid point. If any metric value of the centroid point of a cluster is greater than the threshold, the cluster is labeled as faulty otherwise labeled as non-faulty.

5 Experimental Design

5.1 Data Sets

We conducted experiments on three real data sets to test our algorithm. The data sets are AR3, AR4, and AR5 [12]. These real data sets are collected from PROMISE Software Engineering Repository. These three data sets are related to software fault prediction.

5.2 Metric Threshold

The dimensions and metrics we used in our experiments for AR# data sets are same as [6, 7] and are as follows: lines of code (LoC), cyclomatic complexity (CC), unique operator (UOp), unique operand (UOpnd), total operator (TOp), and total operand (TOpnd). For example, a threshold vector [LoC, CC, UOp, UOpnd, TOp, TOpnd] was chosen as [65, 10, 25, 40, 125, 70].

5.3 Evaluation Parameters

A confusion matrix is formed as shown in Table 1. The actual labels of data items are placed along the rows, while the predicted labels are placed along the columns. For example, a false actual label implies that the module is not faulty. If a not-faulty module (actual label—false) is predicted as non-faulty (predicted label—false), then we get the condition of cell A, which is true negative, and if it is predicted as faulty (predicted label—true), then we get the condition of cell B, which is false positive. Similar definitions hold for false negative and true positive. The FPR is the percentage of not-faulty modules labeled as fault prone by the model and the FNR is the percentage of faulty modules labeled as not fault prone and error is the percentage of mislabeled modules. The following equations are used to calculate these FPR, FNR, error, and accuracy.

Table 1 Confusion matrix

	Predicted labels		
Actual labels		False (Non-faulty)	True (Faulty)
	False (non-faulty)	(True negative) A	(False positive) B
	True (faulty)	(False negative) C	(True positive) D

$$\mathrm{FPR} = \mathrm{B}/(\mathrm{A} + \mathrm{B})$$
$$\mathrm{FNR} = \mathrm{C}/(\mathrm{C} + \mathrm{D})$$
$$\mathrm{Error} = (\mathrm{B} + \mathrm{C})/(\mathrm{A} + \mathrm{B} + \mathrm{C} + \mathrm{D})$$
$$\mathrm{Accuracy} = \frac{\text{No of data set correctly predicted}}{\text{Total Number of data set}}$$

The above performance indicators should be minimized. A high value of FPR would lead to wasted testing effort, while high FNR value means error-prone modules.

5.4 Gain

The optimal number of clusters is said to occur when the intercluster distance is maximized (or intercluster similarity is minimized) and the intracluster distance is minimized (or intracluster similarity is maximized). The clustering gain [10] attains a maximum value at the optimal number of clusters.

The simplified formula for calculation of gain is as follows:

$$\mathrm{Gain} = \sum_{k=1}^{K} (d_k - 1)(z_0 - z_0^k)^2$$

where K is the number of clusters, d_k is the number of data points present in kth cluster, z_0 is global centroid, and z_0^k denotes the centroid of the kth cluster.

5.5 Experimental Setup and Result

Table 2 presents the gain values for all the data sets as obtained by the simple K-means algorithm. Values have been taken for up to 12 clusters. For each cluster, six runs have been executed, and the maximum gain value has been reported. The hyper-quadtree K-means algorithm has four input parameters: MIN, MAX, O, and Δ. The value for MIN has been chosen as 5 %, and for MAX, it is 95 %. In this algorithm, Δ for AR3, AR4, AR5, we give 40, 80, and 40, respectively, and the number of cluster centers obtained was 4, 4, and 3, respectively. To be able to compare our clustering quality with the K-means algorithm, we adjusted the threshold parameter Δ to obtain the same number of clusters (4 for AR3, 4 for AR4, 3 for AR5) which gave maximum gain values for K-means algorithm, we adjusted the threshold parameter Δ to obtain the same number of clusters (4 for AR3, 4 for AR4, 3 for AR5) which gave maximum gain values for K-means algorithm (Table 2).

Table 2 Gain values for various data sets

C∼	AR3	AR4	AR5
1	0000.000	0000.000	0000.000
2	8,482.825	7,629.863	2,663.66
3	9,483.899	9,418.109	5,024.245*
4	9,487.919*	1,0938.666*	4,197.171
5	9,402.292	9,536.031	4,066.295
6	8,679.119	9,466.211	3,928.565
7	8,577.743	9,408.413	3,450.422
8	8,252.662	9,300.438	3,501.186
9	8,213.447	9,218.340	3,633.281
10	8,068.498	9,268.624	3,496.519
11	8,152.776	9,113.923	3,367.876
12	7,818.184	9,028.644	3,160.925

C∼ = Number of clusters, *Maximum gain value

Table 3 Software fault prediction error analyses

Data set	Parameters	HQDK	KM	KMK	KMR
AR3	FPR	21.54	34.54	44.09	43.63
	FNR	18.00	35.00	26.00	28.00
	Error	31.33	55.03	48.77	40.77
AR4	FPR	3.59	39.90	30.00	29,09
	FNR	41.0	51.05	43.89	34.90
	Error	10.15	32.71	31.72	27.40
AR5	FPR	13.28	34.56	30.67	33.45
	FNR	12.50	20.56	19.45	18.00
	Error	13.88	28.44	25.54	20.44

Table 4 Accuracy comparison with different techniques

Number of data	HQDK	KM	KMK	KMR
20	18	15	16	16
40	38	30	32	34
60	57	42	44	46
80	75	54	56	58
100	88	72	74	76

Table 3 presents the error prediction for the HQDK approach compared with other approaches, namely, K-means, K-means with K-d tree, and K-means with R-d tree. Table 3 shows that the error rate obtained by using HQDK is very less when compared with all other approaches. We can also see that the FPR and FNR vales for HQDK is less, when compare with other approaches.

Table 4 presents the accuracy calculation. For the accuracy calculation, we take 100 data from each data set. First, take 20, then we get 18 as the correctly predicted. Take 40 as the number of data use, then we get 36 as the correctly

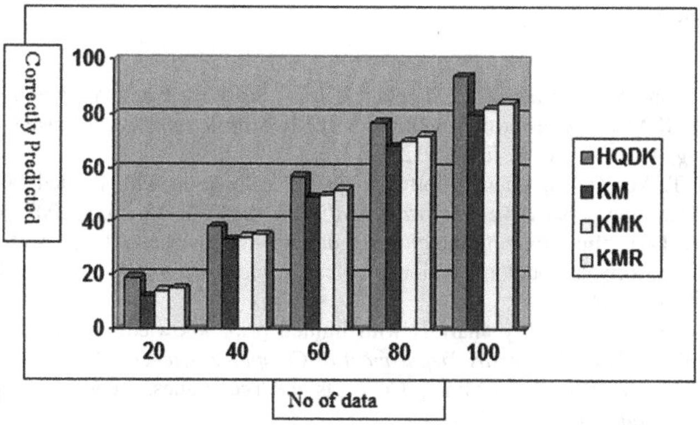

Fig. 3 Accuracy graph

predicted dataset. The result obtained from HQDK is compared with the existing algorithms. The results are presented in Table 4.

The accuracy of result of software fault prediction is shown in Fig. 3. The graph in Fig. 3 represents the values of accuracy with the existing algorithms. The graph shows that hyper-quadtree-based K-means algorithm has high accuracy compared with other techniques for the software fault prediction

6 Conclusion and Future Scope

In this paper, we propose a hyper-quadtree-based K-means algorithm for predicting the faults in the software. The overall objective of this paper is to show how we can use K-means without giving the number of clusters and the initial cluster centers and predict software faults with low error rate. By varying the value of Δ, we get the initial cluster centers and the number of clusters. The proposed algorithm is compared with various existing techniques. The overall error rate of software fault prediction approach by HQDK algorithm is found comparable with other existing algorithms, and HQDK has less error rate. At the same time, accuracy of HQDK is high compared to other techniques.

Here, we labeled the clusters by comparing the value of the centroid with a threshold. In future work, the labeling can be according to the majority of the member. Also, in future work, we can consider evaluating our model for datasets which have noisy instances such as JM1 dataset in PROMISE repository. A preprocessing step is necessary to remove noisy instances before our prediction model is applied or we need to develop a new unsupervised fault prediction model which is insensitive to noisy instances.

References

1. P.S. Bishnu and V. Bhattacherjee, Member, IEEE, "Software Fault Prediction Using Quad Tree-Based K-Means Clustering Algorithm" IEEE *Transactions on Knowledge and Data Engineering,* Vol. 24, No. 6, June 2012.
2. N. Seliya, T. M. Khoshgoftaar, "Software quality estimation with limited fault labels: a supervised learning perspective", *Software quality Journal* Vol. 15, no. 3, 2007, pp. 377–344.
3. C. Catal, B. Diri, "Investigating the effect of dataset size, metrics set, and feature selection techniques on software fault prediction problem," *Information Sciences*, Vol. 179, no. 8, pp. 1040–1058, 2009.
4. N. Seliya, "Software quality analysis with limited prior knowledge of faults," *Graduate Seminar, Wayne State University Department of Computer Science*, 2006.
5. J. Han and M. Kamber, "Data Mining Concepts and Techniques, second ed," pp. 401–404. *Morgan Kaufmann Publisher*, 2007.
6. S. Zhong, T.M. Khoshgoftaar, and N. Seliya, "Unsupervised Learning for Expert- Based Software Quality Estimation," *Proc. IEEE Eighth Int'l Symp. High Assurance Systems Eng.*, pp. 149–155, 2004.
7. P.S. Bishnu and V. Bhattacherjee, "Application of K-Medoids with kd –Tree for software fault prediction," *ACM Software Eng. Note.* Vol. 36, pp. 1–6, Mar. 2011.
8. Yannis Manolopoulos, Alexandros Nanopoulos, Apostolos N. Papadopoulos, Yannis Theodoridis, *"R-Trees: Theory and Application"* pp. 1–6, Dec. 2010.
9. N. Seliya and T.M. Khoshgoftaar, "Software Quality Classification Modelling Using the SPRINT Decision Algorithm," *Proc. IEEE 14th Int'l Conf. Tools with Artificial Intelligence*, pp. 365–374, 2002.
10. C. Catal, U. Sevim, and B. Diri, "Clustering and Metrics Threshold Based Software Fault Prediction of Unlabeled Program Modules," *Proc. Sixth Int'l Conf. Information Technology: New Generations,* pp. 199–204, 2009.
11. H.Samet, *The design and Analysis of Spatial Data structures.* Reading, Mass Addison-Wesley, 2000.
12. http://promisedata.org/, 2012.

Measurement of Volume of Urinary Bladder by Implementing Localizing Region-Based Active Contour Segmentation Method

B. Padmapriya, T. Kesavamurthi, B. Abinaya and P. Hemanthini

Abstract Ultrasound has become increasingly important in medicine and has now taken its place along with X-ray and nuclear medicine as a diagnostic tool. Its main attraction as an imaging modality lies in its non-invasive characteristic and ability to distinguish interfaces between soft tissues. Diagnostic ultrasound can be used to find out the cyst and tumors in the abdominal organs. Considering the importance of measurement of volume of the urinary bladder using diagnostic ultrasound imaging, an image processing technique of edge-based image segmentation has been employed. The technique discussed in this paper deals with a method for automatic edge-based image segmentation of the urinary bladder using localized region-based active contour method from a 2D ultrasound image for finding the area and volume of the urinary bladder accurately. The study of area and volume would provide valuable information about the abnormalities of the bladder and also the extent of abnormality. Experimental results show good performance of the proposed model in segmenting urinary bladder to measure its exact area and volume.

Keywords Diagnostic ultrasound · Urinary bladder · Edge-based segmentation · Localized region-based active contours

B. Padmapriya (✉) · B. Abinaya · P. Hemanthini
Department of Biomedical Engineering, PSG College of Technology, Coimbatore, India
e-mail: priyadhileep@yahoo.co.in

T. Kesavamurthi
Department of Electronics and Communication Engineering, PSG College of Technology, Coimbatore, India

G. S. S. Krishnan et al. (eds.), *Computational Intelligence, Cyber Security and Computational Models*, Advances in Intelligent Systems and Computing 246,
DOI: 10.1007/978-81-322-1680-3_13,

1 Introduction

Bladder ultrasound is used in the acute care, rehabilitation, and long-term care environments. It is a non-invasive alternative to bladder palpation and intermittent catheterization used to assess bladder volume, urinary retention, and post-void residual volume in post-operative patients who may have decreased urine output; in patients with urinary tract infections (UTIs), urinary incontinence, enlarged prostate, urethral stricture, neurogenic bladder, and other lower urinary tract dysfunctions; or in patients with spinal cord injuries, stroke, and diabetes.

Considering the importance of measurement of volume of the urinary bladder, a novel approach using edge-based image segmentation has been employed. The technique discussed in this paper deals with a method for automatic edge-based image segmentation of the urinary bladder from a 2D ultrasound image for finding the area and volume of the urinary bladder accurately. The study of area and volume would provide valuable information about the abnormalities of the bladder and also the extent of abnormality.

Figure 1 shows the bladder ultrasound image of a normal person. The noise present in the image is suppressed by the application of a Gaussian filter.

At present, the area and volume of the bladder are obtained by a trackball arrangement. Here, the operator marks points over the boundary manually. The marked points are approximated into an ellipsoid, as shown in Fig. 2, and then, the area and volume of the ellipsoid are calculated.

This value is considered as the volume of the urinary bladder. The accuracy of this method depends on the tracking points marked by the operator.

The volume can then be calculated using the formula for a prolate ellipse [1]

$$\text{Volume} = \pi/6(W \times D \times H) \quad (1)$$

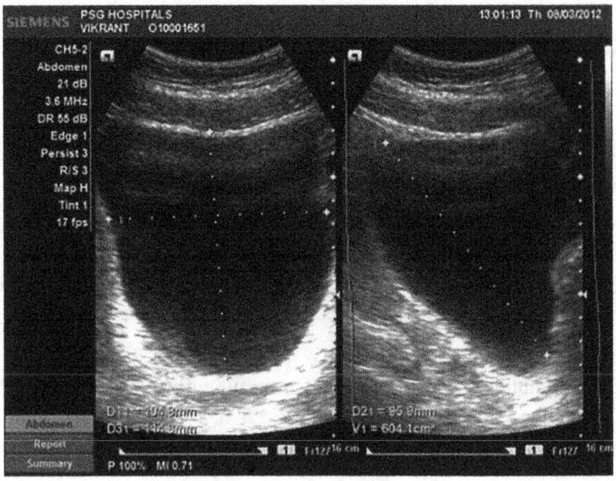

Fig. 1 Bladder ultrasound image

Fig. 2 Ellipsoid method

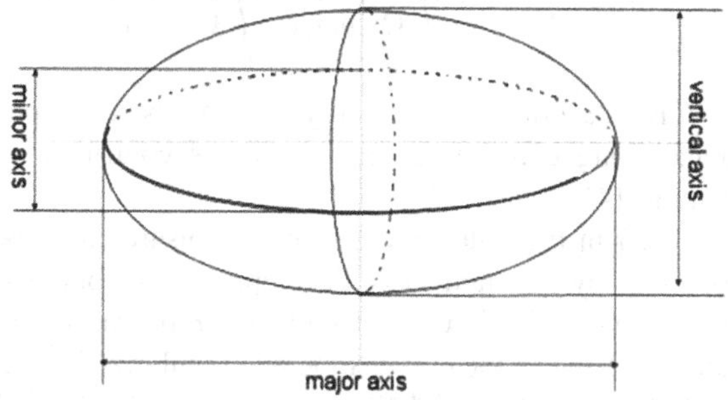

The width (*W*) and the height (*H*) of a bladder are measured from the transversal view of the ultrasound image. The depth (*D*) is the maximum distance in the sagittal view of the ultrasound image [2].

In general, the bladder was assumed an ellipsoid or even spherical configuration when filled with urine [1] in the present clinical studies, and the increase in the urine volume in the bladder was assumed linear. These suppositions are the major factors causing inaccurate estimation of bladder volume.

Here, we propose a method, which helps to accurately calculate the volume of the bladder.

1.1 Localized Region-Based Active Contour

For segmentation method, localized region-based framework for guiding the active contour is used in this study. This framework allows the foreground and background to be described in terms of smaller local regions, removing the assumption that foreground and background regions can be represented with global statistics. Image is defined on the frequency domain, and *C* is a closed contour represented as zero level of a signed distance function φ. First, specify the interior of *C* by following approximation of the smoothed Heaviside function:

$$H\varphi(x) = \begin{cases} 1, & \varphi(x) < -\varepsilon \\ 0, & \varphi(\&x) \geq \varepsilon \end{cases} \tag{2}$$

For the exterior of *C*, it is defined as $(1 - H\varphi(x))$. To specify the area around the curve, the derivative of $H\varphi(x)$, which is a smoothed version of direct delta, is used [3]. The characteristic function in terms of a radius parameter, *r*, is

$$B(x, y) = \begin{cases} 1, & \|x - y\| < r \\ 0, & \text{otherwise} \end{cases} \tag{3}$$

where *r* is used to mask the local regions. Energy function in terms of a generic force function, *F*, can be defined using $\mathcal{B}(x, y)$. The energy function is as follows:

$$E(\varphi) = \int \Omega x\, \delta\varphi(x) \int \Omega y\, \mathcal{B}(x, y) \cdot F(I(y), \varphi(y))\mathrm{d}y\mathrm{d}x \tag{4}$$

F is a generic internal energy measure used to represent local adherence at each point along contour. *E* only considers contributions from the points near the contour [4].

$\delta\varphi(x)$ in the outer integral over x ensures that the curve will not change the topology by spontaneously developing new contours. Every point x selected by $\delta\varphi(x)$, mask with $\mathcal{B}(x, y)$, ensures that *F* operates only on local image information about x. To keep the curve smooth, a regularization term is added. Penalize the arc length of curve and weigh this penalty by a parameter λ. The final energy is as follows:

$$E(\varphi) = \int \Omega x\, \delta\varphi(x) \int \Omega y\, \mathcal{B}(x, y) \cdot F(I(y), \varphi(y))\mathrm{d}y\mathrm{d}x + \lambda \int \Omega x\, \delta\varphi(x)\|\nabla\varphi(x)\|\mathrm{d}x \tag{5}$$

From Eq. (5), the following evolution equation is obtained:

$$\frac{\partial\varphi}{\partial t} = \delta\varphi(x) \int \Omega y\, \mathcal{B}(x, y) \cdot \nabla\varphi(y)\, F(I(y), \varphi(y))\mathrm{d}y\mathrm{d}x \tag{6}$$

Notice that, *F* variation with respect to φ can be computed. All region-based segmentation energies can be put into this framework. The detailed explanation of active contour methods is discussed in [5].

2 Materials and Methods

Numerous data of human urinary bladder ultrasound images were collected. The data had been processed using various image processing techniques such as image enhancement and segmentation and measurement to get the values of height, width, and depth of the bladder. Even though the entire image enhancement is done automatically, the initialization masks still need to be done manually, which finds the region of interest.

In this work, we focused on the technique to improve the quality and information of the content of ultrasound images of the bladder, which includes filtering to suppress speckles and localized region-based active contours to segment the region of interest.

2.1 Ultrasound Image Filtering

Transverse and sagittal views of bladder ultrasound image are used in this study. Therefore, the measurements taken in the study are depth (*D*) and area of bladder region. The ultrasound image is in RGB type, which is an additive color of red,

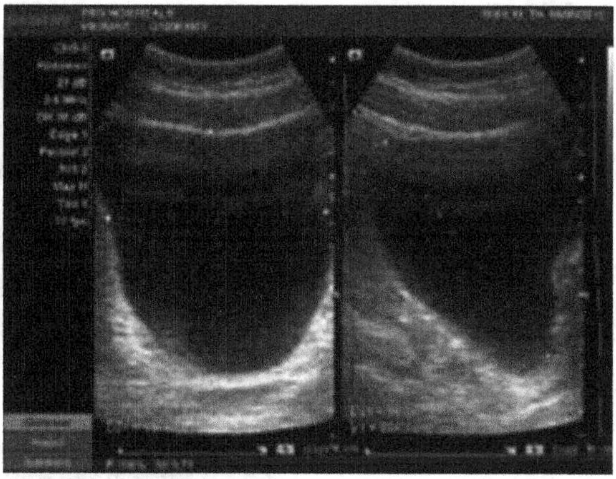

Fig. 3 Bladder image after filtering

green, and blue. The image is first converted into grayscale image for further processing. Image processing toolbox provides image enhancement routines.

The filtering of the image can be observed by applying fspecial [enhances image using Gaussian filter (5 × 5)]. The resulting image after applying fspecial on the original image is shown in Fig. 3. For the segmentation method, traversal view of the bladder is considered. For this purpose, the suitable initialization mask is created for the bladder region, as shown in Figs. 4 and 5.

Resizing the image pixels into the region of interest using initialization mask is significant for efficient image processing. For further processing, the bladder image and initialization mask are segmented using the localized region-based active contour method. The result for the segmentation is shown in Fig. 6. The boundaries or edges of the region behind the mask are detected using suitable operators to find out the region of interest. The data measured are in pixel unit. The numbers of pixels are then converted into its area and displayed. For the sagittal view of bladder region, the methodology is similar to the transversal view, but only

Fig. 4 Initialization mask

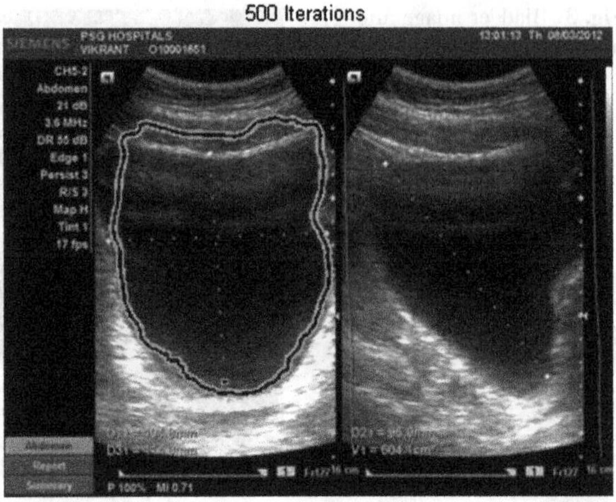

Fig. 5 Localized region-based active contour

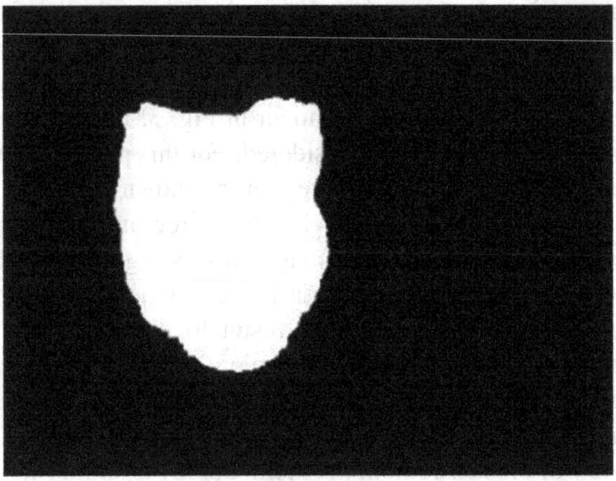

Fig. 6 The binary image from localized region-based active contour

the initialization mask value differs. Finally, the area is displayed in square centimeters. The area is then multiplied by the depth parameter which is obtained in units of centimeters from sagittal ultrasound image to obtain the volume of the bladder in cubic centimeters [6].

3 Results and Discussions

The measurements for depth and area of human urinary bladder are represented in SI unit (centimeter). Experiments involve five samples of human urinary bladder ultrasound image from different people. The number of pixels counted was converted into the area and multiplied with the depth parameter to find the volume of

Table 1 The measurements of calculated volume through image processing technique by ultrasound machine and actual volume of five people

Volume in cm^3 (machine reading)	Calculated volume in cm^3 (Proposed methodology)	Actual (voided) ± Residual volume (5 cm^3)
604	619	616
271	308	305
204	199	197
204	199	197
230	220	217

the urinary bladder. The accuracy of our proposed method is close to the actual voided volume of urine. As per the readings obtained in Table 1, the accuracy is almost 98 %.

4 Conclusion

Ultrasound images are widely used as a tool for clinical diagnosis. Therefore, a convenient system for bladder segmentation, volume measurement, and ultrasound image's noise reduction is of particular interest. The proposed MATLAB method includes image noise reduction using Gaussian filter. The experiment results show that the proposed method can be used to segment the bladder region and give faster convergence compared to the active contour snake method. To conclude, the calculation of depth, area, and volume of bladder is successfully done using MATLAB. In the future, this work could be extended to high-level image processing such as processing in real-time applications.

References

1. Maroulis, D.E., Savelonas, M.A., Karkanis, S.A., Iakovidis, D.K., Dimitropoulos, N, "Variable Background Active Contour Model for Computer-Aided Delineation of Nodules in Thyroid Ultrasound Images" 18th IEEE Symposium of Computer-based Medical System, 2005, pp. 271–276.
2. M. Fa.Gerquist, U. Fa.Gerquist, A.Ode. N and S. G. Blomberg, "Estimation of fetal urinary bladder volume using the sum-of-cylinders method vs. the ellipsoid formula" Ultrasound Obstet Gynecol 2003; 22: 67–73 Wiley InterScience.
3. Slappendel R, Weber EW, "Non-invasive measurement of bladder volume as an indication for bladder catheterisation after orthopaedic surgery and its effect on urinary tract infections" Eur J Anaesthesiol 1999.
4. Lankton, S.; Tannenbaum, A, "Localizing Region-Based Active Contours". IEEE Transaction on Image Processing, 2008, 17(11), pp. 2029–2039.
5. J Dudley et al, "Clinical agreement between automated and calculated ultrasound measurements of bladder volume", pp. 832–834 The British Institute of Radiology, 2003.
6. T. F. Chan and L. A. Vese, "Active contours without edges", IEEE Transactions on Image Processing 2001, pp. 266–277.

Improved Bijective-Soft-Set-Based Classification for Gene Expression Data

S. Udhaya Kumar, H. Hannah Inbarani and S. Senthil Kumar

Abstract One of the important problems in using gene expression profiles to forecast cancer is how to effectively select a few useful genes to build exact models from large amount of genes. Classification is also a major issue in data mining. The classification difficulties in medical area often classify medical dataset based on the outcomes of medical analysis or report of medical action by the medical practitioner. In this study, a prediction model is proposed for the classification of cancer based on gene expression profiles. Feature selection also plays a vital role in cancer classification. Feature selection techniques can be used to extract the marker genes to improve classification accuracy efficiently by removing the unwanted noisy and redundant genes. The proposed study discusses the bijective-soft-set-based classification method for gene expression data of three different cancers, which are breast cancer, lung cancer, and leukemia cancer. The proposed algorithm is also compared with fuzzy-soft-set-based classification algorithms, fuzzy KNN, and k-nearest neighbor approach. Comparative analysis of the proposed approach shows good accuracy over other methods.

Keywords Soft set · Bijective soft set · Classification · Improved bijective soft set

---

S. Udhaya Kumar (✉) · H. Hannah Inbarani · S. Senthil Kumar
Department of Computer Science, Periyar University, Salem 636011, India
e-mail: uk2804@gmail.com

H. Hannah Inbarani
e-mail: hhinba@gmail.com

S. Senthil Kumar
e-mail: ssenthil3@hotmail.com

G. S. S. Krishnan et al. (eds.), *Computational Intelligence, Cyber Security and Computational Models*, Advances in Intelligent Systems and Computing 246,
DOI: 10.1007/978-81-322-1680-3_14,

1 Introduction

Classification and feature reduction are wide areas of research in data mining. Many practical applications of classification involve a large volume of data and/or a large number of features/attributes [1]. Gene expression denotes the activation level of every gene within an organism at an exact point of time. The classification is the process of predicting the classes among the massive amount of dataset using some machine learning algorithms. The classification of different tumor types in gene expression data is of great importance in cancer diagnosis and drug discovery, but it is more complex because of its huge size [2]. There are a lot of methods obtainable to evaluate a gene expression profiles. A common characteristic of these techniques is selecting a subset of genes that is very informative for classification process and to reduce the dimensionality problem of profiles. In this study, entropy filter is used for selecting informative genes [2].

In bijective soft set, every element can be only mapped into one parameter and the union of partition by parameter set is universe [3]. In this study, improved bijective-soft-set-based classification is applied for generating decision rules from the reduced training dataset. The generated decision rules can be used for classification of test data given. In this study, the proposed bijective-soft-set-based approach is compared with fuzzy-soft-set-based classification algorithms, fuzzy KNN, and k-nearest neighbor approach [4]. In comparison with other methods, improved bijective-soft-set-based classification provides more accuracy.

The rest of the paper is structured as follows: Sect. 2 describes the fundamental concepts of soft set and bijective soft set. Section 3 describes overall structure of the proposed work, Sect. 4 discusses experimental analysis, and Sect. 5 describes the conclusion.

2 Basic Concepts of Set Theory and Bijective Soft Set Theory

In this section, we describe the basic notions of soft sets and bijective soft set. Let U be the initial universe of objects and E be a set of parameters in relation to objects in U. Parameters are often attributes, characteristics, or properties of objects [5].

Definition 1 A pair (F, A) is called a soft set over U, where F is a mapping given by $F: A \rightarrow P(U)$.

Definition 2 Let (F, B) be a soft set over a common universe U, where F is a mapping $F: B \rightarrow P(U)$ and B is non-empty parameter set [3].

Definition 3 Let $U = \{x_1, x_2, \ldots, x_n\}$ be a common universe, *X* be a subset of *U*, and (*F*, *E*) be a bijective soft set over *U*. The operation of "(*F*, *E*) restricted AND *X*" denoted by $(F, E)\tilde{\wedge}X$ is defined by $U_{e \in E}\{F(e)\colon F(e) \subseteq X\}$ [3].

Definition 4 Let $U = \{x_1, x_2, \ldots, x_n\}$ be a common universe, *X* be a subset of *U*, and (*F*, *E*) be a bijective soft set over *U*. The operation of "(*F*, *E*) relaxed AND *X*" denoted by $(F, E)\tilde{\wedge}X$ is defined by $U_{e \in E}\{F(e)\colon F(e) \cap X \neq \emptyset\}$ [3].

3 Proposed Work

Based on the bijective soft set theory mentioned in the last section, we begin our experiment. The whole experimental process includes 3 steps: (1) discretization, (2) pretreatment, and (3) classification.

In the first step, datasets are discretized based on class–attribute contingency coefficient discretization algorithm because gene expression datasets include only continuous-valued attributes [6]. In the second step, informative genes are selected using entropy filtering. The efficiency of the genes is considered using entropy filter method. Entropy measures the uncertainty of a random variable. For the measurement of interdependency of two random genes *X* and *Y*, Shannon's information theory is used [2]. Then, improved bijective-soft-set-based classification approach is applied for generating rules.

3.1 Improved Bijective-Soft-Set-Based on Classification: Proposed Approach

In the improved bijective-soft-set-based classification algorithm, two types of rules are generated. First type of rule is deterministic rule (certain rule) and are generated using AND and restricted AND operation [1]. Second type of rule is non-deterministic rule (possible rule) and are achieved by using AND and relaxed AND operations. For each non-deterministic rule, support is computed. Improved bijective-soft-set-based classification approach is presented in Fig. 1.

Table 1 represents a sample of the dataset as an example in order to extract the rules. Let A = {A1, A2} be the set of condition attributes, and D is the decision attribute. Decision attributes {B, D} stand for benign and malignant.

The proposed approach is explained with an example given in Table 1.

Step 1: Construct bijective soft set from conditional attributes [1].

Step 2: Construct bijective soft set from decision attribute.

B (Benign) = {X1, X2, X3, X4, X7, X9, X10} D (Malignant) = {X5, X6, X8}

Step 3: Apply AND operation on the bijective soft set (F_i, A_i).

$(F1, A1) \wedge (F2, A2) = (H, C) = \{\{X1, X3, X9\}, \{X5, X6\}, \{X7, X8\}, \{X2, X4\}\}$

Step 4: Generate deterministic rules by using $U_{e \in E}\{F(e)\colon F(e) \subseteq X\}$

Improved Bijective soft set classification (IBISOCLASS)

Input : Given Dataset with conditional attributes 1, 2 . . . n-1 and Decision attribute n.
Output: A set of Rules R

Step 1: Construct Bijective soft set for all conditional attributes (F_i,E_i) for i=1 to n-1, n is the number of Attributes using Definition **2.**

Step 2: Construct Bijective soft set for decision attribute (G,B) using **Definition 2**.

Step 3: Apply AND on the Bijective soft set (F_i,E_i). Result is stored in (H, C).

Step 4: Generate deterministic rules using (**Definition 3**) $U_{e \in E}$ {F (e):F (e) ⊆X

Step 5: Generate Non-Deterministic rules using (**Definition 4**) $U_{e \in E}\{F(e) : F(e) \cap X \neq \emptyset\}$

Step6: Compute the support value for each non-deterministic rule using $support = \frac{support(A \wedge B)}{support(A)}$ where *A* is the description on condition attributes and *B* the description on decision attributes

Fig. 1 Improved bijective soft set classification

Table 1 Sample dataset

	X1	X2	X3	X4	X5	X6	X7	X8	X9	X10
A1	1	1	1	1	2	2	2	2	2	1
A2	1	2	1	2	1	1	2	2	2	1
D	B	B	B	B	M	M	B	M	M	B

(H, C) Restricted AND B (Benign) = {{X1, X3, X9}, {X2, X4}}
(H, C) Restricted AND D (Malignant) = {X5, X6}
(F, B) = {{X1, X3, X9}, {X2, X4}, {X5, X6}}
If A1 = 1 and A2 = 1 => d = Benign If A1 = 1 and A2 = 2 => d = Benign
If A2 = 2 and A2 = 1 => d = Malignant.
Step 5: Generate non-deterministic rules using $U_{e \in E}\ \{F(e)\text{: } F(e) \cap X \neq \emptyset\}$

$$(F, B) = \{\{X1, X3, X7, X10\}, \{X2, X4\}, \{X5, X6, X8\}\}$$

If A1 = 1 and A2 = 1 => d = Benign If A1 = 1 and A2 = 2 => d = Benign
If A2 = 2 and A2 = 1 => d = Malignant If A1 = 2 and A2 = 2 => d = Benign
If A1 = 2 and A2 = 2 => d = Malignant.
Step 6: Support (Benign, Malignant) = $\left(\frac{1}{10} \wedge \frac{2}{10}\right) = 0.5$.

4 Experimental Analysis

The dataset is collected from public microarray data repository [7]. In this study [8], the three cancer gene expression datasets selected for experiment are lung cancer dataset, leukemia dataset, and breast cancer dataset. The lung cancer dataset consists of 7,219 genes and 64 samples, and it belongs to the class ALL/AML. The leukemia cancer dataset consists of 7,219 genes and 96 samples, and it belongs to the class tumor/normal. The breast cancer dataset consists of 5,400 genes and 34 samples, and it belongs to the class benign/malignant.

4.1 Validation Measures

Validation is the key to making developments in data mining, and it is especially important when the area is still at the early stage of its development. Many validation methods are available. In this paper, we measure accuracy of the proposed algorithm based on precision, recall, and F-measure. The proposed algorithm's accuracy is compared with three existing algorithms such as fuzzy-soft-set-based

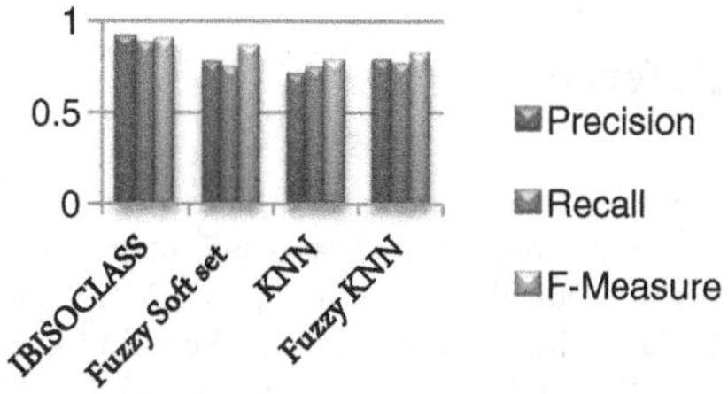

Fig. 2 Comparative analysis of classification algorithms for lung cancer dataset

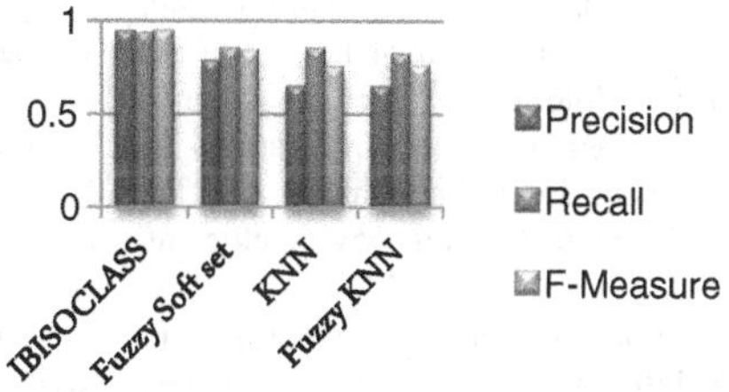

Fig. 3 Comparative analysis of classification algorithms for leukemia cancer dataset

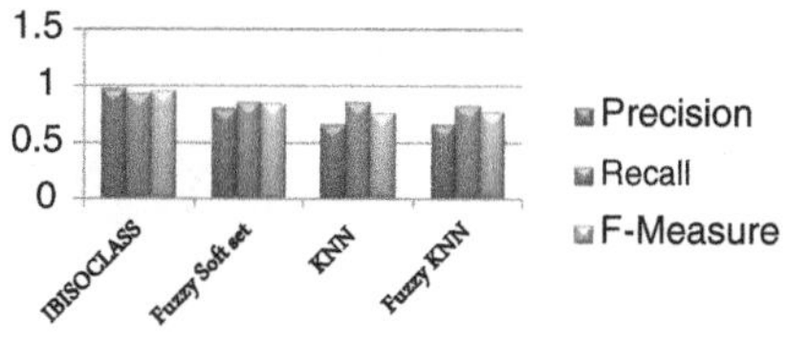

Fig. 4 Comparative analysis of classification algorithms for breast cancer dataset

classification [4], KNN, and fuzzy KNN methods [4]. Figures 2, 3, and 4 show the comparative analysis of classification algorithms for gene expression datasets.

From the above results, it can be easily concluded that the IBISOCLASS algorithm is an effective one for classification of gene expression cancer dataset.

5 Conclusion

Classification is the key method in microarray technology. In this paper, classification technique is applied for predicting the cancer types among the genes in various cancer gene expression datasets such as leukemia cancer gene expression dataset, lung cancer gene expression dataset, and breast cancer gene expression dataset. In this study, improved bijective soft set approach is proposed for classification of gene expression data. The classification accuracy of the proposed approach is compared with the fuzzy soft set, fuzzy KNN, and KNN algorithm. The investigational analysis shows the effectiveness of the proposed improved bijective soft set approach over the other three methods. In future, it can be applied for other datasets also.

References

1. S. Udhaya Kumar, H. Hannah Inbarani, S. Senthil Kumar, "Bijective soft set based classification of Medical data", International Conference on Pattern Recognition, Informatics and Medical Engineering, pp. 517–521, 2013.
2. Hamid Mahmoodian et al., "New Entropy-Based Method for Gene Selection", IETE Journal of Research, vol. 55, no. 4, pp. 162–168, 2009.
3. K. Gong et al., "The Bijective soft set with its operations", An International Journal on Computers & Mathematics with Applications, vol. 60, no. 8, pp. 2270–2278, 2010.
4. N. Kalaiselvi, H. Hannah Inbarani, "Fuzzy Soft Set Based Classification for Gene Expression Data", International Journal of Scientific & Engineering Research, vol. 3, no. 10, 2012.
5. D. Molodtsov, "Soft set theory-first results", An International Journal on Computers & Mathematics with Applications, vol. 37, no. 4–5, pp. 19–31, 1999.
6. Cheng-Jung Tsai, Chien-I Lee, Wei-Pang Yang, "A discretization algorithm based on Class-Attribute Contingency Coefficient", An International Journal on Information Sciences, vol. 178, pp. 714–73, 2008.
7. http://www.broadinstitute.org/cancer/datasets.breastcancer/
8. http://datam.i2r.a.star.edu.sg/datasets/krdb/.

Mammogram Image Classification Using Rough Neural Network

K. T. Rajakeerthana, C. Velayutham and K. Thangavel

Abstract Breast cancer is the second leading cause of cancer deaths in women, and it is the most common type of cancer prevalent among women. Detecting tumor using mammogram is a difficult task because of complexity in the image. This brings the necessity of creating automatic tools to find whether a tumor is present or not. In this paper, rough set theory (RST) is integrated with back-propagation network (BPN) to classify digital mammogram images. Basically, RST is used to handle more uncertain data. Mammogram images are acquired from MIAS database. Artifacts and labels are removed using vertical and horizontal sweeping method. RST has also been used to remove pectoral muscles and segmentation. Features are extracted from the segmented mammogram image using GLCM, GLDM, SRDM, NGLCM, and GLRM. Then, the features are normalized, discretized, and then reduced using RST. After that, the classification is performed using RNN. The experimental results show that the RNN performs better than BPN in terms of classification accuracy.

Keywords Mammogram · BPN · Discretization · Rough neural network (RNN) · Rough set theory

K. T. Rajakeerthana
Department of Electrical and Electronics Engineering, Kongu Engineering College
Perundurai, Erode, Tamil Nadu 638052, India
e-mail: keerthi@yahoo.com

C. Velayutham (✉)
Department of Computer Science, Aditanar College of Arts and Science,
Virapandianpatnam, Tiruchendur, Tamil Nadu 628216, India
e-mail: cvelayutham22@yahoo.com

K. Thangavel
Department of Computer Science, Periyar University, Salem, Tamil Nadu 636011, India
e-mail: drktvelu@yahoo.com

G. S. S. Krishnan et al. (eds.), *Computational Intelligence, Cyber Security and Computational Models*, Advances in Intelligent Systems and Computing 246,
DOI: 10.1007/978-81-322-1680-3_15,

1 Introduction

Breast cancer is the most commonly diagnosed cancer prevalent among women. It is the second leading cause of cancer death in women, exceeded only by lung cancer. The chance that breast cancer may be responsible for a woman's death is about 1 in 36 (about 3 %). In the United Kingdom, every year there are about 45,000 cases diagnosed, and more than 13,200 women die from this cancer, i.e., death rate is 29.3 % (Cancer Research UK, 2009). About 1 in 8 (12 %) women in United States will develop breast cancer during their lifetime.

Mammography is the most common investigation technique used by radiologists in the screening and in the diagnosis of breast cancer. Mammography is a low-dose X-ray procedure that allows visualization of the internal structure of the breast. Mammography is highly accurate, but like most medical tests, it is not perfect. On an average, mammography will detect about 80–90 % of the breast cancers in women without symptoms [1].

Many researchers implemented the rough neural network (RNN) in different domains. Wei Pan [2] have used neural network in intelligent control. Dongbo Zhang and Yaonan Wang [3] have applied fuzzy RNN in vowel recognition. Wei Wang and Hong Mi [4] have applied RNN in customer management system that predicts customer purchasing rule. Gang Wang et al. [5] applied the RNN for classified the counting of knowledge discovery. N. A Setiawan used RNN in predicting missing attributes. Chengxi Dong et al. [6] applied RNN in the decision analysis and obtained good results. Peters et al. [7] applied the approximate neural network based on granularity in the power system, which is made up of rough approximate neuron and decision-making neuron and obtained high accuracy in the test of power system fault classification.

The rest of this paper is organized as follows: Sect. 2 describes RNN classifier. The experimental results are discussed in Sect. 3, and conclusion is presented in Sect. 4.

2 Rough Neural Network Classifier

Rough set and neural networks can solve the complex and high-dimensional problems, which are called RNNs [8, 9]. A rough neuron can be viewed as a pair of neurons, in which one neuron corresponds to the upper boundary and the other corresponds to the lower boundary. Upper and lower neurons exchange information with each other during the calculation of their outputs. The rough set theory (RST) is used as a tool to detect uncertain data, and 'inference decision making' is done based on the information system. Figure 1 shows the algorithm of RNN classifier.

The $\mathrm{BND}_P(Q)$ values are considered as uncertain values, and inference decision making is done based on the similarity measure (1). The similarity measure is

RNN(C, D)
C, Conditional features
D, Decision feature

$[x]_C \leftarrow C$
$[x]_D \leftarrow D$
$\mathrm{IND}(C) \leftarrow [x]_C$
$\overline{C}X \leftarrow \{x \in U \mid [x]_C \cap X \neq \Phi\}$
$\underline{C}X \leftarrow \{x \in U \mid [x]_C \subseteq X\}$
$\mathrm{BND}_C(X) \leftarrow \mathrm{U}\overline{C}X - \mathrm{U}\underline{C}X$
$\mathrm{CEN}_D(X) \leftarrow \underline{C}X$
$\forall \mathrm{BND}_C(X)$
    $\forall \mathrm{CEN}_D(X)$
        $Dij \leftarrow \mathrm{SIM}(\mathrm{BND}_C(D), \mathrm{CEN}_D(X))$
    $ND \leftarrow \min(Dij, \mathrm{CEN}_D(X))$
$Update \;\; D \leftarrow ND$
Call $\mathrm{BPN}(C, D)$

Fig. 1 Algorithm of RNN classifier

evaluated for each element of $\mathrm{BND}_P(Q)$ with the centroid of each class lower approximation $\underset{X \in U/Q}{\cup} \underline{P}X(D)$, and the decision value is updated according to the closest centroid.

$$D_{ij} = \sum_{k=1}^{n} \sqrt{(X_{ik} - X_{jk})^2} \tag{1}$$

3 Experimental Results

This section presents the experimental results of mammogram image classification using BPN and RNN classifiers. The Mammography Image Analysis Society (MIAS) database is used in this paper to study the efficiency of the proposed mammogram image classification using RST. It is a benchmark database available online for research. It consists of 322 images, which belong to three main categories: normal, benign, and malign. There are 209 normal images. Sixty-one benign and fifty malign images are considered abnormal.

There are 12 different data sets that are extracted from the 322 mammogram images and used for machine learning experiment. The data set name described

Table 1 Data set information

Data set index	Data set name	No. of instances	No. of extracted features	No. of classes
1	GLCM_0	322	14	3
2	GLCM_45	322	14	3
3	GLCM_90	322	14	3
4	GLCM_135	322	14	3
5	GLDM	322	14	3
6	SRDM	322	14	3
7	NGLCOM	322	14	3
8	NGLDM	322	14	3
9	GLRLM_0	322	7	3
10	GLRLM_45	322	7	3
11	GLRLM_90	322	7	3
12	GLRLM_135	322	7	3

according to the textural description matrix and orientation (degree) are used to extract features from the mammogram image. The data set index, data set names, number of instances, number of features, and number of classes in the corresponding data set are given in Table 1.

3.1 Texture Description Matrices

The statistical Haralick features are extracted from the segmented mammogram images using the texture description methods GLCM, GLDM, SRDM, NGLCOM, NGLDM, and run-length features using the texture description method GLRLM [10].

3.2 Feature Selection

Six different feature selection algorithms such as quick reduct (QR), entropy-based reduct (EBR), relative reduct (RR), unsupervised quick reduct (USQR) [11], unsupervised entropy-based reduct (UEBR) [12], and unsupervised relative reduct (USRR) [13] are used to select features.

The classification accuracy is evaluated for both BPN classifier and RNN classifier for all the twelve data sets which are given in Table 2. It is observed that the performance of RNN classifier shows some improvement over the performance of BPN classifier in all the data sets. Figure 2 shows the mean classification accuracy of all the data sets for BPN and RNN classifiers. It shows that the classification performance of the proposed RNN classifier gives the highest classification accuracy than the BPN classifier.

Table 2 Classification accuracy

Data set index	QR		EBR		RR		USQR		UEBR		URR	
	BPN	RNN	BPN	RNN	BPN	RNN	BPN	RNN	BPN	RNN	BPN	RNN
1	100.0	100.0	99.7	100.0	97.3	99.8	100.0	100.0	100.0	100.0	99.6	100.0
2	95.6	100.0	94.7	100.0	99.7	100.0	93.7	100.0	96.5	100.0	98.5	99.7
3	98.5	100.0	96.8	99.7	98.5	100.0	98.5	99.8	98.1	99.0	94.5	100.0
4	97.7	100.0	97.4	100.0	95.8	100.0	96.9	100.0	96.7	99.8	96.7	100.0
5	88.2	100.0	92.7	99.7	89.2	97.3	90.5	99.0	89.2	99.5	93.4	99.3
6	94.0	99.8	96.4	99.8	93.7	99.4	95.0	98.3	95.6	100.0	93.0	99.0
7	89.4	97.0	88.5	97.7	91.2	98.7	88.7	99.5	90.6	100.0	90.7	98.6
8	99.7	100.0	98.5	100.0	98.7	100.0	98.0	98.9	97.5	99.6	98.2	100.0
9	62.8	99.6	66.4	99.7	67.4	98.3	67.5	99.5	72.5	98.1	69.7	98.6
10	66.0	98.0	69.7	98.4	69.7	99.2	69.7	97.6	70.3	99.0	68.2	99.5
11	72.5	98.8	73.4	99.6	78.7	98.7	76.5	98.9	78.0	97.0	75.6	96.8
12	58.5	99.5	61.5	98.2	72.5	97.5	70.1	99.0	62.8	98.5	63.5	97.5

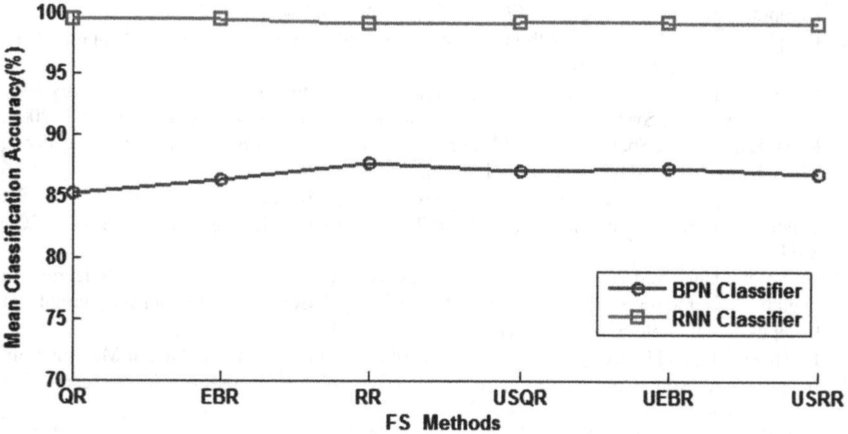

Fig. 2 Mean classification performance of BPN classifier and RNN classifier

4 Conclusion

In this paper, the RNN is proposed for mammogram image classification. It is tested on real database MIAS with different feature extraction methods and different FS methods. The RNN classifier produces 99.2 % for overall classifying accuracy, which is higher than 86.7 % using neural network back-propagation (BPN) classifier.

Acknowledgments The third author gratefully acknowledges the UGC, New Delhi, for partial financial assistance under UGC-SAP(DRS) Grant No. F3-50/2011.

References

1. J. Michaelson, S. Satija, and R. Moore, "The pattern of breast cancer screening utilization and its consequences", vol. 94, no. 1, pp. 37–43, 2002.
2. Wei Pan, "Rough set theory and its application in the intelligent systems", Proceedings of the 7th World Congress on Intelligent Control and Automation, pp. 3076–3081, 2008.
3. Dongbo Zhang, Yaonan Wang, "Fuzzy-rough neural network and its application to vowel recognition", Control and Decision, vol. 21, no. 2, pp. 221–224, 2006.
4. Wei Wang, and Hong Mi, "The application of rough neural network in RMF model", Proceedings of 2nd International Asia Conference on Informatics in Control, Automation and Robotics, pp. 210–213, 2010.
5. Gang Wang, Chenghong Zhang, and Lihua Huang, "A study classification algorithm for data mining based on hybrid intelligent systems", Proceedings of Ninth ACIS International Conference on Software Engineering, Artificial Intelligence, Networking, and Parallel/ Distributed Computing, pp. 371–375, 2008.
6. Chengxi Dong, Dewei Wu, and Jing He, "Decision analysis of combat effectiveness based on rough set neural network", Proceedings of Fourth International Conference on Natural computation, pp. 227–231, 2008.
7. J F Peters, L Han, and S Ramanna, "Rough neural computing in signal analysis", Computational Intelligence, vol. 17, no. 3, pp. 493–513, 2001.
8. Dongbo Zhang, "Integrated methods of rough sets and neural network and their applications in pattern recognition", Hunan university, 2007.
9. Weidong Zhao, and Guohua Chen. "A survey for the integration of rough set theory with neural networks", Systems engineering and electronics, vol. 24, no. 10, pp. 103–107, 2002.
10. R. M. Haralick, K. Shanmugan, and I. Dinstein, "Textural features for image classification", IEEE Trans. Syst., Man, Cybern., vol. 3, pp. 610–621, 1973.
11. C. Velayutham, and K. Thangavel, "Unsupervised Quick Reduct Algorithm Using Rough Set Theory", Journal of Electronic Science and Technology (JEST), vol. 9, no. 3, pp. 193–201, 2011.
12. C. Velayutham, and K. Thangavel, "Entropy Based Unsupervised Feature Selection in Digital Mammogram Image Using Rough Set Theory", International Journal of Computational Biology and Drug Design, vol. 5, no. 1, pp. 16–34, 2012.
13. K. Thangavel, and C. Velayutham, "Unsupervised Feature Selection in Digital Mammogram Image Using Rough Set Theory", International Journal of Bioinformatics Research and Applications, vol. 8, no. 5, pp 436–454, 2012.

Performance Assessment of Kernel-Based Clustering

Meena Tushir and Smriti Srivastava

Abstract Kernel methods are ones that, by replacing the inner product with positive definite function, implicitly perform a nonlinear mapping of input data into a high-dimensional feature space. Various types of kernel-based clustering methods have been studied so far by many researchers, where Gaussian kernel, in particular, has been found to be useful. In this paper, we have investigated the role of kernel function in clustering and incorporated different kernel functions. We discussed numerical results in which different kernel functions are applied to kernel-based hybrid c-means clustering. Various synthetic data sets and real-life data set are used for analysis. Experiments results show that there exist other robust kernel functions which hold like Gaussian kernel.

Keywords Clustering · Kernel function · Gaussian kernel · Hyper-tangent kernel · Log kernel

1 Introduction

Fuzzy clustering has emerged as an important tool for discovering the structure of data. Kernel methods have been applied to fuzzy clustering, and the kernelized version is referred to as kernel-based fuzzy clustering. The kernel-based classification in the feature space not only preserves the inherent structure of groups in the input space, but also simplifies the associated structure of the data [1]. Since

M. Tushir (✉)
Maharaja Surajmal Institute of Technology, New Delhi, India
e-mail: meenatushir@yahoo.com

S. Srivastava
Netaji Subash Institute of Technology, New Delhi, India
e-mail: ssmriti@yahoo.com

G. S. S. Krishnan et al. (eds.), *Computational Intelligence, Cyber Security and Computational Models*, Advances in Intelligent Systems and Computing 246,
DOI: 10.1007/978-81-322-1680-3_16, © Springer India 2014

Girolami first developed the kernel k-means clustering algorithm for unsupervised classification [2], several studies have demonstrated the superiority of kernel clustering algorithms over other approaches to clustering [3–5]. The point raised regarding the kernel-based clustering method of data partitioning is the choice of the type of kernel function chosen in defining the nonlinear mapping. Clearly, the choice of kernel is data specific; however, in the specific case of data partitioning, a kernel which will have universal approximation qualities such as RBF is most appropriate. In [6], we proposed kernel-based hybrid c-means clustering (KPFCM) as an improvement over possibilistic fuzzy c-means clustering [8] using Gaussian kernel function. In most papers, Gaussian kernel function is used as the kernel function. Different kernels will induce different metric measures resulting in new clustering algorithms. Very few papers have studied other kernel functions such as hyper-tangent kernel function [7, 9]. In this paper, we have tried to investigate the effect of different kernel functions on the clustering results. To our knowledge, this is the first such comparison of kernel clustering algorithms using different kernel functions for general purpose clustering. The paper is organized as follows. A background of kernel-based approach is given in Sect. 2. Kernel-based hybrid c-means clustering with different kernel functions incorporated is described in Sect. 3. The experimental results and comparative analysis are given in Sect. 4 followed by the main conclusions presented in Sect. 5.

2 Kernel-Based Approach

A kernel function is a generalization of the distance metric that measures the distance between two data points as the data points are mapped into a high-dimensional space in which they are more clearly separable. By employing a mapping function, $\Phi(x)$, which defines a nonlinear transformation: $x \to \Phi(x)$, the nonlinearly inseparable data structure existing in the original data space can possibly be mapped into a linearly separable case in the higher-dimensional feature space. Given an unlabeled data set $X = \{x_1, \ldots, x_N\}$ in the p-dimensional space R^p, let Φ be a nonlinear mapping function from this input space to a high-dimensional feature space H.

$$\Phi\colon R^p \to H,\ x \to \Phi(x)$$

It is possible to compute Euclidean distances in feature space without knowing explicitly Φ. This can be done using kernel trick in which the computation of distances of vectors in feature space is just a function of the input vectors.

$$\begin{aligned} \|\Phi(x_k) - \Phi(v_i)\|^2 &= (\Phi(x_k) - \Phi(v_i)) \cdot (\Phi(x_k) - \Phi(v_i)) \\ &= \Phi(x_k) \cdot \Phi(x_k) - 2\Phi(x_k)\Phi(v_i) + \Phi(v_i) \cdot \Phi(v_i) \\ &= K(x_k, x_k) - 2K(x_k, v_i) + K(v_i, v_i) \end{aligned} \quad (1)$$

Table 1 List of kernel functions

Name of kernel	Kernel function	Attribute
Gaussian	$\exp\left(-\lVert x_i - x_j\rVert^2/2\sigma^2\right)$	$K(x,x) = 1$
Hyper-tangent	$1 - \tanh\left(\beta\lVert x_i - x_j\rVert^2/2\sigma^2\right)$	$K(x,x) = 1$
Log	$\log\left(1 + \beta\lVert x_i - x_j\rVert^2\right)$	$K(x,x) = 0$

Some examples of robust kernel functions are given in Table 1.

3 Kernel-Based Hybrid c-Means Clustering

We proposed a kernel-based hybrid c-means clustering (KPFCM) in [6] which used Gaussian kernel in the induced distance metric. The KPFCM model minimizes the following objective function:

$$J_{\text{KPFCM}}(U,V,T) = \sum_{k=1}^{N}\sum_{i=1}^{c}\left(\text{au}_{ik}^m + \text{bt}_{ik}^\eta\right)\lVert\Phi(x_k) - \Phi(v_i)\rVert^2 + \sum_{i=1}^{c}\gamma_i\sum_{k=1}^{N}(1 - t_{ik})^\eta \tag{2}$$

where $\lVert\Phi(x_k) - \Phi(v_i)\rVert^2$ is the square of distance between $\Phi(x_k)$ and $\Phi(v_i)$. Also $0 \le u_{ik} \le 1$, $t_{ik} < 1$, $a > 0$, $b > 0$, $m > 1$ and $\eta > 1$. The constant a defines the relative importance of fuzzy membership, whereas b relates to the typicality value in the objective function. If we confine ourselves to the Gaussian kernel function which is used almost exclusively in the literature, then $K(x,x) = 1$. $\lVert\Phi(x_k) - \Phi(v_i)\rVert^2 = 2(1 - K(x_k - v_i))$. Thus, Eq. 2 can be rewritten as

$$J_{\text{KPFCM}}(U,V,T) = 2\sum_{k=1}^{N}\sum_{i=1}^{c}\left(\text{au}_{ik}^m + \text{bt}_{ik}^\eta\right)(1 - K(x_k, v_i)) + \sum_{i=1}^{c}\gamma_i\sum_{k=1}^{N}(1 - t_{ik})^\eta \tag{3}$$

The update of u_{ik}, v_i and t_{ik} is as follows:

$$u_{ik} = (1/(1 - K(x_k, v_i)))^{1/m-1} / \sum_{j=1}^{c}(1/(1 - K(x_k, v_i)))^{1/m-1} \tag{4}$$

$$v_i = \sum_{k=1}^{N}\left(\text{au}_{ik}^m + \text{bt}_{ik}^\eta\right)K(x_k, v_i)x_k / \sum_{k=1}^{N}\left(\text{au}_{ik}^m + \text{bt}_{ik}^\eta\right)K(x_k, v_i) \tag{5}$$

$$t_{ik} = 1/\left(1 + [2b(1 - K(x_k, v_i)/\gamma_i]^{\frac{1}{\eta-1}}\right) \tag{6}$$

We now incorporate different kernel functions to examine and compare with Gaussian kernel. The names given to different clustering algorithms are KPFCM (using Gaussian kernel), KPFCM-H (hyper-tangent kernel), and KPFCM-L (log kernel). Using a hyper-tangent kernel function, the objective function for KPFCM-H clustering is as follows:

$$J_{\text{KPFCM-H}} = 2\sum_{k=1}^{N}\sum_{i=1}^{c}\left(\text{au}_{ik}^{m} + \text{bt}_{ik}^{\eta}\right)\tanh\left(\frac{\|x_k - v_i\|^2}{\sigma^2}\right) + \sum_{i=1}^{c}\gamma_i\sum_{k=1}^{N}(1 - t_{ik})^{\eta} \quad (7)$$

The cluster center v_i for hyper-tangent kernel function is as follows:

$$v_i = \frac{\sum_{k=1}^{N}\left(\text{au}_{ik}^{m} + \text{bt}_{ik}^{\eta}\right)K(x_k, v_i)\left(1 + \tanh\left(\frac{\|x_k - v_i\|^2}{\sigma^2}\right)x_k\right)}{\sum_{k=1}^{N}\left(\text{au}_{ik}^{m} + \text{bt}_{ik}^{\eta}\right)K(x_k, v_i)\left(1 + \tanh\left(\frac{\|x_k - v_i\|^2}{\sigma^2}\right)\right)} \quad (8)$$

The expressions for u_{ik} and t_{ik} remain the same as in Eqs. 4 and 6, respectively. Using a log kernel function, the objective function for KPFCM-L clustering is as follows:

$$J_{\text{KPFCM-L}} = -2\sum_{k=1}^{N}\sum_{i=1}^{c}\left(\text{au}_{ik}^{m} + \text{bt}_{ik}^{\eta}\right)\log\left(1 + \beta\|x_k - v_i\|^2\right) + \sum_{i=1}^{c}\gamma_i\sum_{k=1}^{N}(1 - t_{ik}) \quad (9)$$

The cluster center v_i for log kernel function is as follows:

$$v_i = \frac{\sum_{k=1}^{N}\left(\text{au}_{ik}^{m} + \text{bt}_{ik}^{\eta}\right)\left(1/1 + \beta\|x_k - v_i\|^2\right)x_k}{\sum_{k=1}^{N}\left(\text{au}_{ik}^{m} + \text{bt}_{ik}^{\eta}\right)\left(1/1 + \beta\|x_k - v_i\|^2\right)} \quad (10)$$

The update of u_{ik} and t_{ik} for log kernel function is as follows:

$$u_{ik} = \left(\log\left(1 + \beta\|x_k - v_i\|^2\right)\right)^{-1/(m-1)} / \sum_{i=1}^{c}\left(\log\left(1 + \beta\|x_k - v_i\|^2\right)\right)^{-1/(m-1)} \quad (11)$$

$$t_{ik} = 1/(1 + (-b(\log(1 + \beta\|x_k - v_i\|^2)/\gamma_i)^{-1/(\eta-1)})) \quad (12)$$

4 Experimental Study

A series of experiments were run for a variety of data sets using the KPFCM clustering. The objective of this comprehensive suite of experiments is to come up with a thorough comparison of the performance of the KPFCM clustering using different kernel functions in the objective function. Many two-dimensional

Table 2 Terminal centroids produced by KPFCM, KPFCM-H, and KPFCM-L on X_{12}

Clustering	KPFCM	KPFCM-H	KPFCM-L
Centroids	[−3.33 0.002 3.33 0.002]	[−3.4 0 3.4 0]	[−3.33 0.001 3.33 0.001]
E_*	0.005	0	0.004

synthetic data sets were used with a wide variety in the shape of clusters, number of data points, and count of features of each datum. The real-life data sets used in the experiments are well-known Iris data sets.

4.1 Identical Data with Noise

The first simulation experiment involves the data set X_{12} [8]. The ideal (true) centroids for the X_{12} data set are $V_{\text{ideal}} = \begin{bmatrix} -3.34 & 0 \\ 3.34 & 0 \end{bmatrix}$. Let V^{12}_{KPFCM}, $V^{12}_{\text{KPFCM-H}}$, $V^{12}_{\text{KPFCM-L}}$ be the final centroids identified by their respective algorithms. The results of the final centroids identified by their respective algorithms are tabulated in Table 2. To show the effectiveness of the proposed algorithm, we also compute the error $E_* = \left\| V_{\text{ideal}} - V^{12}_* \right\|^2$, where * corresponds to KPFCM/KPFCM-H/KPFCM-L, respectively. Figure 1 displays the clustering results of three clustering algorithms with different kernel functions. The centroids produced by clustering techniques using different kernel functions are almost identical.

4.2 Dunn Data Set

In general, all clustering techniques perform well for pattern sets that contain patterns of similar volume and similar number of patterns. By changing the volume of clusters in a pattern set, we observe the effectiveness of different clustering

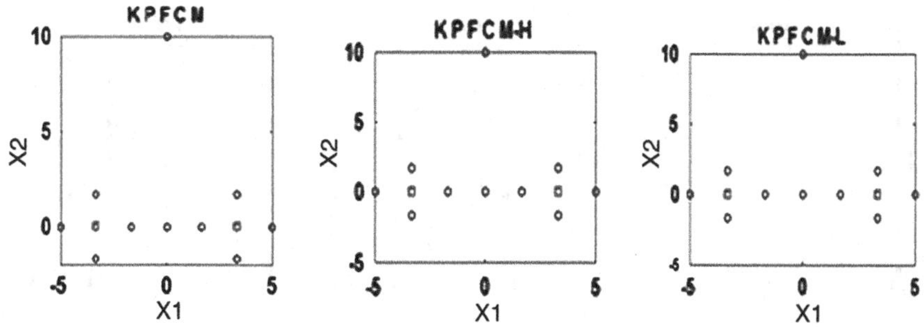

Fig. 1 Clustering results for data X_{12}

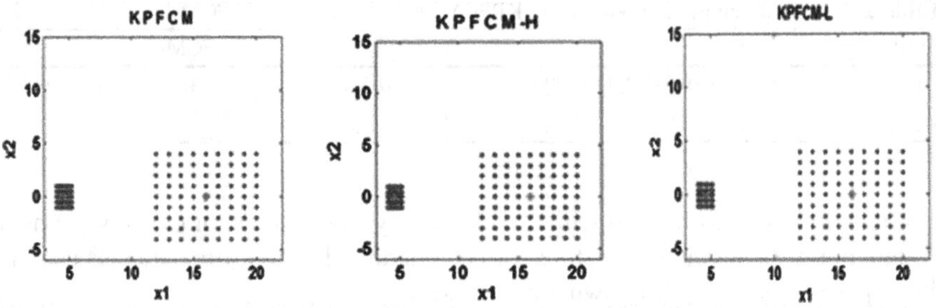

Fig. 2 Clustering results for Dunn data set

algorithms. Figure 2 displays the clustering results of three clustering algorithms. We see that all clustering algorithms correctly partition the two clusters with almost same accuracy.

4.3 Gaussian Random Data

In this example, a Gaussian random number generator was used to create a data set consisting of two clusters. The noise points are then added to the lower cluster to obtain a difference in volume compared to the upper cluster. Figure 3 shows the clustering results of three clustering algorithms. There are only two misclassifications in case of KPFCM-H. The results produced by KPFCM and KPFCM-L are completely identical.

4.4 Iris Data Set

This is a four-dimensional data set containing 50 samples each of three species of Iris flowers. One class is linearly separable from the other two; the latter are not linearly separable from each other. As indicated in Table 3, the typical result of

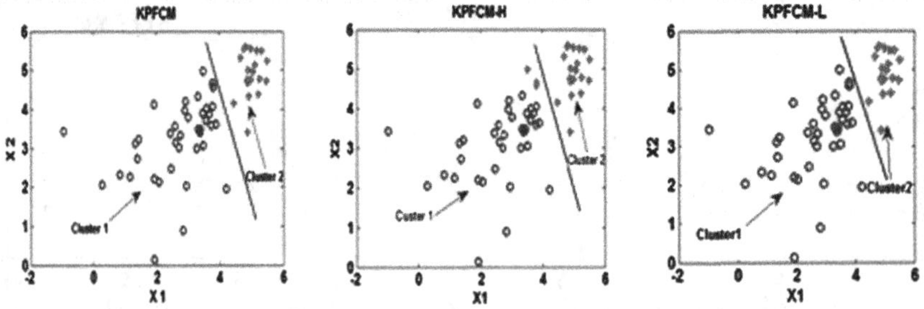

Fig. 3 Clustering results for Gaussian random data set

Table 3 Number of misclassified data and accuracies using KPFCM, KPFCM-H, and KPFCM-L method for the Iris data set

Name of algorithm	Misclassifications	Accuracy (%)
KPFCM ($\sigma = 0.5, a = 1, b = 2, \eta = 2$)	11	92.6
KPFCM-H ($\sigma = 0.25, a = 2, b = 2, \eta = 4$)	11	92.6
KPFCM-L ($\beta = 15, a = 1, b = 2, \eta = 2$)	08	94.6

comparing KPFCM partitions to the physically correct labels of Iris is 11 errors. KPFCM-H also gives 11 errors. The number of misclassified data by KPFCM-L algorithm is 08 with an accuracy of 94.6 %.

5 Conclusions

This paper has presented the detailed analysis of different kernel functions on KPFCM clustering. In literature, most kernel-based clustering algorithms use Gaussian kernel functions. We have studied two nonGaussian kernel functions, namely hyper-tangent and log kernel functions. We incorporated these kernel functions in KPFCM clustering and applied these clustering algorithms on a wide variety of synthetic data as well as real-life data set. From the experiments, we have seen that the two nonGaussian kernel functions have worked as well as Gaussian kernel function. To summarize, we conclude that we have another class of robust kernel functions that have worked well in typical clustering examples of nonlinear classification boundaries.

References

1. K.R. Muller et al., An Introduction to Kernel-based Learning Algorithms, IEEE Trans. onNeural Networks, 12 (2) (2001) 181-202.
2. M. Girolami, Mercer Kernel–based Clustering in Feature Space, IEEE Trans. on Neural Networks, 13 (3) (2002) 780-784.
3. F. Camastra, A. Verri, A Novel Kernel Method for Clustering, IEEE Transaction on Pattern Analysis and Machine Intelligence, 27 (5) (2005), 801-805.
4. Z.d.Wu, W.X.Xie, Fuzzy c-means Clustering Algorithm Based on Kernel Methods in: Proc. Of fifth Intl. Conf. on Computational Intelligence and Multimedia Applications (2003) 47-54.
5. D.Q. Zhang, S.C. Chen, Fuzzy Clustering using Kernel Methods, in: Proc. of Intl. Conf. on Control and Automation, China, (2002) 123-128.
6. M. Tushir, S. Srivastava, A New Kernelized Hybrid c-means Clustering Model with Optimized Parameters, J. Applied Soft computing, 10 (2) (2010) 381-389.
7. D.Q. Zhang, S.C. Chen, Clustering Incomplete Data using Kernel-based Fuzzy c-means Algorithm, Neural Processing Letters, 18 (3) (2003) 155-162.
8. N.R. Pal, K. Pal, J. Keller, J.C. Bezdek, A Possibilistic Fuzzy c-means Clustering Algorithm, IEEE Trans. Fuzzy Syst., 13 (4) (2005) 517–530.
9. S. R. Kannanet al, Robust Kernel FCM in Segmentation of Breast Medical Images, Intl Journal Expert Systems with Applications, 38 (4) (2011), 4382-4389.

Registration of Ultrasound Liver Images Using Mutual Information Technique

R. Suganya, R. Kirubakaran and S. Rajaram

Abstract Registration of medical images is done to investigate the disease process and understand normal development and aging. Image registration is the process of transforming different sets of medical image data into one coordinate system. Mutual information (MI) is a popular similarity measure for medical image registration. The preprocessing step for the registration process is done by means of two successive filters namely speckle reduction by anisotropic diffusion (SRAD) filter and median filter called S-mean filter. This work focuses on the registration of ultrasound liver images, and comparison is done by means of optimization techniques using MI, to bring utmost accuracy in computation time, and is very well suited for clinical applications.

Keywords Mutual information · Image registration · S-mean filter · Rigid body transformation · Ultrasound liver images

1 Introduction

In modern radiology, image registration is important for diagnosis, surgical planning, and treatment control. The geometric alignment or registration of multimodality images is a fundamental task in numerous applications in three-dimensional medical image processing.

R. Suganya (✉) · R. Kirubakaran
Department of Computer Science and Engineering, Thiagarajar College of Engineering, Madurai, India
e-mail: rsuganya@tce.edu

R. Kirubakaran
e-mail: kirubakarancse@gmail.com

S. Rajaram
Department of Electronics and Communication Engineering, Thiagarajar College of Engineering, Madurai, India
e-mail: rajaram_siva@tce.edu

G. S. S. Krishnan et al. (eds.), *Computational Intelligence, Cyber Security and Computational Models*, Advances in Intelligent Systems and Computing 246,
DOI: 10.1007/978-81-322-1680-3_17, © Springer India 2014

The bulk of registration algorithms in medical imaging can be classified as landmark based, surface based, and intensity based. Meyer et al. used mutual information (MI) to match breast ultrasound images [1]. Pluim et al. discussed MI-based image registration and its applications [2]. Mark P. Wachowiak et al. also presented multimodal image registration by optimization techniques [3]. But, this method is limited only to "fine-tune" the initial guess, especially where real-time performance is required. Intensity-based registration methods optimize a functional measuring the similarity of all geometrically corresponding intensity pairs for some feature.

In this paper, we proposed an intensity-based image registration using MI technique, to describe the behavior of the 2D ultrasound liver images. Section 2 discusses preprocessing by S-Mean filter. Section 3 explains theoretical background on MI and optimization techniques; Sect. 4 explains MI-based image registration. Section 5 gives the conclusion of the work.

2 Preprocessing by S-Mean Filter

2.1 Importance of Speckle Reduction

The speckle refers to the variation in the intensity of pixels due to several factors. The presence of speckle in a medical image would hide the necessary pathological information, leading to improper results in diagnosis shown in Fig. 1. For better medical image registration, it is essential to remove the speckles from the images. Some of the major limitations in the existing despeckle filters are difficulty in retaining subtle features, such as small cyst or lesion in ultrasound liver images and lack of edge preservation. After several analyses from Yu and Acton [4] and Abd-Elmoniem [5], it was found that the passing of input medical images into a system of subsequent filters consisting of SRAD and median filter effectively removed the noise factor in those images. In order to accomplish the task, the S-mean filter is proposed.

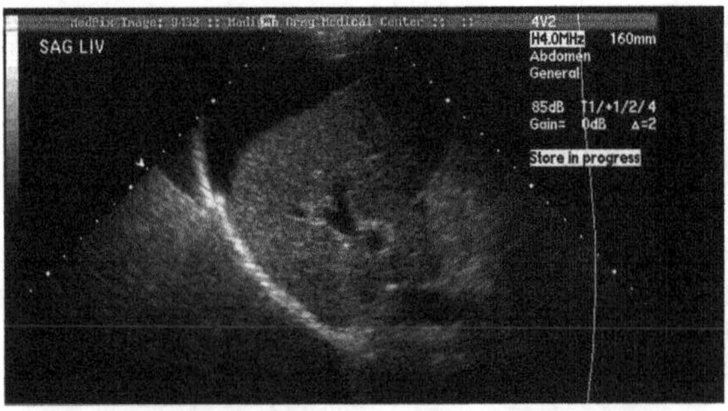

Fig. 1 Image with noise

2.2 Implementation of S-Mean Filter (Speckle Reduction by Anisotropic Diffusion and Median Filter)

Speckle reduction by anisotropic diffusion (SRAD) was introduced by Perona and Malik as a tool to filter images while preserving details and even enhancing edges [6]. The diffusion equation is a partial differential equation given by

$$\frac{\partial \Psi(x,y,t)}{\partial t} = \nabla(w(x,y,t)\nabla\Psi(x,y,t)) \tag{1}$$

$$\Psi(x,y,t) = \Psi(x,y) \tag{2}$$

where Ψ (x, y, t) is the intensity (amplitude) of the input image pixel (seismic event) at locations **x** and **y** and at the diffusion time **t**, and **w** denotes the diffusion coefficients (diffusivity function) which adaptively control the smoothing amount. ∇ denotes the gradient of the image. This method is a monotonically decreasing continuous function of the image gradient magnitude to define the diffusivity function w for their anisotropic nonlinear diffusion filter. This diffusion method can preserve or even enhance prominent edges.

2.3 Median Filter

The output of the SRAD filter is fed into another filter called the median filter. Median filter is a nonlinear filtering method, which is used to remove the speckle noise from an ultrasound image. This filter is popular for reducing the noise without blurring the edges of the image. Thus, the input images have gone through the S-mean filter which is shown in Fig. 2.

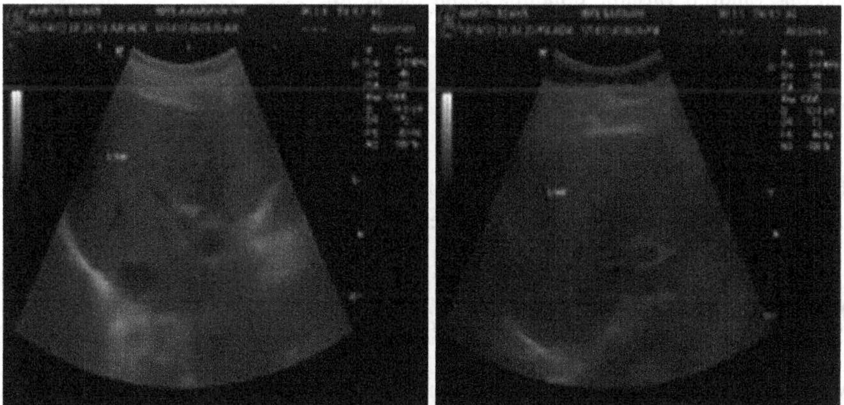

Fig. 2 Liver images after applying both SRAD and median filter

3 Background Theory

3.1 Mutual Information

MI is an automatic, intensity-based metric. Furthermore, it is one of the few intensity-based measures well suited to registration of both monomodal and multimodal images. MI is a statistical measure that finds its roots in information theory [6]. MI is a measure of how much information one random variable contains about another. The Shannon entropy for joint distribution can be defined as

$$-\sum_{i,j} P(i,j) \log P(i,j) \tag{3}$$

where $P(i, j)$ is the joint probability mass function of the random variables i and j, respectively. The MI can also be written in terms of the marginal and joint entropy of the random variables A and B as follows

$$I(A,B) = H(A) + H(B) - H(A,B) \tag{4}$$

where $H(A)$ and $H(B)$ are the entropies of A and B, respectively, and $H(A,B)$ is the joint entropy between the two random variables.

The strength of the MI similarity measure lies in the fact that no assumptions are made regarding the nature of the relationship between the image intensities in both modalities, expect that such a relationship exists. This is not the case for correlation methods, which depend on a linear relationship between image intensities.

3.2 Measures: Entropy

For any probability distribution, entropy is a measure of the amount of information that one random variable contains about the another.

3.3 Transformation: Rigid

Rigid body registration is also used to approximately align images that show small changes in object shape or small changes in object intensity.

3.4 Implementation: Joint Histogram

A joint histogram incorporates additional information from the medical image without sacrificing the robustness of color histograms. This is accomplished through careful selection of a set of local features [8].

3.5 Optimization Techniques

3.5.1 Dividing Rectangles

The algorithm as follows:

1. Normalize the search space to be the unit hypercube. Let C_1 be the center point of this Hypercube and evaluate $f(C_1)$.
2. Identify the set S of the potentially optimal rectangles.
3. For all rectangles j € s: Divide the rectangle containing c into thirds along the dimensions in I, starting with the dimension with the lowest value of $f(c \pm \delta e_i)$ and continuing to the dimension with the highest $f(c \pm \delta e_i)$.

3.5.2 Nelder: Mead Method

A simplex method tries to crawl into global minimum of several variables has been devised by Nelder and Mead. The algorithm is as follows:

1. Evaluate f at the vertices of S and sort the vertices of S so that $f(x_1) \leq f(x_2) \leq \cdots \leq f(x_{N+1})$ holds.
2. Set $f_{\text{count}} = N + 1$.
3. While $f(x_{N+1}) - f(x_1) \leq \tau_{\text{i}}$.

- Compute $x, x = \frac{1}{N} \sum_{i=1}{}^{N} x_i$, $x(\mu_r), x\mu = (1 + \mu)x - \mu x_{N+1}$ and $f_r = f(x(\mu_r))$. $f_{\text{count}} = f_{\text{count}} + 1$
- Reflect: if $f_{\text{count}} = k_{\text{ max}}$ then exit.

  If $f(x_1) \leq f(x_r)) \leq f(x_N)$ replace x_{N+1} with $x(\mu_r)$ and next step

- Shrink if $f_{\text{count}} \geq k_{\max} - N$ then exit.

  For $2 \leq i \leq N + 1$: set $x_i = x_1 - (xi - x1)/2$; compute $f(x_i)$.

- Sort: sort the vertices of S so that

  $f(x_1) \leq f(x_2) \leq \cdots \leq f(x_{N+1})$ holds.

This procedure continues until the stopping criterion is reached. It is effective and computationally compact.

4 Registration of Monomodal Ultrasound Liver Images

4.1 Mutual Information for Ultrasound Liver Images

Ultrasound devices operate with frequencies from 20 kHz up to several gigahertz. The fundamental techniques for image registration involve feature extraction, feature matching, rigid body transformation, and further analyzing for diagnosis. The steps are as follows: (a) preprocessing by S-mean filter on both the target image and reference image, (b) calculating MI with the help of entropy values, and (c) applying a rigid body transformation—"rotation" which helps to relate the target image space to the reference image space. Apply intensity-based image registration for ultrasound-noise-free liver images, finally, a histogram is generated which actually displays the actual output of registration after subsequent transformation. Apply two optimization technique namely DIRECT and Nelder-Mead method in order to bring the computation time to utmost accuracy.

4.1.1 Experimental Results and Discussion

The experimental 2D ultrasound liver dataset used in this work is collected from GEM Hospital, Coimbatore, and Meenakshi mission Hospital, Madurai. Each of these images is a clinical image taken for diagnosis or treatment purposes. The experimental results are as follows: Fig. 3 shows the MI-based image registration. The performances of the two optimization techniques namely DIRECT and Nelder-Mead method for MI-based image registration have been compared.

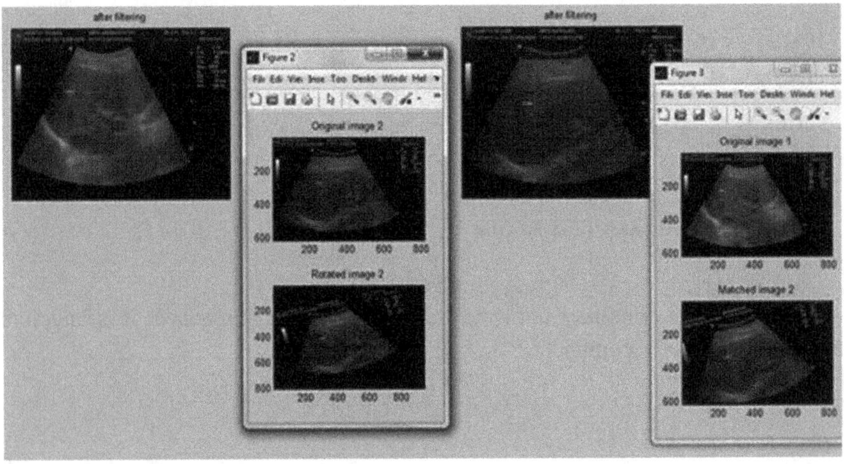

Fig. 3 Mutual information-based image registration

Table 1 Optimization methods using MI

Iteration	Optimization method using MI	Min value
50	DIRECT	0.719
50	Nelder-Mead	0.980

The experimental results indicate DIviding RECTangles (DIRECT) method can quickly produce optimal solutions, as the computation time has been extremely decreased compared to Nelder-Mead method which is shown in Table 1.

5 Conclusion

In this paper, a brief overview of MI and how it can be used to perform ultrasound liver image registration is given. The proposed S-mean filter and further optimization techniques DIRECT and Nelder-Mead are applied to bring utmost accuracy in computation time. The DIRECT method can quickly produce optimal solutions, as the computation time has been extremely decreased compared to Nelder-Mead method. The result demonstrates that our registration technique allows fast, accurate, robust, and completely automatic registration of ultrasound medical images and is very well suited for clinical applications.

References

1. Meyer, C.R., Boes, J. L., Kim, B., and Bland, PH., Lecarpentier, GL., Fowlkes, JB., Roubidoux, MA., and Carson, PL., "Semiautomatic registration of volumetric ultrasound scans," Ultrasound in Medicine & Biology, vol. 25, no. 3, pp. 339–347, 1999.
2. Pluim, J.P.W., Antoine Maintz, J.B., and Viergever, M.A., "Mutual-information-based registration of medical images: a survey", IEEE Transactions on Medical Imaging, vol. 22, pp. 986–1004, 2003.
3. Wachowiak, MP., member, IEEE, and Terry Peters, M., Senior member, IEEE, "High performance Medical image registration using new optimization techniques", IEEE Trans. on Information Technology in Biomedicine, vol. 10, no. 2, Apr. 2006.
4. Y.J. Yu and S.T. Acton, "Speckle reducing anisotropic diffusion," IEEE Trans. Image. Process., vol. 11, no. 11, pp. 1260–1270, Nov. 2002.
5. K. Z. Abd-Elmoniem, A. M. Youssef, and Y. M. Kadah, "Real-time speckle reduction and coherence enhancement in ultrasound imaging via nonlinear anisotropic diffusion," IEEE Trans. Biomed. Eng., vol. 9, no. 9, pp. 997–1014, Sep. 2002.
6. P. Peona and J. Malik, "Scale-space and edge detection using anisotropic diffusion," IEEE Trans. Pattern Anal. Mach. Intell., vol. 12, no. 7, pp. 629–639, Jul. 1990.
7. I. Vajda, Theory of Statistical Inference and Information, Dodrecht, The Netherlands: Kluwer, 1989.
8. Josien P. W. Pluim, J. B. Antoine Maintz, Max A. Viergever, "Mutual information based registration of medical images: a survey", IEEE Trans. Medical Imaging, 2003.

Sentiment Mining Using SVM-Based Hybrid Classification Model

G. Vinodhini and R. M. Chandrasekaran

Abstract With the rapid growth of social networks, opinions expressed in social networks play an influential role in day-to-day life. A need for a sentiment mining model arises, so as to enable the retrieval of opinions for decision making. Though support vector machine (SVM) has been proved to provide a good classification result in sentiment mining, the practically implemented SVM is often far from the theoretically expected level because their implementations are based on the approximated algorithms due to the high complexity of time and space. To improve the limited classification performance of the real SVM, we propose to use the hybrid model of SVM and principal component analysis (PCA). In this paper, we apply the concept of reducing the data dimensionality using PCA to decrease the complexity of an SVM-based sentiment classification task. The experimental results for the product reviews show that the proposed hybrid model of SVM with PCA outperforms a single SVM in terms of classification accuracy and receiver-operating characteristic curve (ROC).

Keywords Sentiment · Opinion · Mining · Hybrid model · PCA

1 Introduction

With the rapid growth of e-commerce and large number of online reviews in digital form, the need to organize them arises. Various machine learning classifiers have been used in sentiment classification [8]. Many studies in machine learning communities have shown that combining individual classifiers is an effective technique for improving classification accuracy. There are different ways in which classifier can be combined to classify new instances. Dimension reduction plays an

G. Vinodhini (✉) · R. M. Chandrasekaran
Department of Computer Science and Engineering, Annamalai University, Annamalai Nagar, Chidambaram 608002, India
e-mail: g.t.vino@gmail.com

G. S. S. Krishnan et al. (eds.), *Computational Intelligence, Cyber Security and Computational Models*, Advances in Intelligent Systems and Computing 246,
DOI: 10.1007/978-81-322-1680-3_18,

important part in optimizing the performance of a classifier by reducing the feature vector size. Principal component analysis (PCA) can transform the original dataset of correlated variables into a smaller dataset of uncorrelated variables that are linear combinations of the original ones. Support vector machines (SVMs) have been recognized as one of the most successful classification methods for many applications including sentiment classification. Even though the learning ability and computational complexity of training in support vector machines may be independent of the dimension of the feature space, reducing computational complexity is an essential issue to efficiently handle a large number of terms in practical applications of text sentiment classification. In this study, we introduce a SVM-based hybrid sentiment classification model with PCA as dimension reduction for online product reviews using the product attributes as features. The results are compared with an individual statistical model, i.e., support vector machine. The rest of the paper is organized as follows. We provide an overview of the related work in Sect. 2. Section 3 presents an overview of the data source used. Methodology of the work is discussed in Sect. 4, which presents the feature reduction and classification methods used. The experimental results are discussed in Sect. 5 and conclusion in Sect. 6.

2 Related Work

Many studies on sentiment classification have used machine learning algorithms, with SVM and naive Bayes (NB) being the most commonly used. In comparison, SVM has outperformed other classifiers such as NB, centroid classifier, k-nearest neighbor, and window classifier [4–7, 9–11, 13]. So in this study, we considered SVM as baseline classifier. Feature selection is the most crucial task in sentiment mining [12]. Tan and Zhang [7] presented a sentiment categorization using four feature selection methods. The experimental results indicate that information gain performs the best for selecting the sentimental terms. Wang et al. [10] presented a hybrid method for feature selection based on the category-distinguishing capability of feature words and IG. Gamon [1] presented a feature reduction technique based on log-likelihood ratio to select the important attributes from a large initial feature vectors. Hatzivassiloglou and Wiebe [2] used words, bigrams, and trigrams, as well as the parts of speech as features in each sentence. In recent years, we witnessed the advance in machine learning methodology, such as SVM and PCA independently. However, the literature does not contribute much to sentiment classification using hybrid classification model with the combination of SVM and PCA. Even though Prabowo and Thelwall [5] used multiple classifiers in a hybrid manner, the hybrid combination of SVM and PCA is not experimented so far.

Our contribution in this work shows the effectiveness of the proposed hybrid classification method in sentiment mining of product reviews. Though SVM can efficiently deal with high-dimensional data, the linear dependence between different variables of the sample influence the generalization of SVM method. On the contradiction, PCA can deal with linear dependence between variables effectively and reduce dimension of the input samples and strengthen the ability of SVM to approximate to a nonlinear function. PCA is a suitable dimension reduction method for SVM classifiers because SVM is invariant under PCA transform. Though SVM has been proved to provide a good classification result in sentiment mining, the practically implemented SVM is often far from the theoretically expected level because their implementations are based on the approximated algorithms due to the high complexity of time and space. To improve the limited classification performance of the real SVM, we propose to use the hybrid model of SVM and PCA. Sentiment analysis is conducted at feature-based sentence level.

3 Data Source

The dataset used contains review sentences of products, which were labeled as positive, negative, or neutral. We collected the review sentences from the publicly available customer review dataset. This dataset can be downloaded from http://www.cs.uic.edu/~liub/FBS/FBS.html. This dataset contains annotated customer reviews of five different products. From those five products, we have selected reviews of digital camera, mobile phone, and music player. For our classification problem, we have considered only positive reviews and negative reviews. The product attributes discussed in the review sentences are collected. Unique unigram product features alone are grouped, which results in a final list of product attributes (features) for each product. In terms of these, the descriptions of review dataset model (model I) to be used in the experiment are given in Table 1.

4 Methodology

The following steps are involved in this work to develop the prediction system.

4.1 Create a Word Vector Model

A word vector representation of review sentences is developed for model using the unigram features. To create the word vector list, the review sentences are

Table 1 Description of dataset

Product	No. of reviews	Feature	No. of features	Positive reviews	Negative reviews
Camera	500	Unigrams only	95	365	135
Mobile phone	456	Unigrams only	86	248	208
Music player	328	Unigrams only	58	174	154

preprocessed. The data preprocessing includes tokenization, stopping word removal, and stemming. After preprocessing, the reviews are represented as unordered collection of words and the features are modeled as a bag of words. A word vector is developed for model using the respective features based on the term binary occurrences. The binary occurrences of the each feature word (n) in the processed review sentences (m) result in a word vector X of size $m \times n$ for model I.

4.2 Dimension Reduction Using PCA

PCA is a common technique for finding patterns in high-dimensional data. PCA algorithm works by calculating the covariance matrix, eigenvalues, and eigenvector. Then, the dimensionality of the data is reduced. A standardized transformation matrix is developed, and finally, the reduced components are obtained. These new components are called principal components (PC). Using Weka, the PC for the model with their unigram features are identified. The PC with variance less than 0.95 are obtained. A word vector model is recreated using review sentences and the reduced PC. The description of principle components obtained for model is shown in Table 2.

4.3 Construct SVM Model as Baseline

Support vector machine belongs to a family of generalized linear classifiers. It is a supervised machine learning approach used for classification to find the hyperplane maximizing the minimum distance between the plane and the training points. The basic idea behind the training procedure for binary classification is to find a hyper plane that separates the document vectors in one class from those in the other. Also the separation margin is as larger as possible. The SVM model is employed using Weka tool. The kernel type chosen is the polynomial kernel with default values for kernel parameters such as cache size and exponent.

Table 2 Description of principle components

Product	Camera	Mobile	Music player
No. of components	PC1–PC57	PC1–PC45	PC1–PC24
Variance	<0.95	<0.95	<0.95
Standard deviation	0.67	0.62	0.07
Proportion of variance	0.003	0.002	0.002
No. of features (original)	95	86	58
No. of principle components (reduced)	57	45	24
No. of reviews	500	456	328
Positive reviews	365	248	174
Negative reviews	135	208	154

4.4 Construct Hybrid SVM Model

A hybrid classifier is a collection of several classifiers whose individual decisions are combined in such a way to classify the test examples. It is known that combined model often shows much better performance than the individual classifiers used. SVM has been known to show a good generalization performance and is easy to learn exact parameters for the global optimum. Due to these advantages, their combination may not be considered as a method for increasing the classification performance. However, when implementing SVM practically, approximated algorithms have been used in order to reduce the computation complexity of time and space. Thus, a single SVM may not learn exact parameters for the global optimum. Sometimes, the support vectors obtained from the learning are not sufficient to classify all unknown test examples completely. So, we cannot guarantee that a single SVM always provides the global optimal classification performance over all test examples. Moreover, the dimension of the feature space does not influence the computational complexity of training or testing due to the use of the kernel function. But the computational complexity of SVM training depends on the dimension of the input space. Therefore, more efficient testing and training is expected from dimension reduction. To overcome these limitations, we propose to use hybrid model of support vector machines with PCA as dimension reduction technique. In hybrid model, each individual SVM is trained independently using the randomly chosen dimension-reduced training samples via a bootstrap technique and then aggregated. Each classifier is trained on a sample of examples taken from the training set and thereby produces a combined model that often performs better than the single model built from the original single training set. The algorithm is shown in Fig 1.

Fig. 1 Hybrid algorithm

Input: Data set $D = \{(x_1,y_1),(x_2,y_2),\cdots,(x_m,y_m)\}$;
Base learning algorithm *B; //SVM*
Number of learning rounds *R*. //10
Process: **for** $i = 1,\cdots,R$:
/* Generate a bootstrap sample from *D**/
Di = Bootstrap (*D*);
/* Train base learner *hi* from the bootstrap sample */
hi = *B*(*Di*);
end.
/* the value of *l*(a) is 1 if a is true and 0 */
Output : $O(x) = \arg\max\, y \in Y \sum_{t=1}^{T} l(y = h_i(x))$

4.5 Aggregating Support Vector Machines

After training, we need to aggregate several independently trained SVMs in an appropriate combined manner. Majority voting is to the simplest method for combining several SVMs. The results of individual SVMs are aggregated by determining the large number of SVMs whose decisions are known. The LSE-based weighting treats several SVMs in the SVM ensemble with different weights. Often, the weights of several SVMs are determined in proportional to their accuracies of classifications. The double-layer hierarchical combining uses another SVM to aggregate the outputs of several SVMs in the SVM ensemble. So, this combination consists of double-layer SVMs hierarchically where the outputs of several SVMs in the lower layer are fed into a super SVM in the upper layer.

5 Results

For hybrid models, we trained groups of 10 individual SVM models, each with different bootstrapped training subsets. The results of the individual classification model in the hybrid were combined using various methods such as the majority voting, the LSE-based weighting, and the double-layer hierarchical combining to produce the final result. We used tenfold cross-validation to compare classification accuracy. The performance evaluation is done using accuracy as a measure (Table 3.). Experiment is conducted on the customer reviews of digital camera, mobile phone, and music player (Sect. 3).

Receiver-operating characteristic (ROC) curve is an alternative technique for selecting classifiers based on their performance in which true-positive rate (TP rate) is plotted along the *y*-axis and false-positive rate (FP rate) is plotted along the *y*-axis. Since SVM is a discrete classifier, it is represented by only one point on an ROC graph. So a very approximate ROC curve is constructed by connecting this point with the points denoting both default classifiers [3].

Figures 2, 3 and 4 show ROC graph to compare the performance of baseline (SVM) method and hybrid method for three different product reviews. The hybrid method performed well and often outperformed the single SVM in terms of ROC curves.

Table 3 Classification accuracy

	Accuracy (%)		
Method	Camera	Mobile phone	Music player
Single SVM	86.99	85.34	88.73
Hybrid SVM—Majority voting	87.55	86.05	90.51
Hybrid SVM—LSE-based weighting	87.82	86.98	90.87
Hybrid SVM—Hierarchical SVM	88.01	87.57	91.08

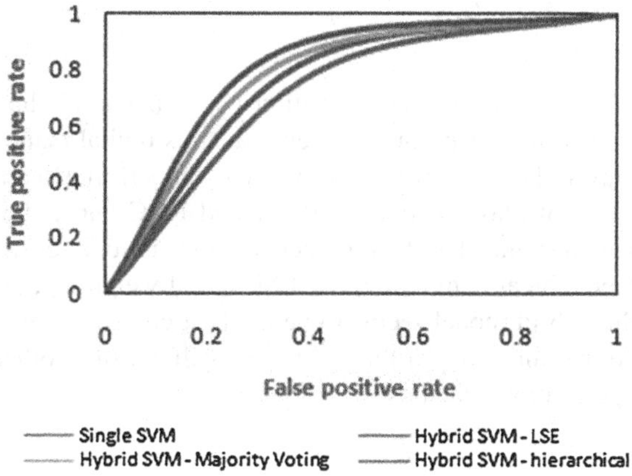

Fig. 2 ROC curve for camera reviews

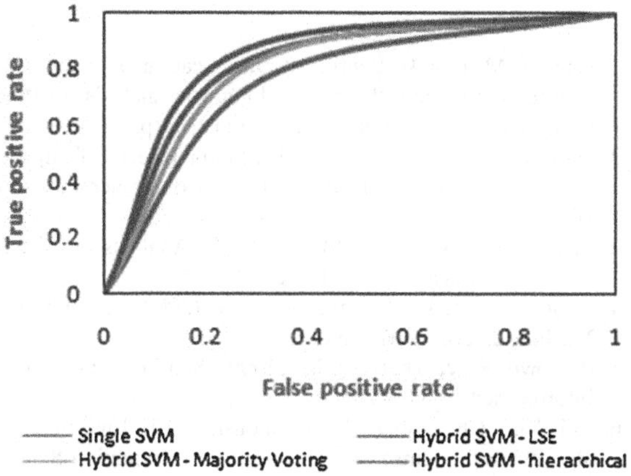

Fig. 3 ROC curve for mobile phone reviews

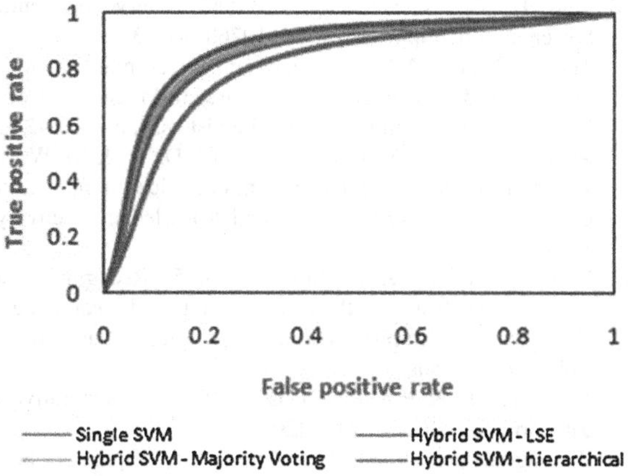

Fig. 4 ROC curve for music player reviews

6 Conclusion

We evaluated the classification performance of the proposed hybrid SVM over three different product reviews such as digital camera, mobile phone, and music player. The hybrid SVM with PCA outperforms a single SVM for all products in terms of classification accuracy and ROC curve. For three different aggregation methods, the classification performance is superior in the order of the double-layer hierarchical combining, the LSE-based weighting, and the majority voting. Thus, the hybrid model seems to be the best choice if time and computational resources are not an issue. In the future, we will consider other classifier combinations and aggregation schemes.

References

1. Gamon, M. (2004). Sentiment classification on customer feedback data: Noisy data, large feature vectors, and the role of linguistic analysis. In Proceedings of the 20th International Conference on Computational Linguistics (p. 841).
2. Hatzivassiloglou, V., Wiebe, J. (2000). Effects of adjective orientation and gradability on sentence subjectivity. In International Conference on Computational Linguistics (COLING-2000).
3. Matjaz Majnik, Zoran Bosni, ROC Analysis of Classifiers in Machine Learning: A Survey,Technical report MM-1,2011.
4. Pang, B., Lee, L., Vaithyanathan, S. (2002). Thumbs up? Sentiment Classification Using Machine Learning Techniques. EMNLP'.
5. Prabowo, R., & Thelwall, M. (2009). Sentiment analysis: A combined approach. Journal of Informetrics, 3, 143–157.
6. Rui Xia, Chengqing Zong, Shoushan Li, Ensemble of feature sets and classification algorithms for sentiment classification, Information Sciences 181 (2011) 1138–1152.
7. Tan, S. B., & Zhang, J. (2008). An Empirical study of sentiment analysis for Chinese documents. Expert Systems with Application, 34(4), 2622–2629.
8. Tang, H., Tan, S., & Cheng, X. (2009). A survey on sentiment detection of reviews. Expert Systems with Applications, 36, 10760–10773.
9. Turney, P. D. (2002). Thumbs up or thumbs down? Semantic orientation applied to unsupervised classification of reviews. In Proceedings of the 40th Annual Meetings of the Association for Computational Linguistics, (pp. 417–424). Philadelphia, PA.
10. Wang, S. G., Wei, Y. J., Zhang, W., Li, D. Y., & Li, W. (2007). A hybrid method of feature selection for chinese text sentiment classification [C]. In Proceedings of the 4th International Conference on Fuzzy Systems and Knowledge Discovery (pp. 435–439). IEEE Computer Society.
11. Wilson,T., Wiebe, J., Hoffman, P. (2005). Recognizing contextual polarity in phrase level sentiment analysis. In Proceedings of the Human Language Technology Conference and Conference on Empirical Methods in Natural Language Processing (pp. 347–354). British Columbia, Canada.
12. Yang, Y. M., & Pedersen, J. O. (1997). A Comparative study on feature selection in text categorization. ICML, 412–420.
13. Ziqiong Zhang, Qiang Ye, Zili Zhang, Yijun Li, Sentiment classification of Internet restaurant reviews written in Cantonese, Expert Systems with Applications 38 (2011) 7674–7682.

Shape Based Image Classification and Retrieval with Multiresolution Enhanced Orthogonal Polynomial Model

S. Sathiya Devi and R. Krishnamoorthi

Abstract This paper proposes a simple edge-based shape representation with multiresolution enhanced orthogonal polynomial model and morphological operations for effective classification and retrieval. The proposed method consists of four phases: (1) orthogonal polynomial computation, (2) edge image construction, (3) approximate shape boundary extraction with morphological operation, and (4) invariant Hu's moment computation. Initially, the orthogonal polynomials are computed and the obtained coefficients are reordered into one-level subband-like structure. Then, the edge image is obtained by utilizing gradient in horizontal and vertical directions from the detailed subbands of the reordered structure. The rough shape boundary is computed with morphological operation. The global invariant shape features such as Hu's moment and eccentricity are extracted in the fourth phase. The obtained features are termed as global shape feature vector and are used for retrieving and classifying similar images with Canberra distance metric and Bayesian classification, respectively. The efficiency of the proposed method is experimented on a subset of standard Corel, Yale, and MPEG-7 databases. The results of the proposed method are compared with those of existing techniques, and the proposed method provides significant results.

Keywords Multiresolution · Orthogonal polynomials · Morphological closing · Shape similarity · Moments · Eccentricity · Bayesian classification

S. S. Devi (✉) · R. Krishnamoorthi
Vision Lab, Department of Computer Science and Engineering, University College of Engineering, Bharathidasan Institute of Technology, Thiruchirappalli, Tamil Nadu, India
e-mail: sathyadevi.2008@gmail.com

R. Krishnamoorthi
e-mail: rkrish07@yahoo.com

G. S. S. Krishnan et al. (eds.), *Computational Intelligence, Cyber Security and Computational Models*, Advances in Intelligent Systems and Computing 246,
DOI: 10.1007/978-81-322-1680-3_19,

1 Introduction

Shape is one of the important low-level features of any CBIR system, and it is more effectively perceived by the human eye than texture and color. Hence, shape-based searching and retrieving has gained much attention in CBIR. The shape representation can be classified into two categories: region based and contour based [1]. Region-based techniques have frequently used moment descriptors to obtain shape representation [2, 3], which considers all the pixels inside the shape to compute the shape features. Contour-based shape representation [1] only exploits shape boundary information and is classified into continuous approach (global) and discrete approach (structural). Both region- and contour-based representation methods compute shape features either in spatial domain or in frequency domain. Spatial domain descriptors are sensitive to noise and are not robust. In addition, it requires intensive computation during similarity calculation, due to the hard normalization of rotation invariance. As a result, these spatial representations need further processing using spectral transform such as Fourier transform and wavelet transform. Though wavelet-based multiscale analysis is vital for edge extraction, the existing wavelet-based methods exploit multiple stages of transformation for extracting edges at the expense of high computational complexity with some sort of degraded retrieval performance. Hence, this paper proposes a new edge-based shape extraction with multiresolution enhanced orthogonal polynomial model and this paper is organized as follows: In Sect. 2, the image retrieval technique with orthogonal polynomial model is presented. The edge extraction and linking with morphological operations are presented in Sect. 3. The invariant shape features such as moments and eccentricity are computed in Sect. 4. Section 5 discusses the performance evolution metric. In Sect. 6, the experimental result of the proposed, Haar wavelet-based methods is presented. Finally, conclusion is given in Sect. 7.

2 Orthogonal Polynomial Model Computation

Based on the work given in [4], the orthogonal polynomial computation and reordering the obtained coefficients into one-level multiresolution subband-like structure are performed.

3 Edge Image Construction

From the reordered logarithmically spaced subband frequencies, the subband representing horizontal and vertical information is selected to compute gradient magnitude because larger values in regions have prominent edges and small values on nearly uniform edges. The gradient image G in terms of the proposed reordered orthogonal polynomial coefficients is defined as follows:

$$G = \sum_{i=0}^{X} \sum_{j=0}^{Y} \left(|\beta_{ij}^{\prime S_2}| + |\beta_{ij}^{\prime S_3}| \right) \tag{1}$$

where X and Y are the size of the subbands and $\beta_{ij}^{\prime S_2}$, $\beta_{ij}^{\prime S_3}$ are the coefficients of subbands S_2 and S_3, respectively. The edge points are extracted from the gradient magnitude G by applying the non-maximum suppression, followed by a dynamic threshold T. The non-maximum suppression is conventionally performed based on the orientation of the gradient points with high computational cost. This paper introduces a simple method for calculating the non-maximum suppression either in x direction or in y direction, depending on which $\beta_{ij}^{\prime S_2}$ and $\beta_{ij}^{\prime S_3}$ have the larger magnitude. If $\beta_{ij}^{\prime S_3}$ has the larger magnitude and is not greater than its neighbors on both sides of the x direction, then $\beta_{ij}^{\prime S_3}$ and $\beta_{ij}^{\prime S_2}$ values are suppressed. Corresponding comparisons and suppressions are performed in the y direction if $\beta_{ij}^{\prime S_2}$ has the larger magnitude. The non-maximum suppression process is summarized below:

for all $G(i,j)$

$\text{if}\left(\left|\beta_{i,j}^{\prime S_3}\right| > \left|\beta_{i,j}^{\prime S_2}\right|\right) \;\; \text{if}\left(\left|\beta_{i,j}^{\prime S_3}\right| | \cdot | \left|\beta_{(i-1),j}^{\prime S_2}\right|\right) \| \left|\beta_{i,j}^{\prime S_3}\right| < \left|\beta_{(i+1),j}^{\prime S_2}\right|\right)$

$\left\{\beta_{i,j}^{\prime S_3} = 0;\; \beta_{i,j}^{\prime S_2} = 0\right\};$

$\text{else if } \left(\left|\beta_{i,j}^{\prime S_2}\right| \leq \left|\beta_{i,(j-1)}^{\prime S_2}\right|\right) \| \left(\left|\beta_{i,j}^{\prime S_2}\right| < \left|\beta_{i,(j+1)}^{\prime S_2}\right|\right) \left\{\beta_{i,j}^{\prime S_2} = 0;\; \beta_{ij}^{\prime S_3} = 0\right\}$

The dynamic threshold T is a sum of the mean of gradient magnitude of S_2 and S_3 subbands and is defined as follows:

$$T = \frac{1}{MN} \left(\sum_{i=0}^{M-1} \sum_{j=0}^{N-1} G_{ij} \right) + \frac{1}{MN} \left(\sum_{i=0}^{M-1} \sum_{j=0}^{N-1} G_{ij} \right) \tag{2}$$

where M and N are the size of the subbands.

Thus, the dynamic threshold is effectively applied to the result of non-maximum suppression, which removes the majority of edge responses, and the edge points $(s_1, s_2, \ldots s_n)$ are extracted.

3.1 Edge Linking (or) Approximated Shape Boundary Extraction

The method that we have discussed in the previous section extracts reasonable edge points. But due to noise, the edge points are not continuous and many false edges also exist. Hence, in this subsection, we utilize the morphological erosion operation in order to obtain the approximate boundary of the image and to detect and remove the false edges. The major part of morphological operations can be defined as a combination of two basic operations: dilation and erosion. Erosion operation reduces the object's geometrical area by setting the contour pixels of an object to the background value. Erosion is defined as follows:

$$E\Theta B = \{p \in \varepsilon^2 : p + b \in E, \quad \forall b \in B\} \tag{3}$$

where p is the element of the image space ε^2 and B is the structuring element.

After performing the erosion operation of the edge image E, the edge linking and removal of false edges are achieved by subtracting the eroded image from the edge image, as given in Eq. (4).

$$E_{\text{link}} = (E - (E\Theta B)) \tag{4}$$

The E_{link} now preserves the global rough shape of an object in the image. The shape feature extraction from E_{link} is discussed below.

4 Shape Feature Extraction

Generally, the real-world images and the images in the database are often geometrically distorted. Hence, it is important to design a feature, which is invariant even the images are geometrically distorted. This section describes the global invariant shape feature vector extraction based on Hu's seven [5] moment and eccentricity. The extracted global feature vector is invariant to scaling, rotation, and translation. The eccentricity is the ratio of the major axis to the minor axis and is defined as follows:

$$\varepsilon = \frac{(\mu_{20} - \mu_{02})^2 + 4\mu_{11}^2}{(\mu_{20} + \mu_{02})^2} \tag{5}$$

Hence, the global feature vector FV of dimension 8 is derived from Hu's seven moments and (6) as given below:

$$\text{FV} = (\varepsilon, \phi_1, \phi_2, \phi_3, \phi_4, \phi_5, \phi_6, \phi_7). \tag{6}$$

5 Similarity and Performance Measure

In the proposed retrieval scheme, similarity between the query image and the images present in the database are calculated using the well-known Canberra distance as shown in Eq. (7).

$$d_C(x, y) = \sum_{i=1}^{d} \frac{|x_i - y_i|}{|x_i| + |y_i|} \quad (7)$$

The performance of the proposed method is measured in terms average retrieval ratio (ARR) that is defined as follows:

$$\text{ARR} = \frac{1}{N} \sum_{i=1}^{N} \frac{m_i}{N} \quad (8)$$

where N is the total number of similar images in one category and m_i is the number of retrieved relevant images. The performance is also measured using popular measure precision and recall rate. The classification accuracy A_i of an individual class i is defined as the number of query image samples correctly classified (true positives plus true negatives).

6 Experiments and Results

The retrieval efficiency of the proposed method is tested with the subset of standard Corel [6] image database, MPEG-7_CE-Shape_Part_B [7] shape database, and Yale face database [8]. The experimental results are given in this section. During experimentation, first preprocessing is performed on the image under analysis. In this step, median filter is applied to smooth the image. The median filter preserves the edges and removes the noises. Then, the image is divided into (2×2) non-overlapping blocks and each block is subjected to the orthogonal polynomial model. The transformed coefficients are reordered into multiresolution-like structure. The same process is repeated for all the blocks in the image. The image is now decomposed into one-level subband-like structure, and each subband is named as S_1, S_2, S_3, and S_4. The S_2 and S_3 subband coefficients possess the first-order derivatives in horizontal and vertical directions and are used for computing the gradient magnitude G. Then, the non-maximal suppression is applied to extract the edge image. The approximated boundary is extracted from the edge image E using the set of binary morphological operators as described in Sect. 3.1. Then, the Hu's seven invariant moments and eccentricity are computed and act as edge-based shape feature vector FV. The FV of dimension 8 is invariant to rotation, scaling, and translation. The obtained features are stored in the feature database. Now, in the query phase, for a given query image, the above steps are repeated and the edge-based invariant shape features are extracted. Then, the similarity measure is computed with Canberra distance metric utilizing the feature vector FV of each pair of query and database images. The distances are then sorted in the ascending order, and the top 10 images are retrieved. For Bayesian classification, 50 % of the images in each database are considered for training and the above-mentioned process is repeated for obtaining the features of an image.

Table 1 Average retrieval rate (ARR) of proposed (PM) and Haar wavelet-based method (HM) for MPEG-7 shape database

Class	Top 20		Top 25		Top 30		Class	Top 20		Top 25		Top 30	
	PM	HM	PM	HM	PM	HM		PM	HM	PM	HM	PM	HM
Apple	75	60	80	75	80	80	Crown	60	50	75	60	75	60
Bell	75	65	75	75	85	80	Elephant	65	65	75	75	90	90
Bird	70	55	75	65	80	75	Dog	60	60	75	65	80	75
Bottle	80	80	85	85	90	90	Fish	85	85	90	90	95	90
Camel	75	75	80	80	85	85	Fly	75	75	80	75	85	90
ARR (%)	72	64	79	74.5	84.5	81.5							

We calculate the average retrieval rate of the proposed and Haar wavelet-based methods using the MPEG-7_CE-Shape-1_Part_B and Yale face databases. ARR is computed by considering the each image in the class as the query image, and the percentage of retrieval is obtained for top 20, top 25, and top 30 images. Similarly, the average retrieval efficiency of all the remaining nine classes is calculated and the obtained results are listed in Table 1. Then, the average retrieval efficiency of all the classes is computed, which gives the overall retrieval efficiency of the proposed and existing methods for the MPEG-7 database. The proposed method gives the average retrieval rate of 82.088 % for top 30 images, whereas the Haar wavelet-based method yields 79.988 %. The proposed method also performs well for top 20 and top 25 images during image retrieval.

Similarly, ARR is computed for Yale face database. ARR is obtained for the 11 classes of face images, and the obtained results are given in Table 2. The experimental results reveal that for some of the face images, our proposed method performs well than Haar wavelet-based method when considering the first 20 retrieved images. For the standard subset of Corel database, the performance is measured with the well-known measure precision and recall. These measures are computed from the top 10 to top 100 successfully retrieved images. We perform the experiment for 7 classes of images, and each class contains 100 images. Seven classes of images are dinosaur, elephant, roses, buses, vintage cars, penguins, and pigeons. The elephant and pigeon classes contain images with textured background and thin edges. The experiment is conducted with these images for both proposed

Table 2 Average retrieval rate (ARR) of proposed (PM) and Haar wavelet-based method (HM) for Yale face database

Class	Top 20		Top 25		Class	Top 20		Top 25	
	PM	HM	PM	HM		PM	HM	PM	HM
Subject1	81.82	72.73	90.91	81.82	Subject6	63.64	72.73	72.73	81.82
Subject2	63.64	63.64	90.91	63.64	Subject7	81.82	81.82	81.82	81.82
Subject3	72.73	54.55	72.73	72.73	Subject8	63.64	63.64	72.73	72.73
Subject4	63.64	72.73	81.82	72.73	Subject9	72.73	72.73	90.91	81.82
Subject5	81.82	63.64	81.82	72.73	Subject10	81.82	72.73	90.91	90.91
ARR (%)	72.731	69.09	82.73	77.28					

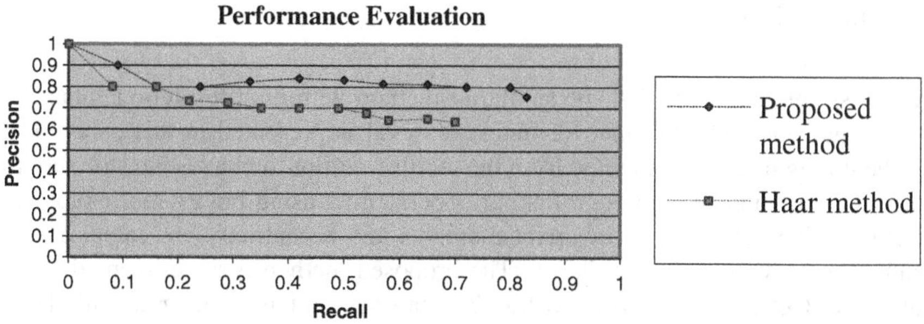

Fig. 1 Precision and recall of the proposed and Haar wavelet-based methods for subset of Corel database

and Haar wavelet-based methods. The precision and recall are computed, and the obtained results are plotted as a graph with recall along x-axis and precision along y-axis and are shown in Fig. 1.

The classification accuracy of the proposed and Haar wavelet-based methods with Bayesian classification for three databases is shown in Table 3. From the results, we infer that the proposed method performs well compared with the existing methods. The proposed method yields high classification rate of 96.71 % for Corel database. The proposed algorithm is tested against the noisy images, and the retrieval performance is evaluated. The Gaussian noise of 10, 20, 25, 30, and 35 % is added to each image in the databases. Then, the performance is evaluated with the average recognition rate. The proposed method maintains the same retrieval rate and classification up to 25 % of the noise, and the retrieval rate is reduced 4 % for 25 % noise. This experimental result reveals that our method is robust to noise for a certain level.

Table 3 Classification accuracy (CA) of the proposed (PM) and Haar wavelet-based method (HM) of MPEG-7, Yale, and Corel databases

MPEG-7 shape database			Yale face database			Subset of Corel database		
Image class	CA (%)		Image class	CA (%)		Image class	CA (%)	
	PM	HM		PM	HM		PM	HM
Apple	96	96	Subject1	91	88	Dinosaur	95	92
Bell	96	94	Subject2	93	89	Sunflower	100	100
Bird	90	90	Subject3	91	90	Roses	98	96
Bottle	86	81	Subject4	92	88	Vintage cars	98	95
Camel	95	93	Subject5	91	88	Buses	97	96
Crown	93	90	Subject6	91	91	Penguin	95	92
Elephant	96	94	Subject7	93	90	Elephant	94	92
Dog	88	89	Subject8	89	86			
Fish	98	95	Subject9	93	90			
Fly	91	98	Subject10	88	87			
Average CA	92.9	92	Average CA	91.2	88.7	Average CA	96.71	94.71

7 Conclusion

A new simple edge-based shape similarity method with multiresolution enhanced orthogonal polynomial model for image retrieval is presented in this paper. The gradient magnitude is computed from the multiresolution subband-like structure to form the edge image, and edge linking is performed using binary morphological operation. The Hu's seven invariant moments and eccentricity are calculated as feature vectors for shape similarity. The proposed method is experimented using subset of COREL, Yale, and MPEG-7 databases and is compared with Haar wavelet-based method. The proposed method yields better retrieval result than the existing methods.

References

1. D. Zhang, G. Lu, "Review of shape representation and description techniques", Pattern Recognition, Vol. 37, pp. 1–19, 2004.
2. J. Flusser and T. Suk, "Rotation Moment Invariant for Recognition of Symmetric Objects", IEEE Transactions on Image Processing, Vol.15, no. 12, pp. 3784–3790, 2006.
3. T. Gevers and A.W.M. Smeulders, "PicToSeek: Combining Color and Shape Invariant Features for Image Retrieval", IEEE Transactions on Image Processing, Vol.9, no.1, pp. 102–119, 2000.
4. R. Krishnamoorthi and S. Sathiya Devi, "A Simple Computational Model for Image Retrieval With Weighted Multi Features Based On Orthogonal Polynomials And Genetic Algorithm", Neurocomputing, Special issue on advanced theory and methodology in intelligent computing, Vol. 116, pp. 165–181, 2013.
5. M. K. Hu, "Visual Pattern Recognition by Moment Invariants", IRE Transactions on Information Theory, Vol. 8, no. 2, pp. 179–187, 1962.
6. www.corel.com.
7. www.Visionlab.uda.edu.
8. www.cvc.yale.edu.

Using Fuzzy Logic for Product Matching

K. Amshakala and R. Nedunchezhian

Abstract Product matching is a special type of entity matching, and it is used to identify similar products and merging products based on their attributes. Product attributes are not always crisp values and may take values from a fuzzy domain. The attributes with fuzzy data values are mapped to fuzzy sets by associating appropriate membership degree to the attribute values. The crisp data values are fuzzified to fuzzy sets based on the linguistic terms associated with the attribute domain. Recently, matching dependencies (MDs) are used to define matching rules for entity matching. In this study, MDs defined with fuzzy attributes are extracted from product offers and are used as matching rules. Matching rules can aid product matching techniques in identifying the key attributes for matching. The proposed solution is applied on a specific problem of product matching, and the results show that the matching rules improve matching accuracy.

Keywords Product matching · Data integration · Fuzzy logic · Matching dependency

1 Introduction

Product matching is a particular case of entity matching that is required to recognize different descriptions and offers referring same product. Although entity matching has received an enormous amount of effort in research [5], only modest work has been dedicated to product matching [7]. Product matching for e-commerce

K. Amshakala (✉)
Department of CSE and IT, Coimbatore Institute of Technology, Coimbatore, India
e-mail: amshakalacse@yahoo.in

R. Nedunchezhian
Sri Ranganathan Institute of Engineering and Technology, Coimbatore, India

G. S. S. Krishnan et al. (eds.), *Computational Intelligence, Cyber Security and Computational Models*, Advances in Intelligent Systems and Computing 246,
DOI: 10.1007/978-81-322-1680-3_20,

Websites introduces several specific challenges that make this problem much harder. In particular, there is a huge degree of heterogeneity in specifying name and descriptions of the same product by different merchants. For example, a desktop PC can be referred in different Websites using different terms like “Data Processor,” “Processor,” “Computer,” “Personal computer.” Similarly, offers on products come from multiple online merchants like jungle.com and pricegrabber.com. When exact string matching is not sufficient, synonym-based similarity matching can be used. WordNet ontology [12] is used to extract synonyms of product names. Not all products are described using same set of attributes. Furthermore, offers often have missing or incorrect values and are typically not well structured but merge different product characteristics in text fields. These challenges highlight the need for efficient entity matching technique that tolerates different data formats, missing attribute values, and imprecise information.

Fuzzy set theory and fuzzy logic proposed by Zadeh (1965) provide mathematical framework to deal with imprecise information. In a fuzzy set, each element of the set has an associated degree of membership. For any set X, a membership function on X is any function from X to the real unit interval [0,1]. The membership function which represents a fuzzy set X' is usually denoted by $\mu(X)$. For an element x of X, the value $\mu_X(x)$ is called the membership degree of x in the fuzzy set X'. The membership degree $\mu_X(x)$ quantifies the grade of membership of the element x to the fuzzy set X'. The value 0 means that x is not a member of the fuzzy set; the value 1 means that x is fully a member of the fuzzy set. The values between 0 and 1 characterize fuzzy members, which belong to the fuzzy set only partially (Fig. 1).

A linguistic variable is a variable that apart from representing a fuzzy number also represents linguistic concepts interpreted in a particular context.

Functional dependencies, conventionally used for schema design and integrity constraints, are in recent times revisited for improving data quality [3, 4, 8]. However, functional dependencies based on equality function, often fall short in entity matching applications, due to a variety of information representations and formats, particularly in the Web data. Several attempts are made to replace equality function of traditional dependencies with similarity metrics. Fuzzy functional dependency (FFD) [10] is a form of FD that uses similarity metrics (membership functions) instead of strict equality function of FDs. FFDs are also used to find dependencies in databases with fuzzy attributes, whose domain has fuzzy values like high, low, small, large, young, old, etc. For example, the attributes size, price, and weight are fuzzy attributes in product database. A typical membership function for the price attribute is shown in Fig. 2.

Fig. 1 Fuzzy membership function

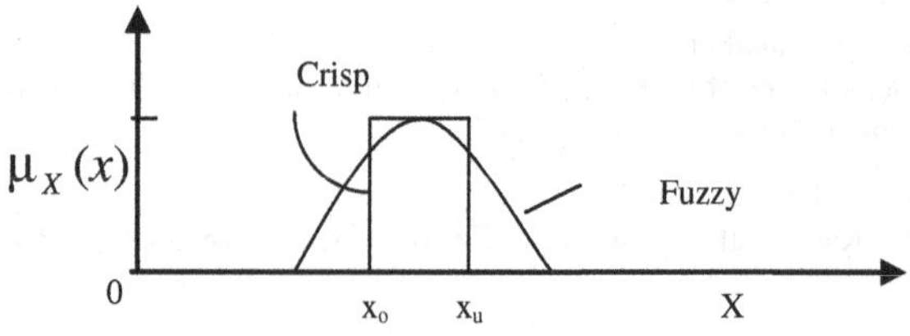

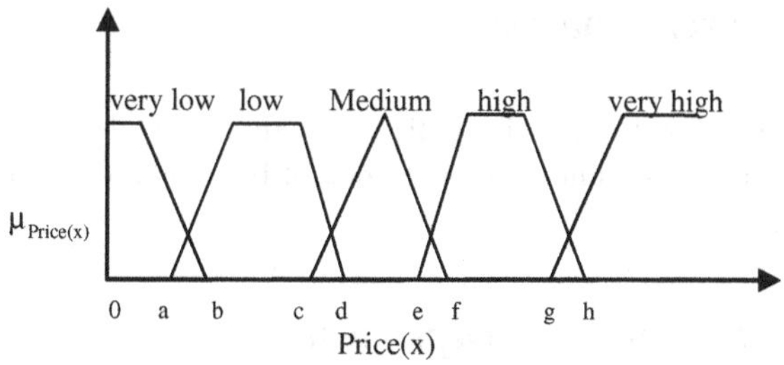

Fig. 2 Fuzzy membership function for price attribute

Recently, matching dependencies (MDs) [4, 8] are used for data quality applications, such as record matching. In order to be tolerant to different information formats, MDs target on dependencies with respect to similarity metrics, as an alternative of equality functions in conventional dependency. An MD expresses, in the form of a rule, that if the values of certain attributes in a pair of tuples are similar, then the values of other attributes in those tuples should be matched (or merged) into a common value. For example, the MD $R1[X] \approx R2[X] \rightarrow R1[Y] = R2[Y]$ says that if R1-tuple and R2-tuple have similar values for attribute X, then their values for attribute Y in R1 and R2 should be made equal. In practice, MDs are often valid in a subset of tuples and not on all the tuples of a relation. Along with FFDs, conditional matching dependencies (CMDs) proposed in [9] are used by the proposed work to infer matching rules that are appropriate for product matching. CMDs, which are variants of conditional functional dependencies (CFDs) [3], declare MDs on a subset of tuples specified by conditions. Approximate functional dependencies are also generalizations of the classical notion of a hard FD, where the value of X completely determines the value of Y not with certainty, but merely with high probability. Conditional functional dependencies (CFDs) [3] and approximate functional dependencies (AFDs) [6] differ with respect to the degree of satisfaction. While AFDs allow a small portion of tuples to violate the FD statement, conditional FDs are satisfied only by the tuples that match the condition pattern.

In this study, an information theory measure called entropy is used to define fuzzy functional dependencies and extensions of conventional FDs. Entropy of an attribute indicates the structuredness of the attribute. The reason behind using entropy as a dependency measure is that it captures probability distribution of the attribute values in a single value. The proposed approach defines MDs as fuzzy conditional matching dependencies (FCMDs). Two types of approximation are used in this approach. One is to compensate for uncertainty in matching similar values using FFDs [10] and the other approximation is to compensate for the fraction of tuples violating FDs using AFDs and CFDs. Experimentally, it is shown that the MDs and matching rules improve both the matching quality and efficiency of various record matching methods.

2 Preliminaries

In this section, we formally define functional dependency and explain how entropy is used to identify the presence of functional dependency between attributes.

2.1 Functional Dependency

A functional dependency (FD) $X \rightarrow Y$ is said to hold over sets of attributes X and Y on R if $\forall i, j$ if $ri[X] = rj[X]$ then $ri[Y] = rj[Y]$, where $r[X]$ denotes the tuple r projected on the attributes X.

2.2 Information Theory Measures

Entropy Let $P(X)$ be the probability distribution of an attribute X, the attribute entropy $H(X)$ is defined as

$$H(X) = -\sum_{x \in X} P(x) \log_2 P(x) \tag{1}$$

The entropy is a non-negative value, $H(X) \geq 0$ always. It may be interpreted as a measure of the information content of, or the uncertainty about, the attribute X. It is also called as marginal entropy of an attribute [2].

The joint entropy $H(X, Y)$ between any two attributes X, Y can be computed using joint probability distribution of the two attributes as follows

$$H(X, Y) = -\sum_{x \in X} \sum_{y \in Y} P(x, y) . \log_2 P(x, y) \tag{2}$$

where $P(x, y)$ is the joint probability distribution of the attributes X and Y. Also, $H(X, X) = H(X)$ and $H(X, Y) = H(Y, X)$.

Theorem 1 *Functional dependency* $X \rightarrow Y$, *holds if and only if* $H(X,Y) = H(X)$.

Theorem 1 indicates that the FD $X \rightarrow Y$ holds, if the joint entropy of X and Y is the same as that of X alone. By computing the joint entropy between attribute pairs and attribute entropies for all the attributes in the given table, all those functional dependencies that are true can be determined. When Theorem 1 is put in other words, for the functional dependency $X \rightarrow Y$ to hold true, the difference between $H(X, Y)$ and $H(X)$ must be equal to zero [11].

$$H(X, Y) - H(X) = 0 \tag{3}$$

2.3 Functional Dependency Extensions

Traditional functional dependencies are used to determine inconsistencies at schema level, which is not sufficient to detect inconsistencies at data level. Fuzzy attributes with crisp domain in a relation have to be fuzzified before applying FD discovery methods on the data. Let us assume that an attribute A with crisp domain when fuzzfied using different fuzzy sets results in fuzzy columns $fA_1, fA_2 \ldots fA_n$. The table is partitioned into equivalence classes that include tuple ids of those tuples that qualify as equal values along different linguistic variables associated with the attribute. The relational table is also partitioned based on crisp data values over the crisp attribute and with each linguistic dimension separately. Compute the marginal entropy of all the crisp attributes (not fuzzified) in the given relation using Eq. 1. From the projected fuzzified table, partition the tuples along the fuzzy columns $fA_1, fA_2 \ldots\ldots fA_n$. Entropy for the fuzzy columns is computed by considering data values with membership degree greater than the membership threshold θ as equal. Joint entropy of fuzzified attribute A and a crisp attribute B is computed as the cumulative sum of joint entropy of the fuzzy columns and crisp attribute B

$$H(AB) = \sum_{i=1}^{n} H(fA_i B) \tag{4}$$

Theorem 1 is applied to check whether any functional dependency exists between crisp and fuzzified attributes. If $H\ (AB) = H(B)$, then $B \rightarrow \theta\ A$ is true. Further, dependencies that apply conditionally appear to be particularly needed when integrating data, since dependencies that hold only in a subset of sources will hold only conditionally in the integrated data. A CFD extends an FD by incorporating a pattern tableau that enforces binding of semantically related values. Unlike its traditional counterpart, the CFD is required to hold only on tuples that satisfy a pattern in the pattern tableau, rather than on the entire relation. The minimum number of tuples that are required to satisfy the pattern is termed as the support of CFD.

When a relation U with m tuples is considered, the support entropy is calculated as follows

$$H_s = -\left(\frac{m_k}{m}\log_2\left(\frac{m_k}{m}\right) + m_r\left(\frac{1}{m}\right)\log_2\left(\frac{1}{m}\right)\right) \tag{5}$$

H_s is nothing but the entropy of candidate that has at least one partition with m_k tuples, where m_k is the minimum number of tuples that should have the same constant value for the CFD to get satisfied. m_r is the remaining number of tuples in the relation, $m_r = m - m_k$. Under certain circumstances where the relational table has uncertain data, the functional dependency $X \rightarrow Y$ may not be strictly satisfied by all the tuples in a relation. When few number of tuples violate the functional dependency, then the difference between the joint entropy *H*(*X*, *Y*) and the marginal entropy *H*(*X*) may be close to 0, but not strictly equal to zero

$$H(X,Y) - H(X) \approx 0 \tag{6}$$

Such dependencies are called as approximate FDs [6].

3 Proposed Approach

3.1 FCMD Discovery Algorithm

Matching entities gathered from multiple heterogeneous data sources place lot of challenges because the attributes describing the entities may have missing values or the values may be represented using different encodings. MDs are a recent proposal for declarative duplicate resolution tolerating uncertainties in data representations. Similar to the level-wise algorithm for discovering FDs [6], the FCMD discovery algorithm also considers the left-hand-side attributes incrementally, i.e., traverse the attributes from a smaller attribute set to larger ones to test the possible left-hand-side attributes of FDs. The following pruning rules are used to reduce the number of attribute sets to be tested for the presence of FCMDs.

3.2 Pruning Rules

3.2.1 Pruning Rule 1: Support Entropy-Based Pruning

Candidates that do not even have one partition with size greater than or equal to k need not be verified for the presence of FCMDs. Candidates with support less than k need not be checked. Supersets of such candidates are also not k-frequent. So, candidates with entropy higher than H_s can be removed from the search space.

3.2.2 Pruning Rule 2: Augmentation Property-Based Pruning

The supersets of the LHS candidate of the FCMDs need not be verified further, once when a FCMD is satisfied by the candidate. This pruning rule prevents discovering redundant FCMDs.

The time complexity of the FCMD discovery algorithm varies exponentially with respect to the number of attributes and varies linearly with respect to the number of tuples (Table 1).

Table 1 FCMD discovery algorithm

Input: Sample training relation

Output: Set of Matching Dependencies (MD)

1. Compute the marginal entropy of all the crisp attributes (not fuzzified) in the given relation using Eq. 1
2. The fuzzy attributes are associated with appropriate membership functions and they are projected as fuzzy columns in the given relation
3. From the projected fuzzified table, partition the tuples along the fuzzy columns $fA1, fA_2 \ldots\ldots fA_n$
4. Entropy for the fuzzy columns are computed by considering data values with membership degree greater than the threshold as equal
5. Joint Entropy of fuzzy attribute A and a crisp attribute B is computed as the cumulative sum of joint entropy of the fuzzy columns and crisp attribute B
$$H(AB) = \sum_{i=1}^{n} H(fA_i B)$$
6. Using the user specified support threshold k, the support entropy H_s is computed using Eq. 6
7. If the RHS and LHS value of Eq. 5 is greater than or equal to H_s then the attribute B is the possible candidate to act as LHS of FCMD (Rule 1)
8. Equation 4 is used to check if any functional dependency exists between crisp and fuzzified attributes
$$\text{If } H(AB) - H(B) \approx 0 \text{ then } B \rightarrow_\theta A \text{ is true}$$
9. Add the discovered Matching dependency to the set MD
10. Repeat through step 5 for all candidates level by level (Pruning rule 2)

As the probability distribution is represented using a single entropy measure, the set closure and set comparison operations are not required to test the presence of FCMDs.

4 Experimental Results

4.1 Experimental Setup

Experiments were carried out on a 2.16-GHz processors with a 2 GB RAM with Windows XP operating system. The implementation of this project is done using Java.

4.2 Dataset

The product catalog for electronic goods was collected from 20 Websites and consolidated as a dataset with 150 entities. On average, the maximum number of duplicates for an entity is 10. The data records are randomly duplicated, and dataset with 10–80 K records were created. The product catalog has four fields including product_ID, Product_Name, Manufacturer_Name, and Price.

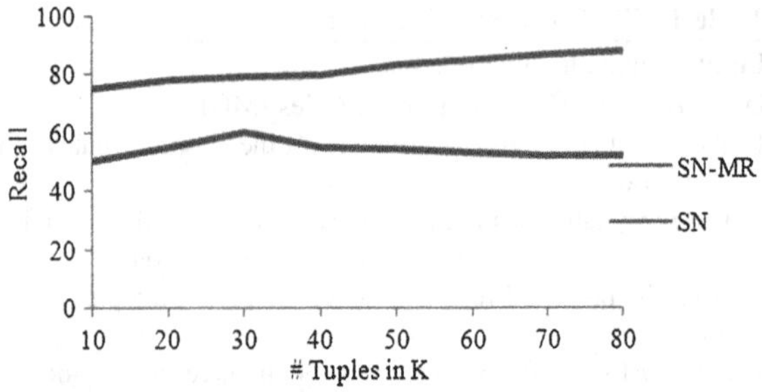

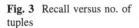

Fig. 3 Recall versus no. of tuples

A record linkage tool called Fine-grained Records Integration and Linkage tool (FRIL) provides a rich set of tools for comparing records [13]. Furthermore, FRIL provides a graphical user interface for configuring the comparison of records. FRIL can be configured to choose different field similarity metrics and different record comparison techniques.

For testing the product catalog integration, FRIL is configured to use Q-grams for comparing Product_Name field, exact string matching for Manufacturer_Name, and numeric approximation for Price field. Sorted neighborhood method is used for record comparisons [1]. The proposed FCMD method first extracts the matching rules and then uses as rules for sorted neighborhood method. Precision and recall are the two measures used to measure the accuracy of the results returned by the record matching algorithms. Precision is used to check whether the results returned are accurate, and recall is used to check whether the results returned are complete. Recall and precision measured for the proposed approach and FRIL are shown in Figs. 3 and 4, respectively.

The matching rules discovered by FCMD method has higher recall than that used by the SN method, because of using dynamic discovery of rules rather than static rule. The precision of the results produced by FCMD method is shown in Fig. 4. The results produced by SN with matching rules discovered by FCMD algorithm includes lesser number of incorrect results, compared to that of SN, because of having matching rules with higher support.

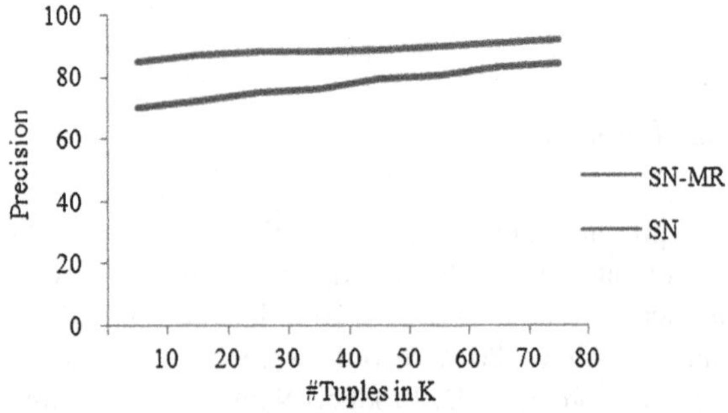

Fig. 4 Precision versus no. of tuples

5 Conclusion

The problem of identifying similar products described used fuzzy attributes is of major concern when integrating product catalogs and matching product offers to products. Most previous work is based on predefined matching rules and supervised way of detecting duplicates using trained datasets. MDs represented using fuzzy dependencies and other FD extensions defined using entropy are proposed in this study. The proposed work also uses synonym-based string field matching, which helps in detecting duplicate records that are missed by exact string matching methods. Fuzzy attributes are modeled using fuzzy functional dependencies. This entity matching technique can be used to identify duplicates in datasets generated on the fly and do not require hand-coded rules to detect duplicate entities, which makes it more suitable for product matching.

References

1. Christen, P. (2012), A Survey of Indexing Techniques for Scalable Record Linkage and Deduplication, IEEE Transactions on, Knowledge and Data Engineering, 24(9), 1537–1555.
2. Divesh S and Suresh V(2010), Information Theory for Data Management, Tutorial in Proceedings of the ACM SIGMOD Conference on Management of Data, 1255–1256.
3. Fan W, Geerts F, (2011) Foundations of Data Quality Management, Synthesis Lectures on Data Management, Morgan & Claypool Publishers.
4. Fan, W., Gao, H., Jia, X., Li, J, and Ma, S. (2011). Dynamic Constraints for Record Matching. The VLDB Journal, 20(4), 495–520.
5. Köpcke, H., and Rahm, E. (2010). Frameworks for Entity Matching: A Comparison. Data & Knowledge Engineering, 69(2),197-210.
6. Liu, J., Li, J., Liu, C., & Chen, Y. (2012). Discover Dependencies from Data-A Review. IEEE Transactions on Knowledge and Data Engineering, 24(2), 251–264.
7. Papadimitriou, P., P. Tsaparas, A. Fuxman and L. Getoor, (2013). TACI: Taxonomy Aware Catalog Integration, IEEE Transactions on Knowledge and Data Engineering, 25: 1643–1655.
8. Song, S. and L. Chen, (2009). Discovering Matching Dependencies, In Proceedings of the 18th ACM Conference on Information and Knowledge Management, pp: 1421–1424.
9. Song, S., L. Chen and J.X. Yu, (2010). Extending Matching Rules with Conditions, Proceedings of the 8th International Workshop on Quality in Databases, 13–17 September.
10. Wang, S. L., Shen, J. W., & Hong, T. P. (2010). Dynamic Discovery of Fuzzy Functional Dependencies Using Partitions. Intelligent Soft Computation and Evolving Data Mining: Integrating Advanced Technologies, 44.
11. Yao, Y.Y., (2003). Information-Theoretic Measures For Knowledge Discovery and Data Mining. Entropy Measures, Maximum Entropy Principle Emerging Applications, 119: 115-136.
12. Zadeh, L.A., (1965). Fuzzy sets. Information and control, 8(3), 338–353. http://wordnet.princeton.edu, 2009
13. http://fril.sourceforge.net.

Part III
Cyber Security

A Novel Non-repudiate Scheme with Voice FeatureMarking

A. R. Remya, M. H. Supriya and A. Sreekumar

Abstract A digital watermark is the type of latent indicator secretly embedded in a noise-tolerant signal such as audio or image data. It is typically used to identify the ownership or copyright of material. "Watermarking" is the process of hiding digital information in a carrier signal in order to confirm the authenticity or integrity of the carrier signal as well as show the identity of its owners. Since a digital copy of data is the same as the original, digital watermarking is a passive protection tool. This technique simply marks the signal with the data neither it degrades nor it controls access to the data, thereby securing the communication. The proposed system introduces a novel non-repudiate scheme to ensure the ownership of every audio communication. This method embeds the prepared watermark in the transform domain of the audio signal using the fast Walsh transforms. Watermark used in this technique is unique for each member, and thus, it provides additional authenticity in every communication compared to state of the art.

Keywords Digital watermarking · Non-repudiation · FeatureMarking · Walsh transforms

A. R. Remya (✉) · A. Sreekumar
Department of Computer Applications, Cochin University of Science and Technology, Kochi, India
e-mail: remyacusat@gmail.com

A. Sreekumar
e-mail: askcusat@gmail.com

M. H. Supriya
Department of Electronics, Cochin University of Science and Technology, Kochi, India
e-mail: supriyadoe@gmail.com

G. S. S. Krishnan et al. (eds.), *Computational Intelligence, Cyber Security and Computational Models*, Advances in Intelligent Systems and Computing 246,
DOI: 10.1007/978-81-322-1680-3_21,

1 Introduction

Advancements in digital technology lead to widespread use of digital communication. This in turn increases the reproduction and retransmission of multimedia data in legal as well as illegal ways. The illegal usage causes a serious threat to the content owner's authority or right. Thus, today's information-based society provides extreme importance in authenticating the information that is mediated on the communication channel.

Digital watermarking is the technique that ensures authorized and legitimate use of digital communication, copyright protection, copy protection, etc. and thus helps to prevent unauthorized manifestations. Watermarking demonstrates the process of embedding copyright or ownership information to the original signal in an unnoticeable way. The embedded watermark should be robust to any signal manipulations and can be unambiguously retrieved at the other end. Audio watermarking is the term coined to represent the insertion of a signal, image, or text of known information in an audio signal in an imperceptible way [1].

Studies on existing watermarking schemes motivate us to develop a non-repudiate scheme which aims to ensure the authenticity of each member participated in the communication. Any member participated in the communication might not be able to deny their presence of participation and thus guarantee an authentic, non-repudiate communication scheme.

The recorded voice signal will be given to the preprocessing module which performs the short-term processing of each voice signal. Features are extracted for each frame using the fast Fourier transformations, and the feature values are given to an online QR code generator to generate the QR code. The extracted features can also be saved in a database. The generated QR code will function as the authentic FeatureMark in the proposed system. Upon receipt of the QR code, FeatureMark embedding module hides the FeatureMark in the transform domain of the audio signal and the FeatureMarked audio is sent to the desired recipient. The receiver confirms the authenticity of the received signal by detecting and extracting the embedded FeatureMark and analysis of the feature values.

Section II of this paper demonstrates some of the definitions and methodologies used in this work. Section III demonstrates the proposed FeatureMarking scheme which helps to avoid the non-repudiate behavior. Section IV shows the experimental results, and the remaining sections include conclusion, future work, acknowledgment, and references. The integral system will thus act as a strong non-repudiate voice scheme to avoid any illegal behavior.

2 Definitions and Methodologies

Voice signal identification can be achieved by extracting the features that can be useful for further processing.

Mel-Frequency Cepstral Coefficients (MFCC) In acoustic signal processing, the short-term power spectrum of a sound is depicted by obtaining its mel-frequency cepstrum (MFC), which is based on a linear cosine transform of a log-power spectrum on a nonlinear mel scale of frequency.

The most commonly used audio features used in the speech or speaker recognition process are mel-frequency cepstral coefficients (MFCCs). MFCCs are treated as the best for speech/speaker recognition because it takes human sensitivity with respect to frequencies into consideration. Reference [2–4] demonstrates audio watermarking as well as audio steganographic techniques based on cepstral coefficients.

Spectral Flux The spectral flux is the term coined to demonstrate the squared difference between the normalized magnitudes of successive spectral distributions corresponding to successive signal frames.

Spectral Centroid The spectral shape of a frequency spectrum is measured with the spectral centroid which makes it usable in voice signal classification activities. The higher the centroid values, the brighter will be the textures with much high frequency.

$$C = \frac{\sum_{n=0}^{N-1} f(n)x(n)}{\sum_{n=0}^{N-1} x(n)} \tag{1}$$

Spectral Roll-Off The spectral roll-off depicts the frequency below which 85 % of the magnitude dissemination of the frequency spectrum is focused. Both the centroid and spectral roll-off are measures of spectral shape, and the spectral roll-off yields higher values for high frequencies. Therefore, it can be stated that there exists a strong correlation between both of these features.

QR Code Nowadays, quick response codes or QR codes are extensively used in various industrial applications. A QR code is represented as a 2D code placed on a white background and consists of black modules arranged in a square pattern. Any kinds of data including binary or alphanumeric can be encoded to an online QR code generator. An online bar code scanner application either on a phone or on PC can decode or display the details hidden in it. A sample QR code is shown in Fig. 1.

Walsh Transforms The Walsh–Hadamard transform (WHT) is defined as a suboptimal, non-sinusoidal, orthogonal transformation that decomposes a signal into a set of orthogonal, rectangular waveforms called Walsh functions. These transformation works in the transform domain of the signal have no multipliers and are real because the amplitude of Walsh (or Hadamard) functions has only two values, +1 or −1 [18].

Fig. 1 A Sample QR code

The forward and inverse Walsh transform pairs for a signal $x(t)$ of length N are

$$y_n = \frac{1}{N}\sum_{i=0}^{N-1} x_i WAL(n,i), n = 1, 2, \ldots, N-1 \tag{2}$$

$$x_i = \frac{1}{N}\sum_{n=0}^{N-1} y_n WAL(n,i), n = 1, 2, \ldots, N-1 \tag{3}$$

Non-Repudiation Non-repudiation allows exchange of digital communication between different groups of individuals in such a way that the people in this communicating group cannot subsequently deny their participation in the communication [6, 7].

2.1 Related Areas

An audio watermarking scheme in the time domain presented in [8] does not require the original signal for watermark detection. An audio watermarking methodology described in [9] exploits the temporal and frequency perceptual masking criteria and constructs its watermark by breaking each audio clip into smaller segments and adding a perceptually shaped pseudo-random noise. A digital audio watermarking in the cepstral domain is presented in [10], which embeds the watermark in the cepstral coefficients of the signal using spread-spectrum techniques. Frequency hopping watermark embedding in the spectral domain is suggested in [11], which embeds the watermark in the power spectrum coefficients of the audio signal. Modified patch work algorithm in transform domain described in [12] is robust to withstand some attacks defined by Secure Digital Music Initiative (SDMI).

Xu et al. [14] demonstrate a method to protect the copyright of digital media over the Internet by introducing a support-vector-regression-based audio watermarking scheme in the wavelet domain, and the watermark information is embedded into randomly selected subaudios. Histogram-based audio watermarking scheme is tested to verify its robustness characteristics by applying different attacks including low-pass filtering, amplification. [15]. Multiple scrambling in the embedding process guarantees unauthorized detection of the embedded watermark, and to enhance the robustness criteria, an adaptive synchronization scheme utilized in the watermark detection process is prescribed in [16]. Employing DCT coefficients in the embedding and extraction procedures of audio watermarking method presented in [17] work with neural network scheme. Many researches have been conducted and being conducted in this area.

3 FeatureMarking System

Each member participated in this communication system is assigned with a unique identification number.

3.1 Preprocessing

Short-term processing of each signal is performed to analyze individual frames and is mandatory in the case of voice signals. Noise reduction is avoided in this work to preserve ambient characteristics of its environment.

Framing The original voice signal is decomposed into a set of overlapping frames for further processing. Recorded voice signals are segmented into frames ranging from 2 to 50 ms.

Once the signal is decomposed into individual frames, calculate the temporal and spectral components by the analysis of its time and frequency components. The time–frequency mapping is usually matched to the perceptual qualities of human auditory system (HAS) [13].

Windowing Windowing function helps to reduce the signal discontinuities toward the beginning and ending of each frame. It is achieved by performing smooth transitions at these regions and hence attenuates the spectral leakage effects. Commonly used windowing techniques include Hamming windowing, Hanning windowing, Bartlett and help to minimize the spectral distortions or sharp discontinuities at these portions.

However, reducing the spectral leakage in the frequency spectrum of each frame by these windowing techniques may result in a modest loss of spectral resolution.

3.2 Feature Extraction

Voice signals are represented as time series in a 2D plane by plotting its amplitude along the *y*-axis and the corresponding time interval along the *x*-axis. The feature extraction module deals with extracting features that help to differentiate each member in the communicating group and store these feature values in the database. Feature extraction is done by performing Fourier transform of the signal, where Fourier analysis decomposes the sampled signal into its fundamental periodic components such as sines and cosines. The fast Fourier transformation such as Cooley–Tukey algorithm maps the given time space into its corresponding frequency space and is important in the context of audio signals [5].

To find the MFCC (as shown in Fig. 2), spectral Flux, Spectral Roll-Off, and Spectral Centroid each chunk of data is windowed with hamming window. MFCC values are assessed by finding the magnitude of the FFT and by taking the cosine transform of its base 10 logarithm. Final step in this process helps to reduce the dimensionality of the results obtained.

3.3 FeatureMark Preparation

Important modules associated with digital audio watermarking scheme include the watermark design, the watermark embedding, and the watermark detection as wells as extraction.

The first two modules are associated with the sender window or the encoding procedure, and the third module is associated with the receiver window or the decoding procedure. The FeatureMark is prepared by inputting feature values in a

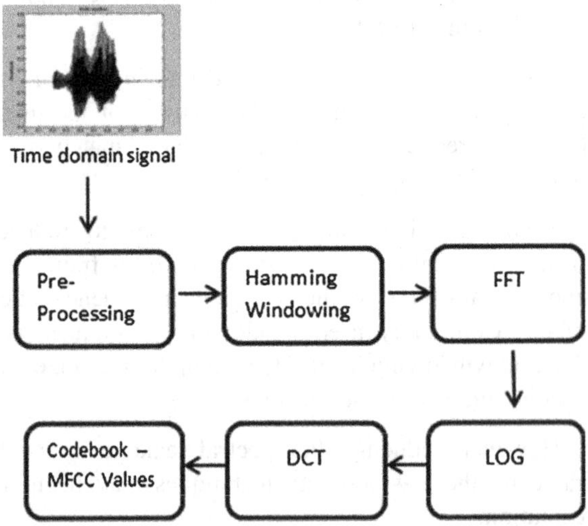

Fig. 2 Extraction of Mel-frequency cepstral coefficients

specific order with the individual unique code to an online QR code generator. According to the input values, the generator outputs signal-specific QR code. The generated QR code is termed as the FeatureMark for the proposed system which depends on the analysis of the features retrieved from the original signal and the unique identification number provided for each participant.

3.4 FeatureMark Embedding

This module embeds the generated FeatureMark in the desired sequence of the original voice signal.

Let

$Y = y_1 y_2 y_3 \ldots y_n$ represent the original voice signal and Wf represent the generated FeatureMark.

Then, the embedding process can be summarized as follows:

$YWf = \rho(Y, Wf)$, where YWf corresponds to the FeatureMarked voice signal.

Watermark embedding procedures are usually being performed in the frequency domain, in the time domain, or in the transform domain of an audio signal in order to exploit the frequency and temporal masking characteristics of the HAS. This system exploits the transform domain of the voice signal and thus inserts the FeatureMark by performing the Walsh transforms of the original signal. Performing the inverse Walsh transform toward the end of the embedding module will revert the changes occurred as part of the transformation. Now, the FeatureMarked voice signal can be sent to the desired recipient. The algorithm behind the watermark embedding module should be available to the receiver so as to detect and extract the FeatureMark efficiently.

3.5 FeatureMark Extraction

At the receiving end, the recipient may expect every voice communication they received as a watermarked one and apply the fast Walsh transforms to detect and extract the embedded FeatureMark. Performing the inverse operations that are applied during the embedding process helps to retrieve the exact replica of the generated FeatureMark. Comparison between the Figs. 3 and 4 will confirm that they are identical in all respects (feature analysis, visual representation, etc.)

According to [13], watermark retrieval process can be grouped in any of the scenarios—when the input audio is employed for decoding and when the input audio is not available for watermark detection:

Case 1: Input audio required for decoding

$$W\prime = \varsigma(Y^w, Y)$$

Fig. 3 Prepared FeatureMark

Fig. 4 Extracted FeatureMark

Case 2: Input audio NOT required for decoding

$$W\prime = \varsigma\,(Y^{w})$$

The proposed non-repudiate scheme behaves like the second category and hence comes under the blind watermarking scheme. The extracted FeatureMark is unambiguous and provides authentic information about voice signal ownership. For a successful digital audio watermarking scheme, the watermark should meet the imperceptibility criteria; it should be perceptually and statistically transparent and should be robust to signal manipulations.

3.6 Non-repudiation

Non-repudiation as the definition shows can achieve a great extent by the performance of this proposed system. Since the watermark prepared for each signal is unique and depends on individual signal, the content owner cannot neglect their presence in later stages.

The extracted FeatureMark will be scanned through an online QR code scanner, the feature values obtained can be compared with the feature values present in the database of the system as well as by performing the feature extraction on the received signal and hence confirm the ownership. By employing Euclidean metric evaluation or direct comparisons (even though a bit time-consuming), the receiver can ensure that they are part of the authentic as well as non-repudiate voice scheme.

Algorithmic steps can be summarized as follows:

Record voice signal

Step 1: Perform the preprocessing

- Framing and windowing

Step 2: Extract feature vectors

- MFC coefficients using fft
- Spectral flux using fft
- Spectral roll-off using fft
- Spectral centroid using fft

Step 3: Feature vectors will be given as input to a QR code generator
Step 4: Using fwht, embed the FeatureMark in the voice signal
Step 5: Perform the inverse fwht and send the FeatureMarked voice signal to the desired recipient
Step 6: Extract out the FeatureMark by applying the fwht on the FeatureMarked voice signal
Step 7: Obtain the feature vectors from the extracted FeatureMark using an online QR code scanner
Step 8: Feature comparison to confirm the authenticity of the voice signal.

4 Experimental Results

The proposed scheme is tested with signal duration up to 10–240 s and found acceptable with around 50 sample signals. Recorded signals are preprocessed with framing and windowing terminology. The input signal is divided into 256 $\times$ 256 frames with a frame rate of 100, and individual data chunks obtained for each voice signal are given as input to the feature extraction module.

MFCCs are restricted to 20 coefficients, which results in a feature vector of 20 $\times$ 256 dimensions. Embedding such a huge volume of audio data will expose the furtiveness of the system hence-forth vector quantization is implemented resulting in a data matrix of 20 $\times$ 1 coefficients.

The saved feature vectors are given to the QR code generator to generate the QR code, which will be used as the FeatureMark in the proposed system. The obtained FeatureMark will be entered in the FeatureMark embedding module, which streams the FeatureMark in order to embed in the original voice signal by transforming it with the fast Walsh transforms. It performs the inverse transform

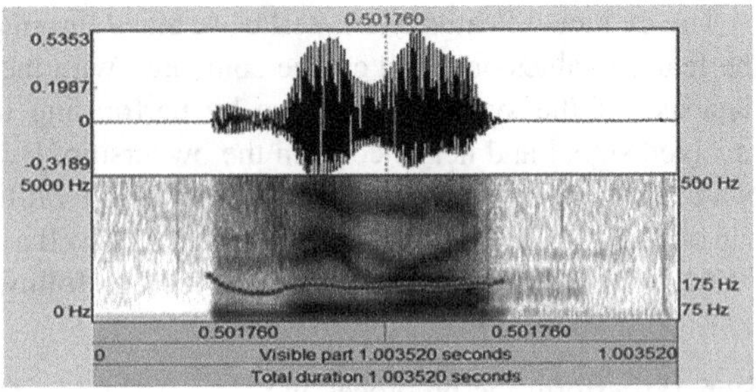

Fig. 5 Original time domain signal

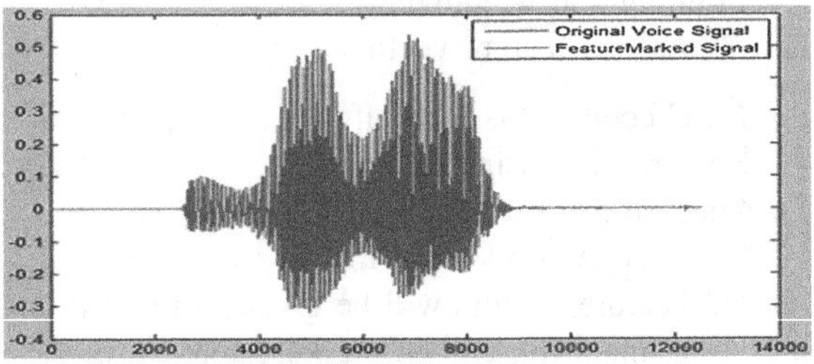

Fig. 6 Comparison of original and FeatureMarked time domain signals

function after the completion of the embedding process. Once the embedding is done, all the processes are reversed to hide the presence of the embedded watermark. The FeatureMarked voice signal is transmitted to the desired recipient. Embedded FeatureMark can be obtained by applying the fast walsh transfoms and is given to an online QR code scanner to retrieve the encoded values.

Authenticity of the voice signal can be confirmed by comparing the extracted features with the database values as well as by performing the feature extraction module at the receiving end, represented in Figs. 5 and 6.

However, as the system can function as a non-repudiate scheme for audio signal communication, it can be incorporated over multiple disciplines as a secure voice authentication system.

5 Conclusion

The proposed method functions as a secure robust voice authentic system with a non-repudiate scheme for voice signal communication. The use of voice signal features, its classification, and FeatureMarking offers an improved scheme for authentic voice communication. The use of an online QR code generator and scanner to create the FeatureMark as well as to extract the feature values, respectively, is vital to this work. The QR code brings in the ease of comparison in the authentication module. Non-repudiation can be avoided by performing the feature extraction module at the receiver end and cross-check with the FeatureMark values in case of denial of participation by any member.

6 Future Works

The proposed system can be enhanced to provide double-layered security and thus can be extended its application to real-time systems as well.

Acknowledgments This work was funded by the Department of Science and Technology, Government of India, under the INSPIRE Fellowship (IF110085).

References

1. Katzenbeisser Stefan and APP Fabien. Information hiding techniques for steganography and digital watermarking. Artech House, London, UK (2000)
2. Kaliappan Gopalan. Audio steganography by cepstrum modification. In Acoustics, Speech, and Signal Processing, 2005. Proceedings.(ICASSP'05). IEEE International Conference on, volume 5, pages v–481. IEEE (2005)
3. Kaliappan Gopalan. Robust watermarking of music signals by cepstrum modification. In Circuits and Systems, 2005. ISCAS 2005. IEEE International Symposium on, pages 4413–4416. IEEE (2005)
4. Christian Kraetzer and Jana Dittmann. Mel-cepstrum based steganalysis for voip-steganography. Proceedings of SPIE, Security, Steganography, and Watermarking of Multimedia Contents IX, 6505:650505–1 (2007)
5. Ingo Mierswa and Katharina Morik. Automatic feature extraction for classifying audio data. Machine learning, 58(2-3):127–149 (2005)
6. Tom Coffey and Puneet Saidha. Non-repudiation with mandatory proof of receipt. ACM SIGCOMM Computer Communication Review, 26(1):6–17 (1996)
7. Steve Kremer, Olivier Markowitch, and Jianying Zhou. An intensive survey of fair non-repudiation protocols.Computer communications, 25(17):1606–1621 (2002)
8. Paraskevi Bassia, Ioannis Pitas, and Nikos Nikolaidis. Robust audio watermarking in the time domain. Multimedia, IEEE Transactions on, 3(2):232–241 (2001)
9. Mitchell D Swanson, Bin Zhu, Ahmed H Tewfik, and Laurence Boney. Robust audio watermarking using perceptual masking. Signal Processing, 66(3):337–355 (1998)
10. Sang-Kwang Lee and Yo-Sung Ho.Digital audio watermarking in the cepstrum domain. Consumer Electronics, IEEE Transactions on, 46(3):744–750 (2000)
11. Nedeljko Cvejic and Tapio Seppänen. Spread spectrum audio watermarking using frequency hopping and attack characterization. Signal processing, 84(1):207–213 (2004)
12. In-Kwon Yeo and Hyoung Joong Kim. Modified patchwork algorithm: A novel audio watermarking scheme. Speech and Audio Processing, IEEE Transactions on, 11(4):381–386 (2003)
13. Andreas Spanias, Ted Painter, and Venkatraman Atti. Audio signal processing and coding. Wiley-Interscience (2006)
14. Xu, Xiaojuan, Hong Peng, and Chengyuan He. "DWT-based audio watermarking using support vector regression and subsampling." Applications of Fuzzy Sets Theory. Springer Berlin Heidelberg, 136–144 (2007)
15. Mali, Manisha D., and S. R. Khot. "Robustness Test Analysis of Histogram Based Audio Watermarking." Wireless Networks and Computational Intelligence. Springer Berlin Heidelberg, 611–620 (2012)
16. Lin, Yiqing, and Waleed H. Abdulla. "Multiple scrambling and adaptive synchronization for audio watermarking." Digital watermarking. Springer Berlin Heidelberg, 440–453 (2008)

17. Chong Wang, Xiaohong Ma, Xiangping Cong, Fuliang Yin, "An Audio Watermarking Scheme with Neural Network", Advances in Neural Networks—ISNN 2005, Lecture Notes in Computer Science Volume 3497, pp 795–800 (2005)
18. MathWorks | Accelerating the pace of engineering and science http://www.mathworks.com/

A Study of Spam Detection Algorithm on Social Media Networks

Jacob Soman Saini

Abstract In the present situation, the issue of identifying spammers has received increasing attention because of its practical relevance in the field of social network analysis. The growing popularity of social networking sites has made them prime targets for spammers. By allowing users to publicize and share their independently generated content, online social networks become susceptible to different types of malicious and opportunistic user actions. Social network community users are fed with irrelevant information while surfing, due to spammer's activity. Spam pervades any information system such as email or Web, social, blog, or reviews platform. Therefore, this paper attempts to review various spam detection frameworks that which deal about the detection and elimination of spams in various sources.

Keywords Spam detection · Spam analysis · Feature extraction

1 Introduction

Social networks such as Facebook, MySpace, LinkedIn, Friendster [1], and Tickle have millions of members who use them for both social and business networking. Due to the astonishing amount of information on Web, users follow the way of searching useful Web pages by querying search engines. Given a query, a search engine identifies the relevant pages on the Web and presents the users with the links to such pages. Spammers try to increase the page rank of the target Web page in search results by search engine optimization (SEO), the injection of artificially created pages into the Web in order to influence the results from search engines to

J. S. Saini (✉)
Department of Computer Science and Engineering, Sree Narayana Gurukulam College of Engineering, Kadayiruppu, Kolenchery, Kerala, India
e-mail: sainijs@rediffmail.com

G. S. S. Krishnan et al. (eds.), *Computational Intelligence, Cyber Security and Computational Models*, Advances in Intelligent Systems and Computing 246,
DOI: 10.1007/978-81-322-1680-3_22,

drive traffic to certain pages for fun or profit. Initially, spams are introduced in mails. Later, this has been extended to social networks.

On the other hand, in email system, spammer sends unsolicited bulk email to users by redirecting them to irrelevant Websites. The success of delivered attacks is dependent almost entirely upon the click-through rate of the email. If the target does not click on the malicious link presented in the email, then the attack usually fails. To improve click-through rates, many techniques exist such as hiding the destination of hyperlinks, falsifying header information, and creative use of images [2, 3], etc.

Email messages also take advantage of some shared context among friends on a social network such as celebrations of birthday functions, residing in the same home town, or common events participation. This shared context dramatically increase email authenticity, filters and increasing the click-through rate for spam that contains advertisements, installs malicious software, or solicits sensitive personal information [4, 5], etc. But, in the content of blog platforms, spammers post irrelevant comments for an already existing post. They focus on several kinds of spam promotion such as, splog (the whole blog is used to promote a product or service), comment spam (comments promoting services with no relation to the blog topic), and trace back spam (spam that takes advantage of the trace back ping feature of popular blogs to get links from them).

This paper primarily focuses on the survey of the literature which deals with the comment spams in blog. Since comments are typically short by nature. Comment spam is essentially link spam originating from comments and responses added to Web pages which support dynamic user editing. As a result of the presence of spammers in a network, there is a decrease in both the quality of the service and the value of the data set as a representation of a real social phenomenon. With the help of the extracted features, it is possible to identify spammers from the legitimate one. Various machine learning, supervised, and unsupervised methods have been used in the literature for the classification of these spams. A study of various spam detection algorithms has been dealt thoroughly in this paper.

2 Related Work

Initially, certain researchers concentrated on the development of honey pots to detect spams. To detect spams, Steve [6] dealt with automatic collection of deceptive spam profiles in social network communities based on anonymous behavior of user by using social honey pots. This created unique user profiles with personal information like age, gender, date of birth, and geographic features like locality and deployed in MySpace community. Spammer follows one of the strategies such as being active on Web for longer time period and sending friend request. The honey profile monitors spammers' behavior by assigning bots. Once the spammers sends friend request, the bots store spammer profile and crawls through the Web pages to identify the target page where advertisements originated.

The spammer places woman's image with a link in the "About Me" section in its profile, and the honey profile bots crawl through the link, parses the profile, extracts its URL, and stores the spammers profile in the spam list. URL does not redirect at times during crawling process, and "Redirection Detection Algorithm" is executed to parse the Web page and extract redirection URL to access it with the motive of finding the source account. Also, he proposed a "Shingling Algorithm" which verifies the collected spam profile for content duplication like URL, image, comments, and to accurately cluster spam and non-spam profile based on the features. In this way, he eliminated spams.

Another researcher named Gianluca [7] used social honey pots to construct honey profiles manually with features like age, gender, DOB, name, surname, etc. Here, the honey profiles have been assigned to three different social network communities (MySpace, Facebook, and Twitter). It considered friend request as well as the message (wall post, status updates) received from spammers and validated with honey pots. The identified spam accounts with the help of spam bots share common traits which have been formalized as features in their honey pots (first feature, URL ratio, message similarity, friend choice, message sent, friend number, etc.). The classifier namely Weka framework with a random forest algorithm has been used to classify spammers for best accuracy. Similarly, during spam campaign, the spam bots were clustered based on spam profiles using naïve Bayesian classifier to advertise same page during message content observation.

Similarly, Kyumin [8] dealt with the social spam detection which has become tedious in social media nowadays. Here, social honey pots are deployed after its construction based on the features such as number of friends, text on profile, age, etc. Here, both legitimate and spam profiles have been used as initial training set and support vector machine has been used for classification. An inspector has been assigned to validate the quality of extracted spam candidates using "Learned classifier" and provide feedback to spam classifier for correct prediction in future. In this paper three research challenges have been addressed. Initially, it validates whether the honey pot approach is capable of collecting profiles with low false positives, next to that, it addresses whether the users are correctly predicted, and finally, it evaluates the effectiveness of fighting against new and emerging spam attacks.

The first challenge is proved using automatic classifier which groups the spammers accurately. The second one considers demographic features for training the classifier using 10-fold cross-validation. It has been tested in MySpace using meta-classifier. In Twitter, it used Bigram model for classification along with the preprocessing steps. Finally, post filters have been used to check the links and remove the spam label by applying "support vector machine" for correct prediction. They also proposed that in future, clique-based social honey pots can be applied with many honey profiles against many social network communities.

Next to honey pots, spammers have been identified in the literature by analyzing content and link-based features of Web pages. In this context, Shrijina [9] has performed spam detection in social media by considering content and link-based features.

Web spam misleads search engines to provide high page rank to pages of no importance by manipulating link and contents of the page. Here, to identify Web spam, Kullback–Leiblerence techniques are used to find the difference between source page features (anchor text, page title, and Meta tags) and target page features (recovery degree, incoming links, broken links, etc.). Therefore, three unsupervised models were considered and compared for Web spam detection. As a first one, hidden Markov model has been used, which captures different browsing patterns, jump between pages by typing URL's, or by opening multiple windows. The features mentioned above were given as input to HMM, and it is not visible to the user. As a result, a link is categorized as spam or non-spam based on how frequently a browser moves from one page to another.

Second method uses "self-organizing maps" a neural model to classify training data (Web links) without human intervention. It classifies each Web link as either spam or non-spam link. One more method called adaptive resonance theory has also been used to clarify a link as either spam or not.

Another work in the literature Karthick et al. [10] has dealt with is the detection of link spam through pages linked by hyperlink that are semantically related. Here, qualified link analysis (QLA) has been performed. The relation existing between the source page and target page is calculated by extracting features of those two pages from Web link and comparing with the contents extracted from these pages. In QLA, the nepotistic links are identified by extracting URL, anchor text, and cached page of the analyzed link stored in the search engines. During query generation, once the page is available with search engines, this result has been compared together with the page features for easy prediction of spam and non-spam links.

In this paper, QLA has been combined with language model detection for better prediction of spams. In language model detection, the KL divergence technique has been used to calculate the difference between the information of the source pages with the content extracted from the link. Once matched, it is clustered as non-spam and vice versa. Here, the result of LM detection, QLA along with pre-trained link, and content features lead to accurate classification and detection.

Qureshi et al. [11] handled the problem of eliminating the existence of irrelevant blogs while searching for a general query in Web. The objective is to promote relevancy in the ordering of blogs and to remove irrelevant blogs from top search results. The presence of irrelevancy is not because of spam, but is due to inappropriate classification for a topic against a query. This approach uses both content and link structure features for detection. The content features calculate the cosine similarity between blog text and blog post title while searching for a particular blog. It has been proved that a co-relation exists between the above two features with which the spammer activity is detected based on the degree of similarity. This detection achieved a precision of 1.0 and recall of 0.87.

The blog link structure feature finds spammer activity by decoupling between two classes (duplicate and unique links) up to three hop counts. Spammers always move within closed group rather than with other blogosphere. The duplicate links are identified and removed.

Jenq-Haur et al. [12] focused on comment spams with hyperlinks. The similarity between the content of page for a post and the link it points to has been compared to identify spam. Here, the collected blogs are preprocessed, which finds the stop word ratio that is found to be less in spammers post. The contents are extracted from the post and are sorted where "Jaccard and Dice's" coefficient is calculated, which provides the degree of overlapping between words. The degree of overlapping is used for calculating inter-comment similarity for a comment with respect to a post. Analysis of content features like inter-comment similarity and post-comment similarity along with the non-content features like link number, comment length, stop words showed better results in identifying spam links.

Next to this, comment-based spams have also been discussed here. Archana et al. [13] has dealt with the spam that gets penetrated in the form of comments in blog. A blog is a type of Web content which contains a sequence of periodic user comments and opinions for a particular topic. Here, spam comment is an irrelevant response received for a blog post in the form of a comment. These comments are analyzed using supervised and semi-supervised methods. Analysis considers various features to identify spams. They are listed below: The post similarity feature has been used to find the relevancy between the post and the comment. Word Net tool has been used to spot out the word duplication features. Word duplication feature identifies the redundant words in comments, and it is found to be higher for spam comments and low for genuine comments. Anchor text feature counts the number of links existing for a comment and predicts that the spammers are the one having higher count. Noun concentration feature has been used to extract comments and part of speech tags from the sentences. In that, the legitimate users have low noun concentration.

Stop word ratio feature considers sentences with a finishing point where spammers have less stop word ratio. Number of sentence feature counts the number of sentences existing in a comment and is found to be higher for spammers. Spam similarity feature checks for the presence of spam words listed and categorizes it. The words identified as spam after preprocessing were assigned a weightage, and the contents that fall above the threshold are detected as spam comments. Here, a supervised learning method (naïve Baye's classifier) has been used along with pre-classified training data for labeling a comment as spam and non-spam. One more unsupervised method directly classifies the comments based on the expert-specified threshold.

Interestingly, in the literature, works have been carried out for book spammers also. Sakakura et al. [14] deal with bookmark spammers who create bookmark entries for the target Web resource which contains advertisements or inappropriate contents, thereby creating hyperlinks to increase search result ranking in a search engine system. Spammer may also create many social bookmark accounts to increase the ranking for that Web resource. Therefore, user accounts must be clustered based on the similarity between set of bookmarks to a particular Website or Web resource and not based on the contents. Here, in this paper, data preprocessing is done by clustering bookmarks by extracting Website URL from the raw URL since spammer may create different bookmark entry for same URL.

Here, the similarity based on raw URL (which is the ratio of number of common URLs to total number of URLs contained in the bookmarks of two accounts) has been considered, and the similarity based on site URL without duplicates (which is the ratio of number of common site URLs to the total number of all URLs in both the accounts) and the similarity based on site URL with duplicates (weight of the sites based on the number of bookmarks common to the user accounts) have been calculated. The agglomerative hierarchical clustering of accounts has been made based on one of the above-mentioned similarities. The cluster which is large and having higher cohesion is categorized as an intensive bookmark account spammer. This work achieves a precision of 100 %.

Yu et al. [15] deals with online detection of SMS spam's using Naïve Bayesian classifier, which considers both content and social network SMS features. The SMS social network is constructed from the historical data collected over a period with the help of telecom operator. The content features are extracted that are presented in vector space, and the weights are assigned to the vector obtained using term frequency function. The feature selection methodologies like information gain and odd ratio have been used for selecting words from SMS with which class dependency and class particularity are found for clustering "content-based features". Features on social network try to extract both the sending behavior of mobile users and closeness for categorizing spammer and legitimate user. Bloom filter is used to test the membership between sender and receiver for removing spammer's relationship. Naïve Bayesian classifier has been used for classifying users as legitimate or spam using the above features.

Ravindran et al. [16] deals with the problem of tag recommendation face which contains popular tags for particular bookmarks based on user feedback and to filter spam posts. In this problem, spammer may increase the frequency of a particular tag, and the system may suggest those tags that have higher frequency to the user. To eliminate this problem, this paper uses "frequency move to set" model to choose a set of tags suggested by user for a bookmark. To find whether a tag is popular or not for placing it in the suggestion set, the tag feature like simple vocabulary similarity has been considered. The suggestion set which is kept updated is measured using the stagnation rate, and unpopular tags are removed randomly from the set. The decision tree classifier has been used here to classify tags as spam and non-spam. The accuracy obtained in this approach is about 93.57 %.

Another work by Ariaeinejad et al. [17] deals with detecting email spam in an email system by considering plain text alone that categorize a mail as spam or ham. The common words in spam and ham emails are eliminated and stored in white list. The collected words are parsed by removing unwanted spaces and other signs among the words. The parsed words are compared with white list, and common words are eliminated. The cleaned words are checked for making decision using "Jaro–Wrinkler" technique. Here, a fuzzy map is constructed as a two dimensional vector using an interval type, and two fuzzy methods have been used– (i) to represent distance of each word in email with closed similarity in dictionaries as a horizontal vector and (ii) to represent weight of the words in dictionary as a vertical vector. Third dimension considers the importance of a word in an email and its frequency which is identified using term frequency inverse document frequency technique. Here, email has been categorized into spam, ham, and uncertain zone using fuzzy-C means clustering. Later, the words are updated consistently for correct prediction.

Another work reported by Ishida et al. [18] deals with detection of spam blogs and keywords mutually by its co-occurrence in the cluster. He employed shared interest algorithm for the blogs collected. This algorithm constructs a bi-partite graph between blogs and low frequency keywords from which clusters of varying size has been formed. The spam score for each cluster is calculated and are ranked by multiplying the number of blogs and keywords in the cluster. The spam blogs with highest score are considered as spam seed and is stored in a list. A threshold is set manually for the ranked spam blogs and keywords. Those that exceed the threshold has been detected as spam blog, and spam keywords are removed from the list. This approach provides mutual detection and thereby reducing the filtering cost and the words are kept updated.

3 Conclusion

This survey has presented various approaches which could identify or detect spams in the social network by extracting necessary information from Web pages. Many researchers worked on honey pot profiles, whereas a few people worked on identifying spam links. Even works have been carried out on email and SMS spams. But, still this area remains in its infant stage, and more number of spam detection algorithms need to be devised for social media networks.

References

1. Danah michele boyd, *"Friendster and publicly articulated social networking"*, proceedings of Conference on Human Factors and Computing Systems (CHI 2004), pp. 1279–1282, 2004.
2. Markus Jakobsson, Jacob Ratkiewicz, *"Designing ethical phishing experiments: a study of (ROT13) rOnl query features"*, Proceedings of the 15th international conference on World Wide Web (2006), pp. 513–522, 2006.
3. Alex Tsow and Markus Jakobsson, *Deceit and Deception: A Large User Study of Phishing*, Technical Report TR649, Indiana University, Bloomington, August 2007.
4. Takeda, T.; Takasu, A., *"A splog filtering method based on string copy detection"*, proceedings of First International Conference on Applications of Digital Information and Web Technologies, pp. 543–548, 2008.
5. Kamaliha, E.; Riahi, F.; Qazvinian, V.; Adibi, J., *"Characterizing Network Motifs to Identify Spam Comments"*, proceedings of IEEE International Conference on Data Mining Workshops, pp. 919–928, 2008.
6. Webb, Steve, Caverlee, James and Pu, Calton, *Social Honeypots: "Making Friends With A Spammer Near You"*, Paper presented at the meeting of the CEAS, 2008.

7. Gianluca Stringhini, Christopher Kruegel, Giovanni Vigna, "*Detecting Spammers on Social Networks*", proceedings of Annual Computer Security Applications Conference (ACSAC) 2010.
8. Kyumin Lee, James Caverlee, Steve Webb, "*Uncovering Social Spammers: Social Honeypots +Machine Learning*", proceedings of ACM-SIGIR 2010.
9. Sreenivasan, Shrijina, Lakshmipathi, B., "*An Unsupervised Model to detect Web Spam based on Qualified Link Analysis and Language Models*", International Journal of Computer Applications, vol. 63, issue 4, pp. 33–37, 2013.
10. K. Karthick, V. Sathiya, J. Pugalendiran, "*Detecting Nepotistic Links Based On Qualified Link Analysis and Language Models*", International Journal of Computer Trends and Technology, pp. 106–109, June 2011.
11. Qureshi, M.A.; Younus, A.; Touheed, N.; Qureshi, M.S.; Saeed, M., "*Discovering Irrelevance in the Blogosphere through Blog Search*", proceedings of International Conference on Advances in Social Networks Analysis and Mining (ASONAM), pp. 457–460, 2011.
12. Jenq-Haur Wang; Ming-Sheng Lin, "*Using Inter-comment Similarity for Comment Spam Detection in Chinese Blogs*", proceedings of International Conference on Advances in Social Networks Analysis and Mining (ASONAM), pp. 189–194, 2011.
13. Archana Bhattarai, Vasile Rus, and Dipankar Dasgupta, "*Characterizing Comment Spam in the Blogosphere through Content Analysis*", proceedings of IEEE Symposium on Computational Intelligence in Cyber Security—CICS, pp. 37–44, 2009.
14. Sakakura, Y.; Amagasa, T.; Kitagawa, H., "*Detecting Social Bookmark Spams Using Multiple User Accounts*", proceedings of IEEE/ACM International Conference on Advances in Social Networks Analysis and Mining (ASONAM), pp. 1153–1158, 2012.
15. Yang Yu; Yuzhong Chen, "*A novel content based and social network aided online spam short message filter*", proceedings of 10th World Congress on Intelligent Control and Automation (WCICA), pp. 444–449, 2012.
16. Ravindran, P.P.; Mishra, A.; Kesavan, P.; Mohanavalli, S., "*Randomized tag recommendation in social networks and classification of spam posts*", proceedings of IEEE International Workshop on Business Applications of Social Network Analysis (BASNA), pp. 1–6, 2010.
17. Ariaeinejad, R.; Sadeghian, A., "*Spam detection system: A new approach based on interval type-2 fuzzy sets*", proceedings of 24th Canadian Conference on Electrical and Computer Engineering (CCECE), pp. 379–384, 2011.
18. Ishida, K., "*Mutual detection between spam blogs and keywords based on cooccurrence cluster seed*", proceedings of First International Conference on Networked Digital Technologies, pp. 8–13, 2009.

Botnets: A Study and Analysis

G. Kirubavathi and R. Anitha

Abstract Botnets are an emerging phenomenon that is becoming one of the most significant threats to cyber security and cyber crimes as they provide a distributed platform for several illegal activities such as launching distributed denial of service (DDoS), malware dissemination, phishing, identity theft, spamming, and click fraud. The characteristic of botnets is the use of command and control (C&C) channels through which they can be updated and directed. We investigate the state-of-art research on recent botnet analysis. This paper aims to provide a concise overview of existing botnets in multiple views. The major advantage of this paper is to identify the nature of the botnet problem and find the specific ways of detecting the botnet.

Keywords Botnets · C&C mechanisms · Botnet analysis · Botnet detection survey

1 Introduction

Explosive growth of the Internet provides much improved accessibility to huge amount of valuable data. However, numerous vulnerabilities are exposed, and the number of incidents is increasing over time. Especially, recent malicious attempts are different from old-fashioned threats, intended to get financial benefits through a large pool of compromised hosts. This horrifying new type of threats endangers millions of people and network infrastructure around the world.

G. Kirubavathi (✉) · R. Anitha
Department of Mathematics and Computational Sciences, PSG College of Technology, Coimbatore, India
e-mail: g.kiruba@gmail.com

R. Anitha
e-mail: anitha_nadarajan@mail.psgtech.ac.in

G. S. S. Krishnan et al. (eds.), *Computational Intelligence, Cyber Security and Computational Models*, Advances in Intelligent Systems and Computing 246,
DOI: 10.1007/978-81-322-1680-3_23,

Botnets are emerging as the most significant threat over Internet [1]. A botnet is a collection of zombie computers, connected to the Internet, called bots which are used for various malicious activities [2, 3]. Bot is a self-propagating application that infects the vulnerable hosts through direct exploitation/Trojan insertion. It also performs user-centric tasks automatically without any interaction from the user. Botnets with a large number of computers have enormous cumulative bandwidth and computing capability. They are exploited by botmaster for initiating various malicious activities, such as email spam, distributed denial-of-service attacks, password cracking, and key logging. The first generation of botnets utilized IRC channels as their command and control (C&C) centers. Since the centralized C&C mechanism of such botnets has made them vulnerable, easily they can be detected and removed. In recent years, botnets use HTTP as communication protocol [4]. HTTP bots request and download commands from Web servers under the control of the botmaster. These Web-based C&C bots try to blend normal HTTP traffic, and hence, detecting such bots is more difficult than the IRC bots. The new generation of botnets which can hide their C&C is Peer-to-Peer (P2P) botnets [5]. P2P botnets do not have the limitation of single point of failure, because they have decentralized C&C servers.

A better analysis and understanding of botnets will help the researchers to develop new technologies to detect and defeat the biggest security threat. Also, the knowledge of existing detection techniques and their limitations will lead to the development of good detection techniques. In this paper, we provide an analysis of some botnets as well as various techniques available for the detection of botnets. The rest of the paper is organized as follows: Sect. 2 describes the botnet lifecycle and classification of botnets. Analysis of botnets is given in Sect. 3. Section 4 briefs botnet detection techniques and their limitations. Finally, Sect. 5 concludes the paper.

2 Botnet Lifecycle

The major reason of remarkable success and increase of botnets is their well-organized and planned formation, generation and propagation. The lifecycle of a botnet from its birth to disastrous spread undergoes the following phases [6] as shown in Fig. 1. The first phase is the Initial Injection; during this phase, the attacker scans a target subnet for known vulnerability such as unpatched vulnerabilities, backdoors left by Trojans, password guessing, and brute force attacks and infects victim machines through different exploitation methods. Once a host is infected and becomes a potential bot, in the second phase, it runs a program known as shell-code that searches for malware binaries in a given network database via FTP, HTTP, or P2P [7, 8]. The bot binary installs itself on the target machine. Once the bot program is installed, the victim host behaves as a real bot or zombie. In connection phase, the bot program establishes a connection with C&C server to receive instructions or updates. This connection process is known as rallying [9]. This phase is scheduled every time the host is restarted to ensure the botmaster that the bot is taking part in the botnet and is able to receive commands to perform

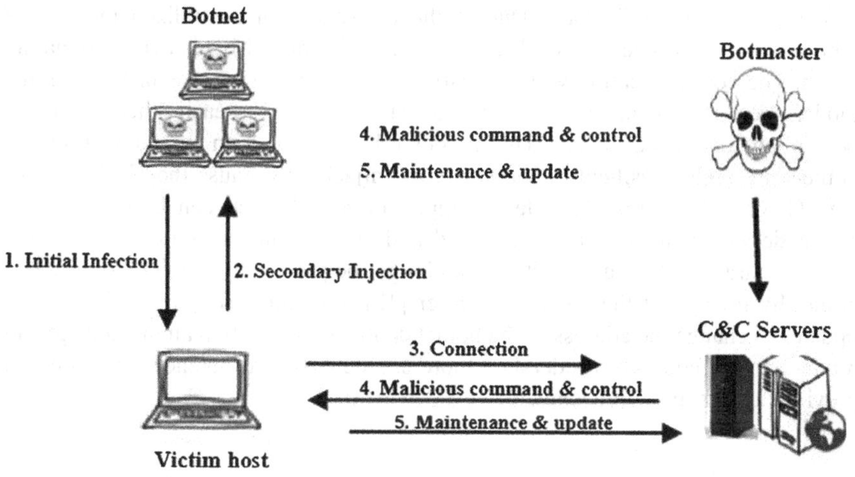

Fig. 1 A typical botnet lifecycle

malicious activities. Therefore, the connection phase is likely to occur several times during the bot lifecycle [10]. After the connection phase, the actual botnet C&C activities will be started. The botmaster uses the C&C to distribute commands to his bot army. Bots receive and execute commands sent by botmaster. The C&C enables the botmaster to remotely control the action of large number of bots to conduct various illicit activities [1]. The last phase of the bot lifecycle is the maintenance and updating of the bots. In this phase, bots are commanded to download an updated binary.

2.1 Command and Control Models

The command and control mechanism is very important and is the major part of the botnet design [11]. This mechanism is used to instruct botnets to operate some tasks such as spamming, phishing, denying services, etc. It directly determines the communication topology of the botnet. Therefore, understanding the C&C mechanisms in botnet has great importance to detect and defend against botnet.

Centralized C&C Model A centralized model is characterized by a central point that forwards messages between clients. In this model, the botmaster selects a host to be the contacting point of all bots; that is, all bots are connected to a centralized server. When the victim is infected, it will connect to the centralized C&C server and wait for commands from the botmaster. The botmaster control thousands of infected bots using centralized model, and this model is easy to construct and efficient for distributing commands. But, this model can be easily detected and disabled [11]. This model uses three different topologies of botnet to connect any zombies in the network namely star topology, multi-server topology, and hierarchical topology.

Decentralized C&C Model Due to the drawbacks of centralized model, the botmaster shifts to decentralized (P2P) model. This model is resilient to dynamic churn. The communication will not disrupt when losing a number of bots. In this model, there is no central server, and bots are connected to each other and act as both client and C&C server. This model follows a random topology. Random botnets are highly resilient to shutdown and hijacking because they lack centralized C&C and employ multiple communication paths between bots. The most recent development in botnet command and control known as fast flux has led some researchers to argue that it may no longer be practically possible to disable a botnet by taking out the command server [12]. Fast flux uses rapid DNS record updates to change the address of the botnet controller very often among a large and redundant set and is considerably more resilient to interference compared to previous command approaches.

2.2 Botnet Communication Mechanisms

Communication mechanisms are the part of the botnet design. The type of communication used between a bot and its C&C server or between any two bots can be classified into two types: push-based mechanism and pull-based mechanism.

The push-based communication mechanism is also known as command forwarding, which means a botmaster issues a command to some bots, and these bots will actively forward the command to others. In this way, bots can avoid periodically requesting or checking for a new command, and hence, this reduces the risk of them being detected. However, the inherent disadvantage of this method is the amount of traffic that can be observed leaving the server, or the tight timing correlation between various monitored nodes that are part of the same botnet, leading to easier detection of infected hosts.

The pull-based communication mechanism is also known as command publishing/subscribing. The bots retrieve commands actively from a place where botmaster publishes commands. This helps not only to avoid flash crowds at a C&C server, but also the injection of random delays in command retrieval from bot to server makes it more difficult to trace a C&C server. This allows the server to hide behind traditional Web traffic.

2.3 Attacks by Botnets

Using botnets, botmaster can launch various attacks like DDoS, phishing, spamming, click fraud, and spyware. DDoS attacks have emerged as a prevalent way to disrupt network services and impose financial losses to organizations or government sectors. DDoS attacks are actions that are intended to destroy or to degrade a service provided by an information system. A DDoS network attack occurs when the victim server receives malicious stream of packets that prevent the legitimate communication from taking place. This can be achieved either by saturating the server's network connection or by using weaknesses in the communication protocols that typically allow the attacker to generate high server resource usage by multiple agents simultaneously. The DDoS attack that disrupted Website operations at Bank of America and at least five other major banks in October 2012 used compromised Websites and flooded the bank's routers, servers, and server applications—layers 3, 4, and 7 of the networking stack with junk traffic [13].

Zombies can help scan and identify vulnerable servers that can be hijacked to host phishing sites, which impersonate legitimate services (e.g., PayPal or banking Websites) in order to steal passwords and other identity data. The Asprox botnet was used for phishing scams. Most of today's email spam is sent by botnet zombies. Botnet-infected computers are said to constitute about 80 % of the total spam. Srizbi is one such bot widely used for spamming. Click fraud is a type of Internet crime that occurs in pay-per-click online advertising. Bots are commonly used to execute click fraud because they can easily be directed to send Web requests that represent "clicks" on the Internet ads for certain affiliates. One such chameleon botnet takes $6 million a month in ad money using click fraud. Adware exists to advertise some commercial entity actively and without the user's permission or awareness. Spyware is a software which sends sensitive information such as credit card no, password, cd keys, etc. to its creator. One such example is the Aurora botnet. Zeus v3 botnet raids on UK bank accounts in 2010. The Citadel Botnets targeted firms to steal $500 million from bank accounts across the world varying from tiny credit unions to big global banks such as BOA, HSBC, Credit Suisse, and the Royal Bank of Canada.

3 Analysis of Botnets

We have analyzed several botnets available in the open source. The details of botnets analyzed and their characteristics are given in Table 1.

Based on the exploration of Barford's research [14], we summarize the binary analysis of botnet into control mechanism, host control mechanism, propagation mechanism, and attack mechanism. Table 2 gives the binary analysis of the botnets BlackEnergy, TDL-4, and Aldi bot.

Table 1 Botnet analysis

Botnets	Year	Prot	Comm.	C&C	Topology	Attacks
Slapper	2002	P2P	Push	Unstructured p2p	Random	Spam, DDoS
Rbot	2003	IRC	Push	Centralized	Multi-server	DDoS, data theft
Phat Bot	2004	P2P	Push	Unstructured p2p	Random	DDoS
Spam Thru	2006	P2P	Push	Custom P2P	Random	Spam
Zeus	2007	HTTP	Pull	Centralized	Multi-server	Steal banking information
Black Energy	2008	HTTP	Pull	Centralized	Multi-server	DDoS
Festi	2009	HTTP	Pull	Centralized	Multi-server	Spam and DDoS
Kelihos	2010	P2P	Push	Unstructured P2P	Random	Bitcoins, Spam
Spyeye	2010	HTTP	Pull	Centralized	Multi-server	Steal banking information
Zero access	2011	P2P	Push	Unstructured P2P	Random	Bitcoin, click fraud
Chameleon	2012	HTTP	Pull	Centralized	Multi-server	Click fraud

Table 2 Botnet binary analysis

Botnets	Mechanisms	Purpose
BlackEnergy	*Botnet Control Mechanisms*	
	auth.php, index.php	To authenticate, botnet control and reporting
	icmp, syn, udp, http, dns	A basic ICMP, TCP syn, UDP, HTTP, DNS traffic flooder
	Host control Mechanisms	
	syssrv.sys, bot.exe	Used to hide bot processes and files, the encrypted bot binary
Aldi	*Botnet Control Mechanisms*	
	DownloadEx, CreateSocks, Update	Download and execute malware code, creates a SOCKS5 proxy, bot update
	Attack Mechanisms	
	StartHTTP, StartTCP	Starts an HTTP DDoS attack, starts a TCP DDoS attack
TDL-4	*Botnet Control Mechanisms*	
	SearchCfg, LoadExe, Search	Search Kad for a new ktzerules file, download and run the executable file, search Kad for a file

4 Botnet Detection

Botnet analysis and detection have been a major research topic in recent years due to increase in the malicious activity. Moreover, the individual bots are not physically owned by the botmaster and may be located in several locations spanning the globe. Differences in time zones, languages, and laws make it difficult to track malicious botnet activities across international boundaries [2, 5]. Many solutions have been proposed in botnet detection and tracking. Mainly, there are two approaches for botnet detection and tracking. Early approach is using honeypots

Table 3 Various botnet detection techniques and their limitations

Different approaches		Level	Features	Technique used	Prot. and struct independent	Limitation
Signature based	Tang and Chen [15]	n/w	In–out bound traffic features	Expectation maximization algorithm, Gibbs sampling algorithm	Yes	In presence of noise, the accuracy of PADS suffers
Anomaly based	Lu et al. [16]	n/w	n-gram feature selection, Traffic payload features,	K-means clustering and X-means clustering	Yes	This approach is payload aware and hard to execute on large scale network.
	Al-Hammadi and Aickelin [17]	Host	API function calls send(), sendto(), recv(), recvfrom() and connect()	Correlation algorithm	IRC	Detection of IRC botnets only
	Liu et al. [18]	n/w	Macroscopic feature of the network streams: characteristics of paroxysm and distribution	P2P k-mean clustering	P2P	Method was unreliable or non-functional even if a single infected machine is present on the network
	Binkley and Singh [19]	n/w	TCP work weight	Combined TCP-based anomaly detection with IRC tokenization and IRC msg statistics	IRC	It could be easily crushed by simply using a minor cipher to encode the IRC commands
	Lin et al. [20]	n/w	Flow (src ip, src port, dst ip, dst port)	Flow correlation, scoring techniques	IRC	It is hard to take down botnets when legitimate IRC servers are used to command the bots
	Yin and Ghorbani [21]	n/w &host	Network traffic features and host data features	Host and network-based detection, correlation algorithm	P2P	It cannot detect bots that are using encrypted messages
	Al-Hammadi and Aickelin [22]	Host	Communication functions, file access functions, registry access function, keyboard status function	Correlation algorithm	P2P	The value of threshold to detect malicious processes is undefined
	Huang et al. [23]	n/w	Peer states	Contact tracing chain	P2P	Once the contact tracing chain is attacked by the botmaster, the number of detected bots will decrease greatly.

(continued)

Table 3 (continued)

Different approaches		Level	Features	Technique used	Prot. and struct independent	Limitation
Mining based	Goebel and Holz [24]	n/w	Unusual or suspicious IRC nicknames, IRC servers, and uncommon server ports	N–gram analysis and scoring system	IRC	Limited to only detecting only IRC-based C&C protocol
	Venkatesh et al. [25]	n/w	TCP connection features: One-way connection of TCP packets, ratio of in/out going TCP packets, ratio of TCP packets, Flags	MLFF NN, bold driver back propagation learning algorithm	HTTP	The proposed method is offline network traffic monitoring
	Gu et al. [26]	n/w	Network traffic flows and packets	Clustering and cross-cluster correlation	Yes	This approach is ineffective, Since current botnets are sending spam through large popular web mail services. Such activities are very hard to detect through network flow analysis due to encryption and overlap with legitimate web mail patterns.
Flow based	Chen et al. [27]	n/w	Time interval, duration, packets, byets, bpp	Mutual authentication and clustering analysis	HTTP	Detects only centralized bots
Modeling based	Wang et al. [28]	n/w	Statistical features-connection level-req, res bytes, duration time, packet level—no. of pkts, avg len of pkts, inter arrival of pkts	X-means clustering, circulation auto correlation	HTTP	Detects only web-based bots
Graph based	Nagaraja et al. [29]	n/w	Unique communication patterns from overlay topologies	Inference algorithm and random walk	P2P	Bootstrapping the detection algorithm may be very challenging and makes the detection results heavily dependent on other systems, thus limiting the real world applicability of BotGrep.
DNS based	Choi et al. [30]	n/w	Src ip addr, domain name of the query, timestamp	Botnet DNS query detection algorithm	Yes	Limited coverage because they use a MAC addr as an identifier of a host rather than an IP addr

(continued)

Table 3 (continued)

Different approaches		Level	Features	Technique used	Prot. and struct independent	Limitation
	Villamarín-Salomón and Brustoloni [31]	n/w	DNS traffic similarity	Bayesian method	Yes	May generate false positive when a domain name is queried by one infected host and a few uninfected host
	Ramachandran and Dagon [32]	n/w	DNSBL queries	Single-host, or third-party, reconnaissance; self-reconnaissance and reconnaissance using other bots	Yes	Approach is useful for just certain types of spam botnets
Symptoms based	Morales et al. [33]	n/w	Symptoms: TCP connection attempts, DNS activities, digital signatures, unauthorized process tampering, process hiding	Decision tree	Yes	Some of the advanced rootkit techniques might be suppressed by the bot to evade detection

[8, 11, 12]. However, honeypots are useful in understanding the botnet characteristics, but they cannot detect bot infection always. The other approach is based on network traffic analysis. Analysis of network traffic can be further classified into signature based, anomaly based, mining based, DNS based, flow based, modeling based, symptoms based, and graph based. Table 3 briefs various attack detection techniques and their limitations.

5 Conclusion

Botnets pose a significant and growing threat against cyber security. Since botnets remain as a large-scale problem that affects the entire Internet community, understanding of botnet characteristics is very important for efficient detection. This paper focuses on the analysis of botnets to understand the behavior on their mechanism which would be helpful for future study of thwarting botnet communications. Diversity of botnet protocols, structures, and encrypted communications makes botnet detection a challenging task. In this paper, a brief survey of some of the existing botnet detection techniques and their limitations is presented. However, most of the current botnet detection techniques are designed only for specific botnet C&C communication protocols and structures. Consequently, when botnets change their C&C architecture, these methods will not be effective in detecting them. But, some of the detection approaches based on DNS and data mining can detect botnets irrespective of the C&C architecture. So, developing techniques to detect botnets regardless of the C&C architecture and with encrypted communications will be challenging, and it can combat botnet threat against computer assets.

References

1. N. Ianelli, A. Hackworth, *Botnets as a Vehicle for Online Crime*—Coordination Center, CERT cMellon University, Carnegie CERT, 2005.
2. T. Micro, *Taxonomy of Botnet Threats*, Technical Report, Trend Micro White Paper, 2006.
3. C. Li, W. Jiang, X. Zou, *Botnet: survey and case study*, in: 4th International Conference on ICIC-2009, pp. 1184–87.
4. Lee, J. S., et.al, *The activity analysis of malicious http-based botnets using degree of periodic repeatability*. In proceedings of International Conference on IEEE Security Technology, 2008, pp. 83–86.
5. J. B. Grizzard, et.al. *Peer-to-peer botnets: Overview and case study*. In USENIX HotBots'07, 2007.
6. Govil, Jivesh, and G. Jivika. "*Criminology of botnets and their detection and defense methods*." In proceedings of IEEE International Conference on Electro/Information Technology, 2007, pp. 215–220

7. Stinson, Elizabeth, and John C. Mitchell. "*Characterizing bots' remote control behavior.*" In Detection of Intrusions and Malware, and Vulnerability Assessment, pp. 89–108. Springer Berlin Heidelberg, 2007.
8. M.A Rajab, et.al., *A multifaceted approach to understanding the botnet phenomenon*, In proceedings of the 6th ACM SIGCOMM Conference on Internet Measurement, IMC'06, ACM, New York, NY, USA, 2006, pp. 41–52.
9. C. Schiller, J. Binkley, *Botnets: The Killer Web Applications*, Syngress Publishing, 2007.
10. Liu, lei, et.al., *BotTracer: Execution-Based Bot- Like Malware Detection*, Information security, Springer, Berlin Heidelberg, 2008, pp. 97–113.
11. Cooke, et.al., *The zombie roundup: Understanding, detecting, and disrupting botnets.* In Proceedings of the USENIX SRUTI Workshop, vol. 39, p. 44. 2005.
12. Lemos, Robert. "*Fast flux foils bot-net takedown.*" 2007.
13. http://arstechnica.com/security/2012/10/ddos-attacks-against-major-us-banks-no-stuxnet/
14. Barford, Paul, and Vinod Yegneswaran. "*An inside look at botnets.*" In Malware Detection, pp. 171–191. Springer US, 2007.
15. Y. Tang and S. Chen. Defending against internet worms: A signature-based approach. In Proc. of Infocom, 2003. vol. 2, pp. 1384–1394. IEEE, 2005.
16. Lu, et.al., "*Clustering botnet communication traffic based on n-gram feature selection.*" Computer Communications 34, no. 3 (2011): 502–514.
17. Al-Hammadi, Y & Aickelin. U, '*Detecting Botnets through Log Correlation*', Proc. IEEE/ IST Workshop on Monitoring, Attack Detection and Mitigation, 2006, pp. 97–100.
18. Liu, et.al., "*A P2P-botnet detection model and algorithms based on network streams analysis.*" In International Conference on Future Information Technology and Management Engineering (FITME), 2010 vol. 1, pp. 55–58.
19. J.R. Binkley and S.Singh, "*An algorithm for anomaly-based botnet detection,*" in Proc. USENIX Steps to Reducing Unwanted Traffic on the Internet Workshop (SRUTI'06), 2006, pp 43–48.
20. Lin, Hsiao-Chung, Chia-Mei Chen, and Jui-Yu Tzeng. "*Flow based botnet detection.*" In Innovative Computing, Information and Control (ICICIC), 2009 Fourth International Conference on, pp. 1538–1541. IEEE, 2009.
21. Yin, Chunyong, and Ali A. Ghorbani. "*P2P botnet detection based on association between common network behaviors and host behaviors.*" In proceedings of IEEE international conference on Multimedia Technology (ICMT), 2011, pp. 5010–5012, 2011.
22. Al-Hammadi, Yousof, and Uwe Aickelin. "*Behavioural Correlation for Detecting P2P Bots.*" In proceedings of IEEE 2nd international conference on Future Networks ICFN'10, 2010. pp. 323–327.
23. Huang, Zhiyong, X. Zeng, and Y. Liu. "*Detecting and blocking P2P botnets through contact tracing chains.*" International Journal of Internet Protocol Technology 5, no. 1, 2010, pp. 44–54.
24. J. Goebel and T. Holz. *Rishi: Identify bot contaminated hosts by irc nickname evaluation.* In USENIX Workshop on Hot Topics in Understanding Botnets (HotBots'07), 2007.
25. Venkatesh, G. Kirubavathi, and R. Anitha Nadarajan. "*HTTP botnet detection using adaptive learning rate multilayer feed-forward neural network.*" In Proceedings of the Springer international workshop in information security theory and practice—WISTP'12, UK, 2012. pp. 38–48.
26. Gu, G et.al., *Bot-Miner: Clustering Analysis of Network Traffic for Protocol-and Structure Independent Botnet Detection.* In 17th Usenix Security Symposium (2008).
27. Chen, Chia-Mei, Ya-Hui Ou, and Yu-Chou Tsai. "*Web botnet detection based on flow information.*" In proceedings of IEEE international conference on Computer Symposium (ICS), 2010, pp. 381–384.
28. Wang, et.al., "*Modeling Connections Behavior for Web-based Bots Detection.*" In proceedings of IEEE 2nd international conference on e-Business and Information System Security (EBISS), pp. 1–4, 2010.

29. Nagaraja, et.al., "*BotGrep: Finding P2P Bots with Structured Graph Analysis.*" In USENIX Security Symposium, pp. 95–110. 2010.
30. H. Choi, H. Lee, and H. Kim, "*Botnet Detection by Monitoring Group Activities in DNS Traffic,*" in proceedings of IEEE 7th international conference on Computer and Information Technology (CIT 2007), 2007, pp. 715–720.
31. Villamarín-Salomón, Ricardo, and J C Brustoloni. "*Bayesian bot detection based on DNS traffic similarity.*" In Proceedings of ACM symposium on Applied Computing, pp. 2035–2041. ACM, 2009.
32. N. F. A. Ramachandran and D. Dagon, "*Revealing botnet membership using dnsbl counter-intelligence,*" in Proc. 2nd Workshop on Steps to Reducing Unwanted Traffic on the Internet (SRUTI'06), 2006.
33. Morales, et.al., "*Symptoms-based detection of bot processes.*" In Computer Network Security, pp. 229–241. Springer Berlin Heidelberg, 2010.

Comparative Study of Two- and Multi-Class-Classification-Based Detection of Malicious Executables Using Soft Computing Techniques on Exhaustive Feature Set

Shina Sheen, R. Karthik and R. Anitha

Abstract Detection of malware using soft computing methods has been explored extensively by many malware researchers to enable fast and infallible detection of newly released malware. In this work, we did a comparative study of two- and multi-class-classification-based detection of malicious executables using soft computing techniques on exhaustive feature set. During this comparative study, a rigorous analysis of static features, extracted from benign and malicious files, was conducted. For the analysis purpose, a generic framework was devised and is presented in this paper. Reference dataset (RDS) from National software reference library (NSRL) was explored in this study as a mean for filtering out benign files during analysis. Finally, through well-corroborated experiments, it is shown that AdaBoost, when combined with algorithms such as C4.5 and random forest with two-class classification, outperforms many other soft-computing-based techniques.

Keywords Malware · Portable executable features · Static analysis · NSRL · Classification

1 Introduction

Contemporary computer and communication infrastructure are exceedingly vulnerable to various types of attacks. The sheer amount and range of known and unknown malware is part of the reason why detecting malware is a complicated

S. Sheen (✉) · R. Karthik · R. Anitha
PSG College of Technology, Coimbatore, India
e-mail: shina_np12@yahoo.com

R. Karthik
e-mail: karthi1998@gmail.com

R. Anitha
e-mail: anitha_nadarajan@mail.psgtech.ac.in

G. S. S. Krishnan et al. (eds.), *Computational Intelligence, Cyber Security and Computational Models*, Advances in Intelligent Systems and Computing 246,
DOI: 10.1007/978-81-322-1680-3_24,

problem. A widespread way of inducing these attacks is using malwares such as worms, viruses, trojans, or spywares. Christodorescu and Jha [1] describe a malware instance as a program whose objective is malevolent. McGraw and Morrisett [2] describe malicious code as "any code added, changed, or removed from a software system in order to deliberately cause harm or subvert the intended function of the system." Vasudevan and Yerraballi [3] describe malware as "a generic term that encompasses viruses, trojans, spywares and other intrusive code."

In this work, we have considered Windows-based malware for analysis. The popularity of Windows based OS among average computer users is a huge reason why Windows has the most viruses of any operating system. It is also true that Microsoft's apparent lack of concern for security in the early days made the problem much worse than it had to be. 2012 saw numerous attacks that were devised for Windows vulnerabilities with Windows 7's malware infection rate as much as 182 %.

According to Sophos security report, many of the malwares found on Macs are also Windows based malware. (http://www.sophos.com/en-us/medialibrary/PDFs/other/sophossecuritythreatreport2013.pdf). Mac users who need occasional access to a Windows program sometimes decide to download it from third parties and may illegally create a license key using a downloadable generator. By doing so, they often encounter malware such as Mal/KeyGen-M, a family of trojanized license key generators that has been identified on approximately 7 % of the Macs running Sophos anti-virus software. Another common source of Windows malware on Macs today is fake Windows Media movie or TV files. These files contain auto-forwarding Web links promising the codec needed to view the video, but deliver zero-day malware instead. Moreover, the Windows' partitions of dual-boot Macs can indeed be infected, as can virtualized Windows' sessions running under Parallels, VMware, VirtualBox, or even the open source WINE program. Windows Media files generally would not run on Macs, but Mac users often torrent these files to improve their "ratios" on private tracker sites, without realizing that the contents are malicious .

Windows' users then attempt to play the videos and become infected. The top Windows malwares found on Macs are shown in Fig. 1. For detecting malwares, a number of non-signature based malware detection methods have been proposed in recent times. These methods typically use heuristic analysis, behavior analysis, or a combination of both to identify malware. Such methods are being robustly explored because of their ability to detect zero-day malware without any a priori knowledge. Some of them have been incorporated into the existing commercial off-the-shelf anti-virus products, but have attained only limited success [4, 5]. The most imperative deficiency of these methods is that they are not real-time deployable. Also, McGraw and Morrisett [2] note that classifying malicious code has become more and more complex as newer versions appear to be combinations of those that belong to existing categories.

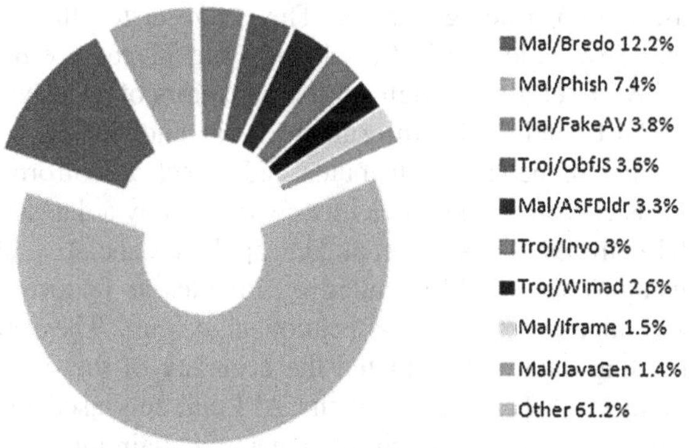

Fig. 1 Sophos 7-day snapshot of 100,000 Macs (April 2012)

2 Related Works

In this paper, we focus on comparative study of many supervised-learning-based soft computing techniques for malware detections by extracting the portable executable features extracted from certain parts of EXE files stored in Win32 PE binaries (EXE or DLL). These are meaningful features that might indicate that the file was created or infected to perform malicious activity. Among the features to be extracted are data extracted from the PE header that describes physical structure of a PE binary (e.g., creation/modification time, machine type, file size), optional PE header information describing the logical structure of a PE binary (e.g., linker version, section alignment, code size, debug flags), import section details, export section details, resources used by a given file, and the version information. String features are based on plain text strings that are encoded in program files (such as "windows," "kernel," "reloc") and possibly can also be used to represent files similar to text categorization problem. Entropy of interpretable strings that is a discriminating feature was also considered in this work.

Shultz et al. [6] extracted DLL information inside PE executables. They retrieved the list of DLLs, the list of DLL function calls, and the number of different function calls within each DLL. RIPPER, an inductive rule learning algorithm, is used on top of every feature vector for classification. They have done experiments on a dataset that consists of 206 benign and 38 malicious executables in the PE file format.

Shafiq et al. [7] presented the "PE-Miner" framework where 189 structural features are extracted from executables to detect new malwares in real time. They evaluated their system using the VX Heavens and Malfease datasets and obtained a detection rate of 99 %. Ye et al. [8, 9] analyzed the Windows APIs called by PE files to develop the intelligent malware detection system using object-oriented association-mining-based classification. The association among the APIs captures the underlying semantics for the data which are essential for malware detection. They first constructed the API execution calls, followed by extracting OOA rules using OOA fast FP-growth algorithm. Finally, classification is based on the association rules generated. They have tested their scheme using 29,580 executables of which 12,214 are benign and 17,366 are malicious executables.

Wang et al. [10] defined the behaviors of an executable by observing its usage of DLL and APIs. Information gain and SVMs were applied to filter out the redundant behavior attributes and select the informative features for training a virus classifier. The model was evaluated by a dataset containing 1,758 benign and 846 viruses. Shaorong et al. [11] applied associative classification based on FP tree using Windows APIs called by PE files as feature set. They proposed an incremental associative classification algorithm. They mined 20,000 executables as training data and validated the accuracy of the classifier on 10,000 executables. Sami et al. [12] extracted the API call sets used in a collection of PE files and generated a set of discriminative and domain interpretable features. These features are then used to train a classifier to detect unseen malwares. They have achieved detection rate of 99.7 %. But the problem with API call feature extraction is it is very time-consuming.

3 Proposed Framework for Comparative Study

A generic framework for malware detection has been proposed, which is elaborately discussed in this section.

3.1 National Software Reference Library and Reference Dataset

The National Software Reference Library (NSRL) is a project supported by the US Department of Justice's National Institute of Justice (NIJ), federal, state, and local law enforcement, and the National Institute of Standards and Technology (NIST) to promote efficient and effective use of computer technology in the investigation of crimes involving computers. It was designed to gather software from different sources and incorporate file profiles into a reference dataset (RDS) of information. The RDS can be used by law enforcement, government, and industry organizations to review files on a computer by matching file profiles in the RDS. This will help alleviate much of the effort involved in determining which files are important as evidence on computers or file systems that have been seized as part of criminal investigations. The RDS is a collection of digital signatures of known, traceable software applications. There are application hash values in the hash set, which may be considered malicious, i.e., steganography tools and hacking scripts.

3.2 Generic Malware Detection Framework

Figure 2 depicts the generic framework for malware detection studied and evaluated in this work.

There are two work flows shown in Fig. 1: Path 1 (* marked) and an auxiliary, Path 2. In Path 2, mainly followed for forensic analysis, all the files in a system are hashed and subsequently compared with the file hash in RDS library. If a match is found, a file is directly classified into a known category, which is harmless. This preprocessing reduces the overall collection of files, which needs to be analyzed with a soft-computing-based classifier. Path 1 can be followed for real time detection of executable files by directly evaluating against the classifier for categorizing into malware and benign files, without doing any hashing.

3.3 Three-Stage Malware and Benign File Classifier

The classification described in the generic malware detection framework is divided into three-stage process as shown in Fig. 3.

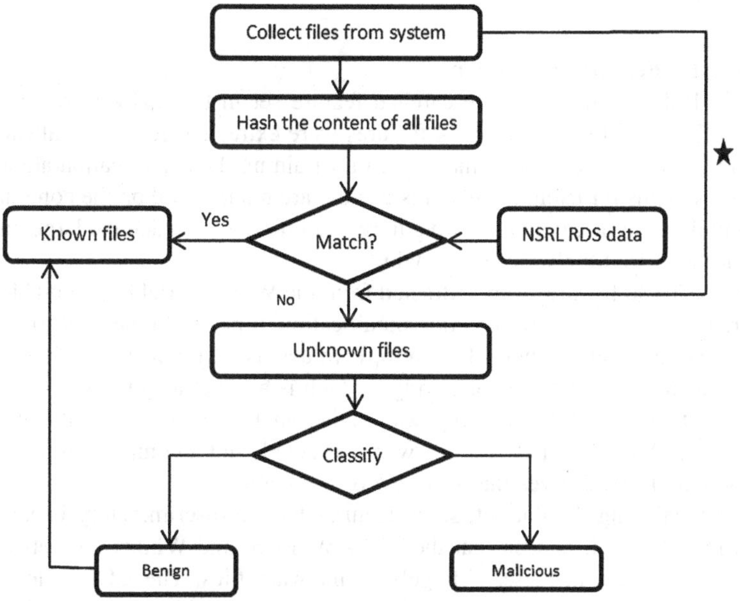

Fig. 2 Generic malware detection framework

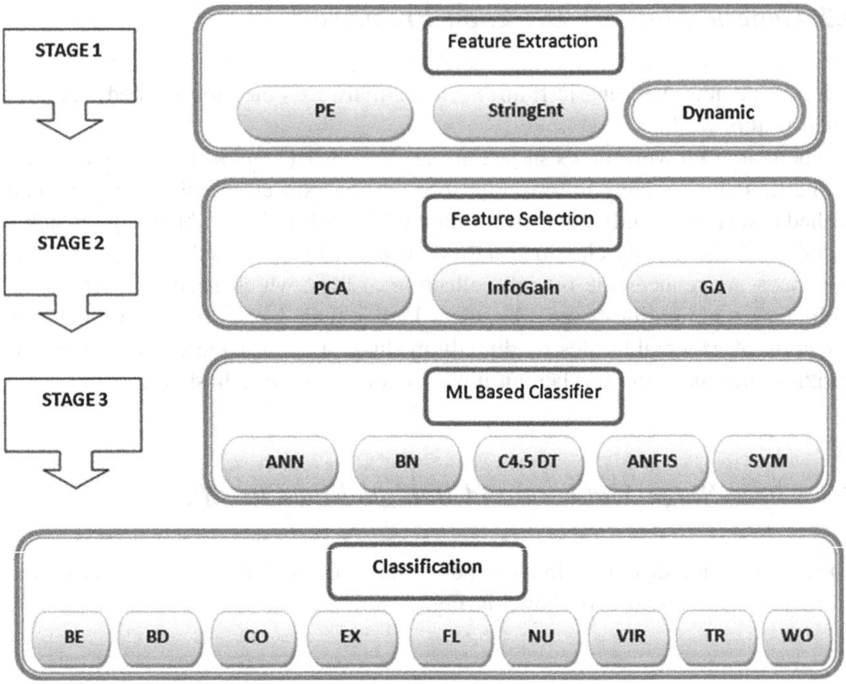

Fig. 3 Three-stage malware and benign file classifier

Stage 1—Feature Extraction

Table 1 shows the widely researched feature set in the malware research literatures. For our study, 2, 3, and 4 features were extracted from the malware and benign files and 2 and 3 were finally used for training the classification algorithm.

The decisions for using the features 2 and 3 are made based on the computation time and the complexity in extracting these features and also due the large number of Windows-based malware in the wild.

A total of 188 features were extracted from the Win32 portable executable (PE) header, to mention a few: creation/modification time, machine type, file size, linker version. Then, a total of 4 entropy values as mentioned in Table 2 were calculated from the interpretable strings, which is a novel idea used in this work. Analysis showed that the entropy was higher in malicious files rather than in benign files. A total of 191 features were extracted, and a comprehensive dataset was formed. Table 2 gives the overview of the features.

When analyzing the dataset, some features have a discriminatory behavior as shown in Table 3. It is seen that the DLLs Winsock and Winnet are networking DLLs, which are made use of largely in malware files. The COFF file header contains important information such as the type of the machine for which the file is intended, the nature of the file (DLL, EXE, or OBJ), the number of sections, and the number of symbols. It is interesting to note in Table 3 that a reasonable number

Table 1 Widely used features in malware detection

Features
1. N-grams from the binary representation of a file
2. Win32 portable executable (PE) header information
3. Entropy of uppercase characters, lowercase characters, digits, and other characters of the interpretable string (StringEnt)
4. Dynamic instruction sequence (opcode)
5. Some of the features that were collected from the dynamic execution "defined file type copied to Windows folder," "defined file type created," etc.

of symbols are present in benign executables. The malicious executables, however, either contain too many or too few symbols. The interesting information in the standard fields of the optional header includes the linker version used to create an executable, the size of the code, the size of the initialized data, the size of the uninitialized data, and the address of the entry point.

In Table 3, the values of major linker version and the size of the initialized data have a significant difference in the benign and malicious executables. The size of the initialized data in benign executables is usually significantly higher compared to those of the malicious executables. The Windows-specific fields of the optional header include information about the operating system version, the image version,

Table 2 Feature description

Portable executables (PE)	Dynamic link library	73	Binary
	COFF file header	7	Numerical
	Optional header (standard fields)	31	Numerical
	Optional header (data directory)	32	Numerical
	Section header (only .text, .data, .rscs)	27	Numerical
	Resource directory table	6	Numerical
	Resource directory resources	11	Numerical
	Checksum	1	Binary
String entropy (StringEnt)	Uppercase characters	1	Numerical
	Lowercase characters	1	Numerical
	Digits	1	Numerical
	Junk and special characters (newline) etc.	1	Numerical

Table 3 Discriminatory PE features

Feature	Benign	Malware
Winsock.DLL	0.003	0.160
Winnet.DLL	0.005	0.103
Number of symbols	384.2	0.002
Major linker version	8.334	4.419
Major image version	160.03	9.93
Data raw size	387,152	182,748
Debug size	27.673	46.7312
COFF header char	6,696.94	13,488.83

Table 4 Efficiency matrix—two class

	TP	FP	F-Measure	ROC	Accuracy
MultiBoost AB + C4.5	0.995	0.011	0.995	0.998	99.5432
C4.5	0.993	0.013	0.993	0.989	99.3434
Naïve Bayes	0.967	0.042	0.967	0.982	96.7314
Random forest	0.995	0.01	0.995	0.999	99.5147
AdaBoostM1 + C4.5	0.996	0.008	0.996	0.998	99.6146
Bagging + C4.5	0.994	0.013	0.994	0.997	99.372
AdaBoostM1 + random forest	0.997	0.008	0.997	0.997	**99.6717**
Multi-layer perceptron	0.99	0.022	0.99	0.996	98.9723
Radial-basis-function-based neural network (RBNet)	0.948	0.029	0.949	0.993	94.7759
Bayes network	0.985	0.022	0.985	0.977	98.487

the checksum, the size of the stack and the heap. It can be seen that their values are significantly higher in the benign executables.

Stage 2—Feature Selection

For building robust learning models for any soft-computing-based classifier, the foremost step is to have a strong and meaningful feature set. By removing most irrelevant and redundant features from the data, feature selection helps improve the performance of learning models. Feature selection was done using PCA, information gain, and genetic search. The above-mentioned feature selection methods were applied to the 191 features extracted from each file, and the corresponding dataset was formed. Further, these datasets were exhaustively used with many soft-computing-based classifiers, which are discussed below. After few trial runs with different combinations of the feature set, 34 features were finally selected.

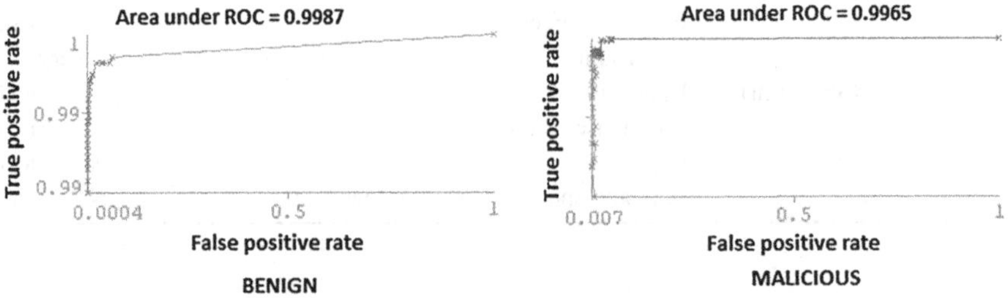

Fig. 4 ROC curve for two-class classification

Table 5 Comparison with related works

Paper	Features used	TP rate (%)	FP rate (%)
Shafiq et al. [7]	189	99	0.5
Siddiqui et al. [13]	84	92.4	9.2
Khan et al. [14]	42	78.1	6.7
Proposed work	34	99.67	0.8

Table 6 Efficiency matrix—9 class

	Class	TP	FP	F-Measure	ROC	Accuracy
AdaBoostM1 + C4.5	BE	0.989	0.004	0.989	0.999	83.2858
	BD + SN	0.825	0.067	0.828	0.959	
	CO + VT	0.573	0.006	0.578	0.957	
	DO + NU	0.55	0.003	0.626	0.954	
	FL	0.542	0.009	0.577	0.94	
	EX + HT	0.615	0.005	0.604	0.945	
	WO	0.754	0.038	0.663	0.965	
	TR	0.771	0.067	0.795	0.95	
	VIR	0.75	0.01	0.423	0.955	
	Avg	0.833	0.043	0.835	0.967	
AdaBoostM1 + random forest	BE	0.99	0.002	0.992	0.998	**84.8273**
	BD + SN	0.854	0.081	0.831	0.959	
	CO + VT	0.612	0.003	0.685	0.947	
	DO + NU	0.577	0.003	0.663	0.928	
	FL	0.526	0.006	0.601	0.931	
	EX + HT	0.59	0.003	0.63	0.913	
	WO	0.783	0.028	0.724	0.961	
	TR	0.786	0.063	0.809	0.953	
	VIR	0.8	0.007	0.529	0.961	
	Avg	0.848	0.044	0.848	0.967	
Bagging + C4.5	BE	0.985	0.003	0.988	0.996	82.3294
	BD + SN	0.818	0.073	0.818	0.944	
	CO + VT	0.505	0.005	0.55	0.907	
	DO + NU	0.486	0.004	0.551	0.894	
	FL	0.495	0.011	0.525	0.908	
	EX + HT	0.462	0.002	0.55	0.843	
	WO	0.764	0.04	0.659	0.942	
	TR	0.764	0.071	0.786	0.931	
	VIR	0.75	0.012	0.395	0.951	
	Avg	0.823	0.045	0.825	0.952	

Stage 3—Classifiers

Pattern recognition and classification refer to an algorithmic procedure for assigning a given piece of input data into one of the given number of categories. An algorithm that implements classification, especially in a concrete implementation, is known as a classifier. In this work, there are many soft-computing-based classifiers used such as artificial neural network, Bayesian network, decision trees along with ensemble of classifiers with boosting.

4 Experimental Results

A total of 66,713 malicious files were collected from a very extensive repository of malwares in the Internet know as VX Heavens [15]. In that, around 26,713 files had a valid PE header. After the removal of outlier data, a comprehensive dataset of 20,154 entries consisting of PE feature values and string entropy of benign and malicious files was collected, from which 13,148 entries were randomly chosen to be used as training dataset and remaining 7,006 entries were used for testing.

In particular, for the two-class classification, AdaBoostM1 + random forest give the best classification accuracy of **99.6717 %,** as shown in Table 4 and the receiver-operating characteristic (ROC) curve, as shown in Fig. 4.

We have also compared the performances of our best classifier with results obtained from related researches, as shown in Table 5. We have achieved a better classification performance with a lesser number of features

The efficiency matrix for nine-class classification is shown in Table 6.

5 Conclusion and Future Direction

In this work, a comparative study of two- and multi-class classification for malicious executable detection was conducted and evaluated with well-corroborated experiments. Also, a generic framework for detecting malicious executables and a three-stage classifier was proposed. Apart from the PE header features, an entropy-based interpretable string feature was used.

As further future direction, dynamic features along with static features could be combined together to form a hybrid feature set for a more robust detection mechanism. Also, feasibility study could be conducted in order to extent this work to other platforms other than the Windows operating system, which is one of the widely used and exploited operating system.

References

1. M. Christodorescu and S. Jha. Testing malware detectors. In Proceedings of the International Symposium on Software Testing and Analysis, July 2004.
2. G. McGraw and G. Morrisett. Attacking malicious code: A report to the infosec research council. IEEE Software, 17(5):33–44, 2000.
3. A. Vasudevan and R. Yerraballi. Spike: Engineering malware analysis tools using unobtrusive binary-instrumentation. In Proceedings of the 29th Australasian Computer Science Conference, pages 311–320, 2006.
4. F. Veldman, "Heuristic Anti-Virus Technology", International Virus Bulletin Conference, pp.67–76, USA, 1993.
5. J. Munro, "Antivirus Research and Detection Techniques", Antivirus Research and Detection Techniques, ExtremeTech, 2002, available at http://www.extremetech.com/article2/0,2845,367051,00.asp.
6. M. G. Schultz, E. Eskin, E. Zadok, and S. J. Stolfo. Data mining methods for detection of new malicious executables. In Proceedings of the 2001 IEEE Symposium on Security and Privacy (S&P'01), pages 38–49, May 2001
7. M. Zubair Shafiq, S. Momina Tabish, Fauzan Mirza, Muddassar Farooq. PE-Miner: Mining Structural Information to Detect Malicious Executables in Realtime. In Proceedings of the 2009 Recent Advances in Intrusion Detection (RAID) Symposium-Springer.

8. YanfangYe, D. Wang, T. Li, and D. Ye. IMDS: Intelligent Malware Detection System. In KDD '07: Proceedings of the 13th ACM SIGKDD international conference on Knowledge discovery and Data Mining
9. Yanfang Ye, Dingding Wang, Tao Li, Dongyi Ye, Qingshan Jiang: An intelligent PE-malware detection system based on association mining. Journal in Computer Virology 4(4): 323–334 (2008)
10. Tzu-Yen Wang, Chin-Hsiung Wu, Chu-Cheng Hsieh, A Virus Prevention Model Based on Static Analysis and Data Mining Methods, IEEE 8th International Conference on Computer and Information Technology Workshops, 2008.
11. Feng Shaorong, Han Zhixue, An Incremental Associative Classification algorithm used for Malware Detection, 2nd International Conference on Future Computer and Communication (ICFCC), 2010.
12. A Sami, B Yadegari, H Rahimi, N Peiravian, S Hashemi and A Hamze, Malware Detection based on Mining API Calls, In Proceedings of the 2010 ACM Symposium on Applied Computing.
13. M. Siddiqui, M. C. Wang, and J. Lee, "Detecting trojans using data mining techniques." in IMTIC, ser. Communications in Computer and Information Science, D. M. A. Hussain, A. Q. K. Rajput, B. S. Chowdhry, and Q. Gee, Eds., vol. 20. Springer, 2008, pp. 400–411
14. H. Khan, F. Mirza, and S. Khayam, "Determining malicious executable distinguishing attributes and low-complexity detection," Journal in Computer Virology, pp. 1–11, 2010, 10.1007/s11416-010-0140-6. [Online]. Available: http://dx.doi.org/10.1007/s11416-010-0140-6
15. VX Heaven http://vx.netlux.org

CRHA: An Efficient HA Revitalization to Sustain Binding Information of Home Agent in MIPv6 Network

A. Avelin Diana, V. Ragavinodhini, K. Sundarakantham and S. Mercy Shalinie

Abstract Home agents (HAs) maintain the binding information of mobile node (MN). The binding cache of HA stores the associated data of MN. It represents a single point of failure in Mobile IPv6 networks. An efficient fault-tolerant method is essential to defend these information without any loss. This paper focuses on the revitalization of HA at the time of failure. The standby HAs are formed as clusters within the redundant HA set. Every home agent synchronize its bindings to the next highest preference value home agent. In this paper, we propose clustered redundant home agent (CRHA) protocol to maintain the binding association within the cluster. The simulation results show that our approach is better than the existing recovery schemes.

Keywords MIPv6 · Fault tolerance · Home agent · Binding update · Preference value

1 Introduction

In MIPv6 network, when the Mobile node (MN) gets away from the home and changes its point of attachment to the Internet, home agents (HA) maintain current location (IP address) information of MN [1]. Correspondent node (CN) is the

A. A. Diana (✉) · V. Ragavinodhini · K. Sundarakantham · S. M. Shalinie
Department of Computer Science and Engineering, Thiagarajar College of Engineering, Madurai, India
e-mail: dianaavelin@gmail.com

V. Ragavinodhini
e-mail: raga.cse.34@gmail.com

K. Sundarakantham
e-mail: kskcse@tce.edu

S. M. Shalinie
e-mail: shalinie@tce.edu

G. S. S. Krishnan et al. (eds.), *Computational Intelligence, Cyber Security and Computational Models*, Advances in Intelligent Systems and Computing 246,
DOI: 10.1007/978-81-322-1680-3_25, © Springer India 2014

network entity on another end for communication, i.e., any node that communicates with MN is called CN. HA is a router on MN's home network, which tunnels datagram for delivery to MN when it is away from home network. In Mobile IPv6, MN should assign three IPv6 addresses to their network interfaces, when they are roaming away from their home network. First is its home address (HoA), which is a stable IP address assigned to the MN. It is used for two reasons: (1) allows a MN which is having a stable entry in the DNS and (2) to hide the IP layer mobility from upper layers. The second is MN's current link, i.e., local address, and the third address is care-of-address (CoA) which is related with MN only when it visits foreign network. The association between the MN's HoA and its CoA along with the remaining life time is known as binding. The central data structure used in MIPv6 is binding cache (BC), a volatile memory consisting of number of bindings for one or more MNs. BC is maintained by both CN and HA. Each entry contains the MN's HoA, CoA, and life time. The life time is valid, if the MN does not refresh the BC entry; the entry is deleted after the lifetime expiry. After configuring its CoA, the MN has to register its binding with HA to determine the current location of MN.

The remainder of this paper is organized as follows: Sect. 2 comprises of related work. The MIPv6 network architecture is described in Sect. 3. Section 4 briefly illustrates the proposed clustered redundant home agent (CRHA) scheme, and its performance analysis with the existing approach is described in Sect. 5. The conclusion is provided in Sect. 6.

2 Related Work

The HA in MIPv6 network transmits packet via tunneling to the MN. The network comprises of active home agent (AHA) and redundant standby home agents (SHAs). The HAs are identified via DHAAD mechanism [2], and the selection of HA is based on the highest preference value [3]. Every HA must maintain a separate HA list [4]. The binding update (BU) is retained to hold this list between MN and HA [5]. This list contains the binding information of MN, and if the AHA gets failed, it is transferred to one of the SHA.

In Mobile IPv6, fault-tolerant methods can be classified into three categories. Central management method is difficult to deploy and least expandable [6, 7]. Passive failure detection and recovery method MN detects HA failure during registration and produces long service break time, and the signaling cost is high. In binding backup, AHA backup its bindings with SHA. At the time of HA failure, one of the SHA will take over the service from AHA. This method deploys different solutions. Full backup method [8] deploys every home agent in a network to maintain all MN bindings. The stable storage method [9] is used to keep all MN bindings in the network. In partial backup method [10], each home agent in a network selects a SHA. All the HAs in virtual home agent method [11] share one global address, and only one home agent is active. IETF draft proposes redundant home agent set (RHAS) method [12, 13] in which every AHA has a SHA from

RHAS. If the AHA fails, the SHAs in the RHAS are responsible for failure detection and service takeover. This method comprises of two switch methods, namely RHAS virtual and RHAS hard which works with HARP protocol [14].

In all the discussed approaches, the recovery duration of a HA at the time of AHA failure is high. Even though it has a selected SHA, the continuous failure of AHAs will lead to loss of binding information. If the replaced new AHA gets failed suddenly without selecting a new SHA, the information in such failed HA will be lost and cannot be recovered. Our approach provides an efficient solution to overcome such issue, and we have also compared our method with other existing approaches.

3 Network Architecture

In Internet topology, the movement of MN can be identified by unique prefix of each network. If the network has one or more prefix, it can be advertised by the neighbor discovery. The global address of the router in *R*-flag depicts that the network has different routers with same prefix and the MN is connected to one router among them. Too many MNs for a single HA cause overload problem. To mitigate this problem, MIPv6 provides DHAAD mechanism that determines the HA address and permits the HA to distribute the load between multiple HAs in a same network by using higher preference value. The network architecture is described in Fig. 1.

4 Proposed Clustered Redundant Home Agent Scheme

4.1 HA Access and Handover Mechanism

The flow chart Fig. 2 illustrates the HA access mechanism in our network. Initially, the incoming BU message from MN is authenticated by its respective AHA. If the BU message is successfully authenticated, AHA sends the registration reply

Fig. 1 Network architecture

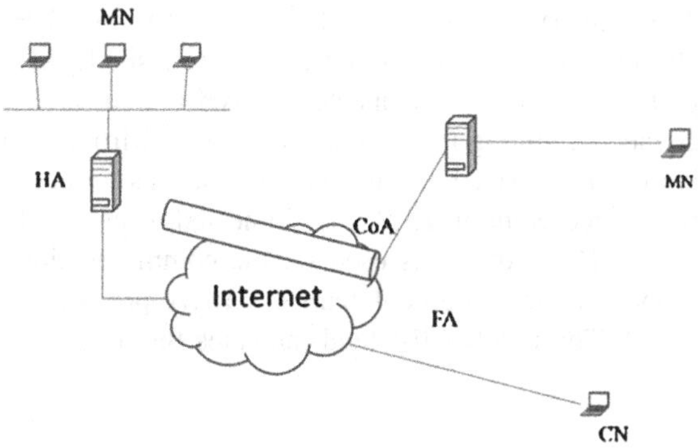

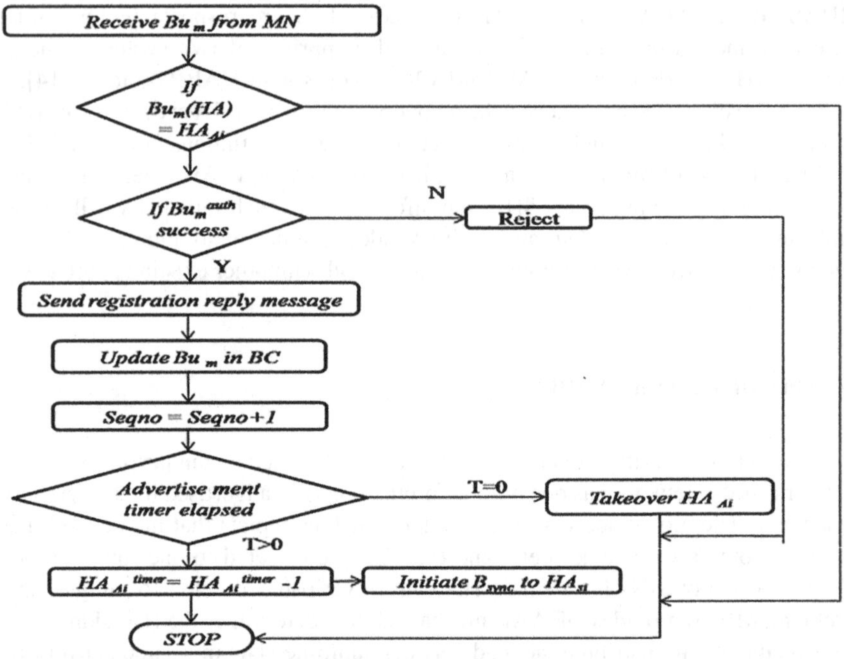

Fig. 2 HA access and handover in MIPv6 network

message to MN else it rejects the MN. Then, the MN updates BU message in BC. The sequence number is incremented by one based on the incoming BU. If the advertisement timer of AHA has elapsed, two cases are to be noticed

- Case 1 ($T = 0$): SHA takeover the AHA.
- Case 2 ($T > 0$): Timer of AHA gets decrement by one and the binding synchronization to SHA is initiated.

4.2 CRHA Mechanism

The network consists of a cluster of SHAs in a RHAS as shown in Fig. 3. Each cluster of virtual SHAs in RHAS serves a single AHA. A single SHA with highest preference value is connected to AHA, and other SHAs within the cluster are connected among themselves and updated from connected SHA. When the SHA knows that AHA fails, it converts itself as an active one and SHA with highest preference value in RHAS is connected to new AHA.

HA–HA protocol is used for the communication between HAs. In RHAS, we propose a new protocol named CRHA protocol for HA communication as in Fig. 4. The number list field indicates the total number of HAs in a cluster. The

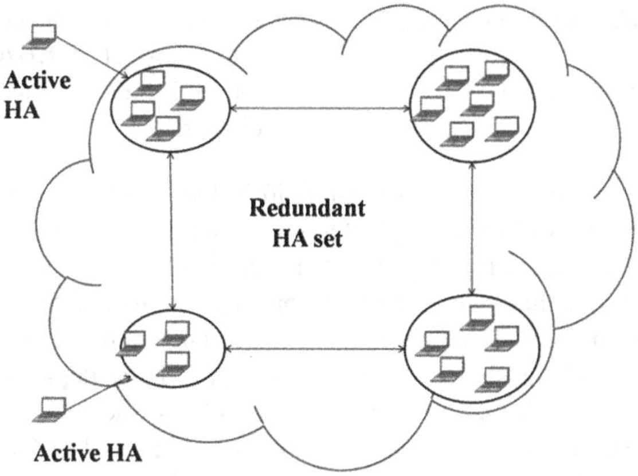

Fig. 3 Clustered redundant HA set

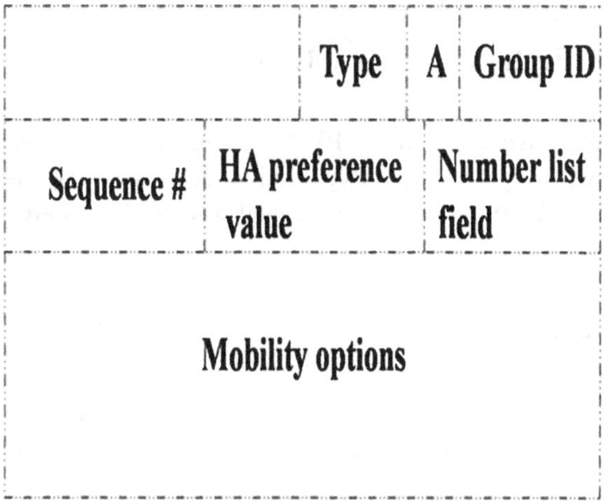

Fig. 4 CRHA protocol

preference value of SHA is based on the nearest distance from AHA. If a SHA from the cluster is removed, the number list field gets decremented.

The type field is to identify the type of HA in a network. Mobility options field consists of any option other than the specific section of message precised to mobility.

Type 0	Assigned for SHA which is directly connected with the AHA
Type 1	Denotes the cluster HA in a redundant HA set
Group ID	Determines the ID for each cluster
A (acknowledgement flag)	Verifies the sender is AHA through the cluster hello message
Sequence	Confirms that the incoming cluster hello message is a most recent one

HA preference value	Based on the highest preference value, SHA from the cluster is set as AHA at the time of AHA failure
Number list	This field indicates the total number of HAs in a cluster.

When the number list field value of cluster gets reduced to ≤2, the AHA relieves from older cluster and binds with a new cluster in RHAS. The left HAs batch up with a cluster constituting least number of HAs which is nearer. Figure 5a shows that two AHAs are batched up with two clusters, i.e., serving clusters. Consider AHA2 with cluster preference value 3 as in Fig. 5a. It gets failed, and one of the SHAs in cluster serves as AHA as depicted in Fig. 5b. Now, the cluster consists of only two standby virtual HAs, and so, it finds a new non-serving cluster with number list 6 and joins it as in Fig. 5c. Finally, the remaining two HAs get attached to the serving cluster with 4 HAs to make it as 6 as shown in Fig. 5d.

5 Performance Analysis

We have compared CRHA scheme with the existing RHAS hard and virtual switching methods. Registration delay of a HA means the delay variation in registration between before and after AHA failure. In all the other approaches,

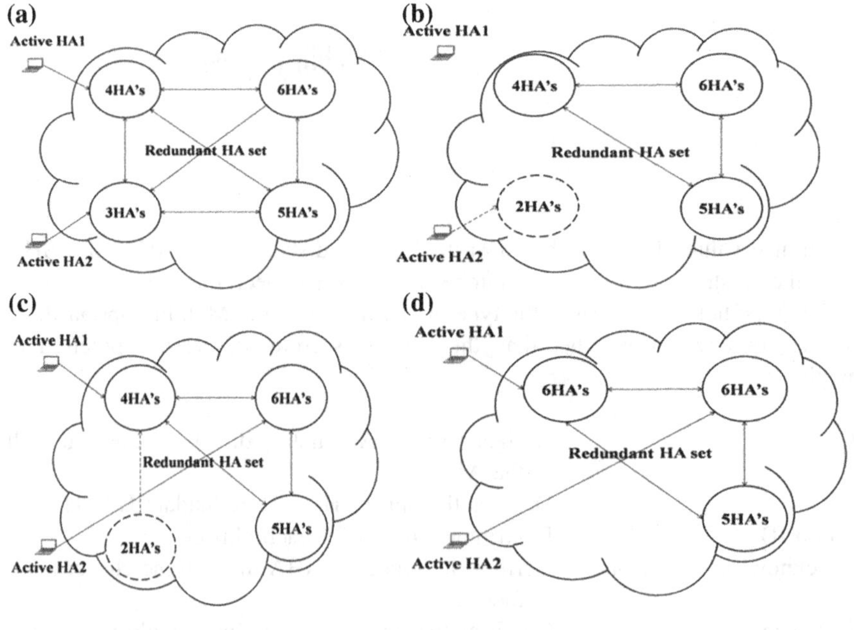

Fig. 5 HA handoff in CRHA method **a** CRHA cluster **b** Active HA2 cluster with one HA decremented **c** Active HA2 joins new cluster **d** Newly formed CRHA cluster

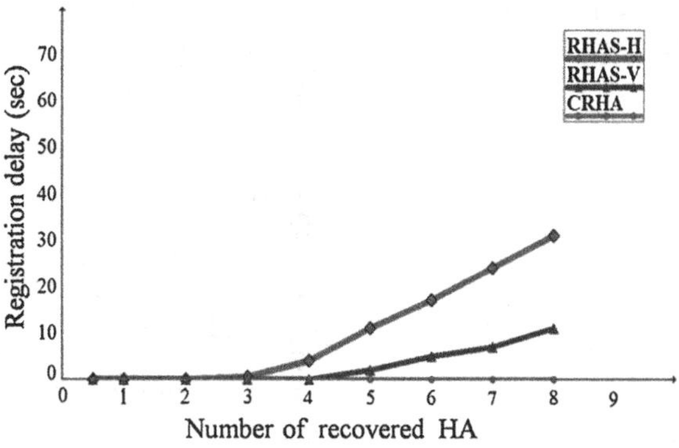

Fig. 6 BU registration delay at HA

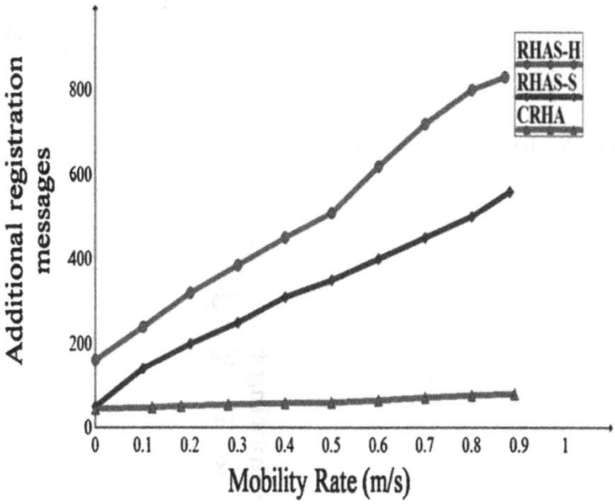

Fig. 7 Additional registration messages

SHAs take time to switch position as AHA at failure, but in CRHA, the failure is not transparent to MN since we use virtual address, and hence, there is no registration delay variation. Figure 6 illustrates the registration delay with recovery HAs of different approach. Figure 7 depicts the additional registration message with the mobility rate of MN. The registration delay is mitigated to negligible count in CRHA approach. Figure 8 shows variations in registration delay with the increased mobility rate. There is no need of exploring time for SHA in CRHA method as it is within the cluster. Figure 9 shows the exploring time of SHA to the number of SHAs explored. Since the AHA failure is not transparent and the exploring time of SHA is negligible, the registration delay variation is constant in CRHA.

Fig. 8 Registration delay variations

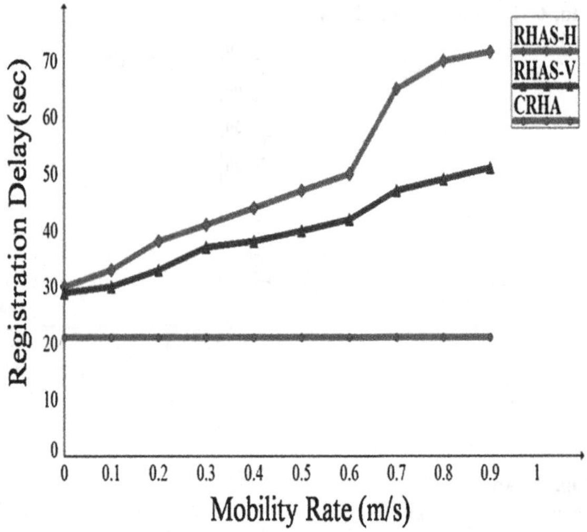

Fig. 9 SHA exploring time

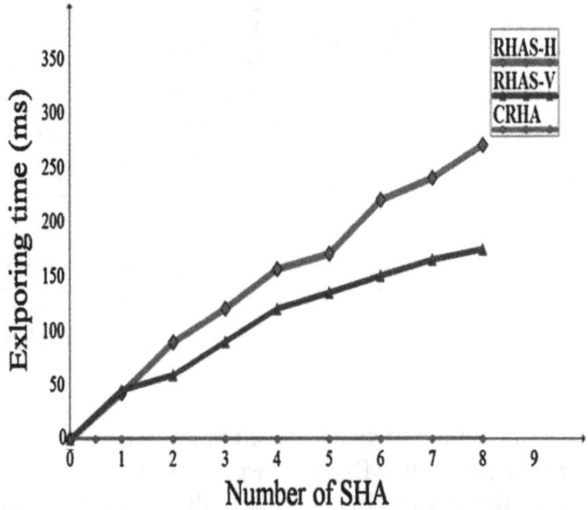

6 Conclusion

This paper proposes CRHA scheme with a protocol for an effective fault-tolerant mechanism. The key point is to reduce the exploring time of SHA and registration delay. Since the SHAs share binding information in the form of a chain within cluster, the exploring time is unnecessary during AHA failure. In CRHA method, registration delay is negligible, and hence, the binding information remains secure. The simulation results reveal that CRHA method shows a better performance when compared with the existing RHAS switching approaches.

References

1. D. Johnson, C. Perkins, J. Arkko. Mobility Support in IPv6. IETF RFC 3775. 2004
2. Qian Sun., et al., "Security Issues in Dynamic Home Agent Address Discovery", draft-sun- mipv6-dhaadsecurity-00.txt, November 2004.
3. Cisco IOS IPv6 Command Reference, ipv6 mobile home-agent (global configuration) through ipv6 ospf database-filter all out. Available FTP: http://www.cisco.com/en/US/docs/ios/ipv6/command/reference/ipv6_07.html
4. IP Mobility: Mobile IP Configuration Guide, Cisco IOS Release 15 M&T, Available FTP: http://www.cisco.com/en/US/docs/ios-xml/ios/mob_ip/configuration/15-mt/ip6-mobile-home-agent.html#GUID-9156DE37-F1F6-489C-BF4D-65B366A2D782
5. Yujun Zhang, Hanwen Zhang, A Mobile Agent Fault-Tolerant Method Based on the Ring
6. JW. Lin, J. Arul. An Efficient Fault-Tolerant Approach for Mobile IP in Wireless Systems. IEEE Transactions on Mobile Computing. 2(3):207–220. 2003.
7. S. Bose, C. Hota. Efficient Fault Tolerant Mobile IP in Wireless Networks Using Load Balancing Approach. Information Technology Journal. 6(3):463–468. 2007.
8. J. Lee, TM. Chung. Performance Evaluation of Distributed Multiple Home Agents with HAHA Protocol. International Journal of Network Management. 17:107–115, 2007.
9. JH. Ahn, CS. Hwang. Efficient Fault-Tolerant Protocol for Mobility Agents in Mobile IP. International Parallel and Distributed Processing Symposium (IPDPS). Pp.1273–1280. 2001.
10. YS. Chen, CH. Chen, HY. Fang. An Efficient Quorum-Based Fault-Tolerant Approach for Mobility Agents in Wireless Mobile Networks. IEEE International Conference on Sensor Networks, Ubiquitous and Trustworthy Computing (SUTC). Pp.373–378. 2008.
11. J. Faizan, HE. Rewini, et al. Introducing Reliability and Load Balancing in Mobile IPv6 Based Networks. Journal of Wireless Communications and Mobile Computing. 6:1–19. 2006.
12. R. Wakikawa. Home Agent Reliability Protocol. IETF Draft of MEXT Working Group, July 2009.
13. R. Wakikawa. Home Agent Reliability Protocol. IETF Draft of MEXT Working Group, May 2011
14. S. Rathi, and K. Thanushkodi, Design and Performance Evaluation of an Efficient Home Agent Reliability Protocol, International Journal of Recent Trends in Engineering, Vol 2, No. 1, November 2009

Interaction Coupling: A Modern Coupling Extractor

S. Gomathi and P. Edith Linda

Abstract Software development plays a vital role in interaction between the methods, classes, and attributes. Coupling is one of the most vibrant internal qualities to measure the design performance. Many object-oriented metrics have been proposed to evaluate different aspects of object-oriented program using coupling. This paper presents a new modern approach, which depicts the concept of interaction coupling, and a prototype is developed to measure the interaction coupling. Three types of metrics response for class (RFC), message-passing coupling (MPC), and method invocation coupling (MIC) that may invoke methods are analyzed, measured, and summarized.

Keywords Interaction coupling · Class loader · Extractor · Reliability · Efficiency

1 Introduction

Object-oriented development has proved its value for systems that must be maintained, reused, and modified. Coupling has been defined as one of the most important qualitative attributes to measure the performance of software at design or implementation phase [1]. Coupling can be categorized into three types: component coupling, interaction coupling, and inheritance coupling. This research is mainly focused on interaction couplings such as MPC, RFC, and MIC. Interaction coupling occurs when the methods of a class invoke methods of another class. In this, the main thing that is to be understood is about message. A message

S. Gomathi (✉)
Sri Krishna Arts and Science College, Coimbatore, India
e-mail: gomathisrinivasan88@gmail.com

P. E. Linda
Dr. G. R. Damodaran College of Science, Coimbatore, India
e-mail: p.lindavinod@gmail.com

G. S. S. Krishnan et al. (eds.), *Computational Intelligence, Cyber Security and Computational Models*, Advances in Intelligent Systems and Computing 246,
DOI: 10.1007/978-81-322-1680-3_26, © Springer India 2014

is a request that an object makes of another object to perform an operation [2]. The operation executed as a result of receiving a message is called a method.

The rest of the paper is organized as follows. Section 2 summarizes about the metrics used in the previous papers. Section 3 highlights about the problem with existing coupling parameters. Section 4 highlights the types of interaction coupling, and Sect. 5 depicts the framework. Section 6 summarizes a new algorithm design to measure those coupling parameters. Sections 7 and 8 offer the results and the conclusion.

2 Literature Survey

Li and Henry [3] identified a number of metrics that can predict the maintainability of a design. They define message-passing coupling (MPC), defined as the number of send statements defined in a class. The number of send statements that are sent out from a class may indicate how dependent the implementation of the local methods is on the methods in other classes. MPC only counts invocations of methods of other classes, not its own. Chidamber and Kemerer [2] proposed and validated a set of six software metrics for object-oriented systems, including two measures for coupling RFC and CBO. The response set (RS) of a class is a set of methods that can potentially be executed in response to a message received by an object of that class. A given method is counted only once. RFC includes methods called from outside the class and also a measure of the communication between that class and other classes. Vijaya Saradhi and Sastry [4] proposed a unique new approach for metric, which delivers the system quality based on cohesion and coupling between classes. The proposed metric shows the relationship between the classes based on the flow in control and the number of occurrences of class, which is mainly based on the existing systems' input. Arisholm et al. [5] stated that the relationships between coupling and external quality factors of object-oriented software have been studied extensively for the past few years. The authors concluded about the empirical relationships between class-level coupling, class fault- proneness and to measure coupling is through structural properties and static code analysis.

3 Problem Specifications

The current research on modeling and measuring the relationships between object-oriented programs through coupling analysis is insufficient. Coupling measures are incomplete in their precision of definition and quantitative computation. Moreover, some existing coupling measures do not reflect the differences in and the connections between design-level relationships and implementation-level connections. Hence, the way the coupling is used to solve problems is not satisfactory. Measuring various types of coupling manually is not possible, and tools fail to measure some important coupling parameters.

4 Proposed Metrics

Different types of coupling have evolved over time. But single type of coupling is inadequate to reduce the complexity of the code. This paper compared various couplings and finally selected the best coupling parameters to evaluate the complexity, quality factor, and reliability of object-oriented program [6]. We propose three metrics MPC, RFC, and MIC that help to detect the reusability and efficiency in design of object-oriented programs at the early stage.

4.1 Message-Passing Coupling

MPC is the count of total number of functions and procedure calls made to external units [3]. The MPC measures the dependency of local methods to methods implemented by other classes.

Viewpoints: This allows for conclusions on reusability, maintenance, and testing effort [7]. Message passing is calculated at the class level.

$$\mathrm{MPC} = \sum_{j=1}^{n} \mathrm{MC}_e \tag{1}$$

where MC_e is method call to external class and j is the number of classes.

4.2 Method Invocation Coupling

It is defined as the relative number of classes that receive the message from the particular class [8].

Viewpoints: The number of methods invoked implies the program reliability and efficiency [9]. The methods invoked should be minimum so as to maintain the system throughput.

$$\mathrm{MIC} = \sum_{j=1}^{n} \mathrm{MI}_i \tag{2}$$

where MI_i is method invoked from other classes and j is the number of classes.

4.3 Response for Class

RFC is the number of functions and procedures that can be potentially be executed in a class. Specifically, RFC is the number of operations directly invoked by member operations in a class plus the number of operations themselves [2].

Viewpoints: If the larger number of methods can be invoked in response to a message, the testing and debugging of class becomes more complicated [10].

$$\mathrm{RFC} = \sum_{j=1}^{n} \mathrm{MC}_e + \mathrm{MC}_I \qquad (3)$$

where MC_e is method call to external class, MC_I is methods of its own class, and j is the number of classes.

5 Framework of the Proposed System

The proposed framework is named as interaction coupling extractor (ICE), which is depicted in Fig. 1.

The framework has been partitioned into three phases. Each phase will perform some unique and important tasks.

5.1 Input Phase

This is the initial phase which is used to get input from the user. Programmer should give the jar file as input. Java ARchive (JAR) file is a collection of text, images, packages, and class files of java. A JAR file is essentially a zip file that contains an optional META-INF directory. But jar file cannot be measured directly. It must be extracted into individual class files. Jar files of variable size are given as input to interaction coupling extractor.

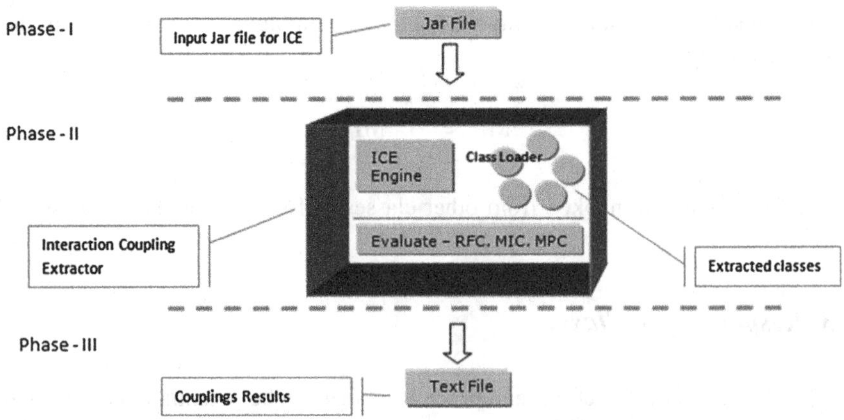

Fig. 1 Interaction coupling extractor

5.2 Processing Phase

This is the second level of extraction. Once a jar file is accepted, the file will be sent to interaction coupling extractor engine where the engine will separate the jar file into individual class files. That class files will be loaded into class loader in order to evaluate the interaction coupling parameters (RFC, MPC, and MIC).

5.3 Result Phase

The measured class will be displayed in the text file, which shows the number of RFC, MIC, and MPC measures of individual class. Then, the results are analyzed.

6 Algorithm Design

This design algorithm shows how to design the interaction coupling parameters MPC, RFC, and MIC and also shows how to measure those parameters.

Algorithm - Interaction Coupling Extractor (ICE)

Task: *To measure the relationship between methods of a jar file*

Input: *Jar file with variable size*

Output: *the Measured value of MPC, MIC and RFC*

Begin

Terminologies: *MC – Methods called, MI – Methods Invoked,* MI_i *– Methods Invoked from other class,* MC_e *– Methods call to external class,* MC_I *– Methods of its own class*

Calculate Message Passing Coupling using

$$MPC = \sum_{j=1}^{n} MC_e$$

Calculate Response for Class

$$RFC = \sum_{j=1}^{n} MC_e + MC_I$$

Calculate Method Invocation Coupling

$$MIC = \sum_{j=1}^{n} MI_i$$

The measured parameters are tabulated in a text file

End

Table 1 Results of JavaCSV.jar

Class name	RFC	MPC	MIC
com.csvreader.CsvReader$HeadersHolder	2	1	1
com.csvreader.CsvReader$RawRecordBuffer	1	0	1
com.csvreader.CsvWriter	58	30	0
com.csvreader.CsvWriter$UserSettings	1	0	1
com.csvreader.CsvReader$UserSettings	1	0	1
com.csvreader.CsvReader	86	36	0
com.csvreader.CsvWriter$Letters	1	0	0
com.csvreader.CsvReader$DataBuffer	1	0	1
com.csvreader.CsvReader$ColumnBuffer	1	0	1
com.csvreader.CsvReader$Letters	1	0	0
com.csvreader.CsvReader$StaticSettings	1	0	0
com.csvreader.CsvReader$ComplexEscape	1	0	0

7 Results

Two jar files, namely JavaSCV and JEdit, are given as input to interaction coupling extractor (ICE), and the results are shown below.

File Name: JavaCSV.jar
Size: 14.0 Kb
Source: findjar.com/jar/net/sourceforge/javacsv/javacsv/2.0/javacsv-2.0.jar.htm
Number of classes: 12

From Table 1, it is clearly shown that the RFC and MPC for class com.csvreader.CsvReader are higher. Hence, that class must be reprogrammed in order to make the jar file efficient.

File Name: JEdit.jar
Size: 114 K
Source: http://www.java2s.com/Code/Jar/j/Downloadjeditsyntaxjar.htm
Number of classes: 46

From Table 2, it is clearly shown that the RFC and MPC for class installer.SwingInstall are higher. MIC of installer.OperatingSystem class is high. Hence, both the classes must be reprogrammed in order to make the jar file efficient.

Other jar files, namely JUnit.jar, HSQL.jar, with more than 200 class files, are given as input, and the results are analyzed. The efficiencies of all the classes are measured. Maintainability of the program is improved based on the measurements. The classes that exceed the maximum ranges must be reprogrammed in order to make the program efficient, to reduce complexity, and to make it more flexible.

Table 2 Results of JEdit.jar

Class name	RFC	MPC	MIC
installer.CBZip2InputStream	36	6	1
installer.CBZip2OutputStream$1	−1	−2	0
installer.CBZip2OutputStream$StackElem	2	1	1
installer.CBZip2OutputStream	43	10	0
installer.CRC	5	1	2
installer.ConsoleInstall	36	34	1
installer.ConsoleProgress	10	4	2
installer.Install	45	38	9
installer.InstallThread	33	30	3
installer.InvalidHeaderException	3	2	2
installer.NonInteractiveInstall	26	25	1
installer.OperatingSystem$HalfAnOS	5	3	1
installer.OperatingSystem$MacOS	6	3	1
installer.OperatingSystem$OSTask	13	4	8
installer.OperatingSystem$Unix$ManPageOSTask	16	13	1
installer.OperatingSystem$Unix$ScriptOSTask	20	17	1
installer.OperatingSystem$Unix	25	19	4
installer.OperatingSystem$VMS	6	4	1
installer.OperatingSystem$Windows$JEditLauncherOSTask	14	11	0
installer.OperatingSystem$Windows	7	4	1
installer.OperatingSystem	15	10	10
installer.ServerKiller	25	22	1
installer.SwingInstall$ActionHandler	5	3	1
installer.SwingInstall$ChooseDirectory$1	4	2	1
installer.SwingInstall$ChooseDirectory$ActionHandler	14	12	1
installer.SwingInstall$ChooseDirectory	30	27	2
installer.SwingInstall$DirVerifier$1	7	5	1
installer.SwingInstall$DirVerifier$2	8	6	1
installer.SwingInstall$DirVerifier	35	25	3
installer.SwingInstall$SelectComponents	37	33	1
installer.SwingInstall$SwingProgress$1	3	1	1
installer.SwingInstall$SwingProgress$2	4	2	1
installer.SwingInstall$SwingProgress$3	3	1	1
installer.SwingInstall$SwingProgress$4	5	3	1
installer.SwingInstall$SwingProgress$5	3	1	1
installer.SwingInstall$SwingProgress	19	12	1
installer.SwingInstall$TextPanel	16	15	1
installer.SwingInstall$WindowHandler	3	1	1
installer.SwingInstall$WizardLayout	17	11	1
installer.SwingInstall	70	66	4
installer.TarBuffer	34	18	2
installer.TarEntry	63	32	4
installer.TarHeader	22	13	1
installer.TarInputStream$EntryAdapter	7	3	0
installer.TarInputStream	41	25	1
installer.TarOutputStream	33	21	0

8 Conclusion and Future Scope

This paper introduced a framework for interaction coupling for object-oriented systems. The interaction coupling is best suited to find the reusability and efficiency of the object-oriented systems. The algorithm used to implement the concept of RFC, MPC, and MIC is easy to understand. The detailed result sets show how the coupling parameters are measured and evaluated. The future work is to measure the component and inheritance coupling. The component coupling show how the arguments passed from one method to another method are measured. Inheritance coupling is used to measure the coupling between the inherited classes. The proposed framework will measure both types of coupling, and the measured parameters are displayed as chart.

References

1. Amjan Shaik et al., "Metrics for Object Oriented Design Software Systems: A Survey", Journal of Emerging Trends in Engineering and Applied Sciences ISSN: 21417016 (2010).
2. Chidamber, Shyam and Kemerer, Chris, "A Metrics Suite for Object-Oriented Design", IEEE Transactions on Software Engineering, June, (1994), pp. 476–492.
3. Li and Henry, "Object Oriented Metrics which predict maintainability", Journal of Systems and Software, Volume 23 Issue 2, Nov (1993).
4. L.C. Briand, J.W. Daly and J.K. Wust, "A Unified Framework for coupling measurement in Object Oriented Systems", IEEE Transaction on Software Engineering (1999), Vol. 25, Issue.
5. E. Arisholm et al., "Dynamic coupling measurement for object-oriented software", IEEE Transactions on Software Engineering, vol. 30, pp. 491–506, August 2004.
6. M.V. Vijaya Saradhi and B.R. Sastry "ESPQ: A New Object Oriented Design Metric for Software Quality Measurement" International Journal of Engineering Science and Technology (2010), Vol. 2, Issue. 3.
7. Harrison. R et al, "Coupling metrics for object-oriented design" proceedings for the Fifth International Software Metrics Symposium, (1998), pages 150–157.
8. Ramanath Subramanyam et al., "Empirical analysis of CK metrics for Object-Oriented design complexity: Implications for Software Defects", IEEE transactions on software engineering (2003), vol. 29, No. 4.
9. K K Agarwal, Yogesh Singh, ArvinderKaur and Ruchika Malhotra, "Empirical Study of Object-Oriented Metrics", (2006), vol. 5, No. 8.
10. Magnus Andersson, Patrik Vestergren, "Object-Oriented Design Quality Metrics" http://citeseerx.ist.psu.edu/viewdoc/summary?doi=10.1.1.1.5047.

Secure Multicasting Protocols in Wireless Mesh Networks—A Survey

Seetha Surlees and Sharmila Anand John Francis

Abstract Security is considered as one of the most significant constraint for the recognition of any wireless networking technology. However, security in wireless mesh networks (WMN) is still in its infancy as little attention has been given to this topic by the research society. WMN is a budding technology that provides low-cost high-quality service to users as the "last mile" of the Internet. Multicasting is one of the major communication technologies primarily designed for bandwidth (BW) conservation and an efficient way of transferring data to a group of receivers in wireless mesh network. The goal of secured group communication is to ensure the group secrecy property such that it is computationally infeasible for an unauthorized member node to discover the group data. In this article, the comparative study on existing approaches has been carried out; in addition to it, the fundamental security requirements and the various security attacks in the field of secure multicasting in WMN have also been discussed.

Keywords Wireless mesh networks · Multicasting · Security

1 Introduction

Wireless mesh networks (WMNs) are multihop, dynamically self-organized and self-configured network, with the nodes in the network automatically establishing an ad hoc network and maintaining the mesh connectivity. Fig. 1 illustrates the

S. Surlees (✉)
Department of IT, Karunya University, Coimbatore, India
e-mail: sitakaru@karunya.edu

S. A. John Francis
Department of MCA, Karunya University, Coimbatore, India
e-mail: sharmila@karunya.edu

G. S. S. Krishnan et al. (eds.), *Computational Intelligence, Cyber Security and Computational Models*, Advances in Intelligent Systems and Computing 246,
DOI: 10.1007/978-81-322-1680-3_27, © Springer India 2014

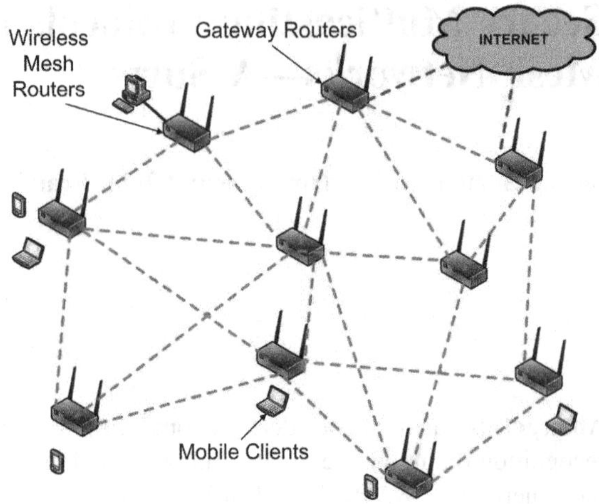

Fig. 1 Wireless mesh architecture—MR, mesh clients, and gateway node

architecture of WMN. WMNs are consisting of two types of nodes: mesh routers (MR) (wireless access points) and mesh clients. A mesh router is usually equipped with multiple wireless interfaces built on either the same or different wireless access technologies [1]. Routers in WMNs are usually stationary and form the mesh backbone for mesh clients. The additional gateway/bridge functionalities in MR enable the integration of WMNs with various other networks. Unlike in MANET, there is no power or computational constraints in WMN and the MR are stationary in most cases where as in MANET the nodes are mobile always [2]. WMN offers enormous applications like broadband home networking, community networking, transportation systems, public safety, and disaster recovery.

Group communication based on multicasting is considered to be a well-known communication paradigm in WMN due to the broadcasting nature of wireless communications. Multicasting is a bandwidth (BW)-conserving technology that helps at reducing the consumption of many applications. More internet users like to watch football matches and TV dramas on the internet as a substitute of traditional TV. Plentiful multicasting applications were foreseen to be deployed in WMNs, such as video on demand, Webcast, distance education, online games, video conferencing, and multimedia broadcasting [3]. These multicasting applications have one or more sources that distribute data to a group of changing receivers. The messages are delivered only once and duplicated only at branch points where links to the destination split.

The multicasting applications use multicast routing protocols that effectively deliver data from a source to multiple destinations organized in a multicast group. Multicasting is especially useful in wireless environments where there is scarce BW and many users are sharing the same wireless channels [4]. A major goal for multicasting is providing data confidentiality among the group members [5]. Based

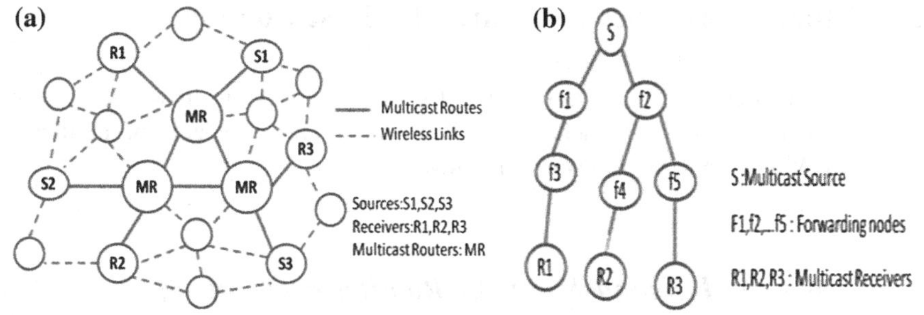

Fig. 2 **a** Mesh-based multicasting, **b** tree-based multicasting

Fig. 3 Classification of multicasting protocol

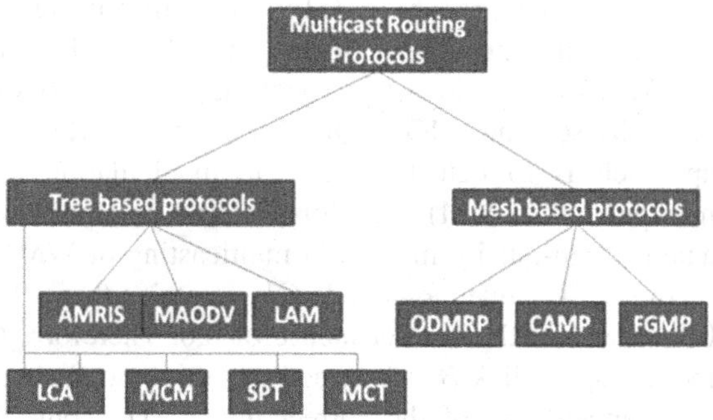

on the way of creating the routes among the members of the multicast group, the multicasting routing protocols categorized into mesh-based and tree-based protocols that are shown in Fig. 2a and b. The tree-based protocol does not always offer sufficient robustness where as a mesh-based protocol addresses robustness, reliability requirements with path redundancy. However, tree-based protocols have been widely used in WMN because of its resilience characteristics that withstand the failure of the nodes in the network. Fig. 3 shows the different kinds of protocols under tree-based and mesh-based approaches.

The features of wireless medium, dynamic changing topology, and cooperative routing protocols of the WMN demand the security measures for authenticating the members in the multicast group [6]. To ensure secured group communication, the forward secrecy and backward secrecy [7] should be followed for the newly joined members and revoked members in a group.

The rest of the paper is organized as follows: In Sect. 2, we briefly review the related works for secure multicast routing in WMN. In Sect. 3, the performance analysis of different secure multicasting protocols is being compared. The article concludes with Sect. 4.

2 Existing Approaches to Secure Multicast Routing

Very few researchers have focused toward secure multicasting in wireless mesh network. The existing approaches that deal with security aspects of multicast routing in WMN have been discussed below.

2.1 Secure On-Demand Multicast Routing Protocol [8]

The recently developed secure multicast protocol focused on selecting a path based on high-quality metric such as expected transmission count (ETX) and success product probability (SPP) rather than using traditional hop-count metric to maximize the throughput of the network. On-demand multicast routing protocol (ODMRP) is a multicast protocol that source periodically recreates the multicast group by sending a JOIN QUERY and JOIN REPLY messages. The aim of this approach is to detect the metric manipulation attacks, namely local metric manipulation (LMM) and global metric manipulation attack (GMM) that results against high-quality metrics in multicasting of WMN.

In Fig. 4, a malicious node C1 maintains that SPP metric value of $B_1 \rightarrow C_1 = 0.9$ instead of the correct metric of 0.6. Therefore, C_1 gathers an incorrect local metric for the link $B_1 \rightarrow C_1$ and advertises to R about the metric $S \rightarrow C_1 = 0:9$ as a replacement of the correct metric. The route $S \rightarrow A_1 \rightarrow B_1 \rightarrow C_1 \rightarrow R$ is highly preferred than the correct route $S \rightarrow A_3 \rightarrow B_3 \rightarrow C_3 \rightarrow R$. The limitation of this approach is that the node is detected as an attacker only when the specified threshold value is met. This leads to some attackers being unnoticed if the difference between expected PDR (ePDR) and perceived PDR (pPDR) is less than threshold value. In addition to it, the approach restricts to accuse only one node at a time. Therefore, it is difficult to secure a network despite of the fact that the majority of the nodes are attackers.

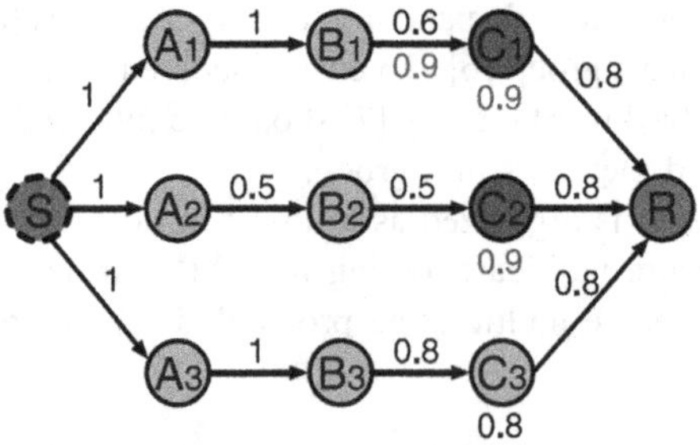

Fig. 4 Metric manipulation attack [8]

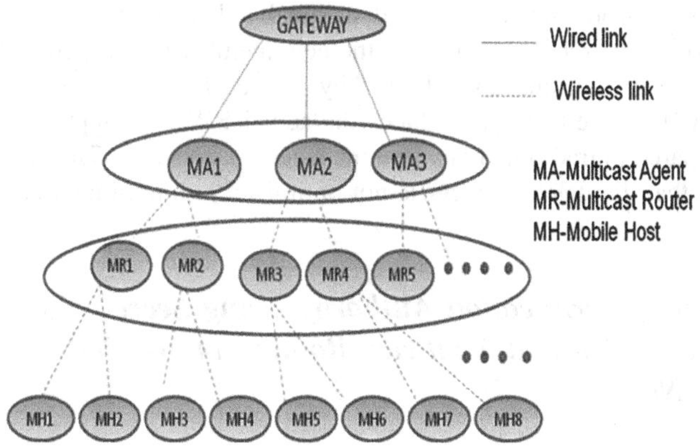

Fig. 5 Two-level hierarchical system model

2.2 Hierarchical Agent-Based Secure Multicast HASM [9]

This approach deals with the secure mobile multicast by ensuring only the authenticated mobile users to access the multicast data that are exchanged among the members of the multicast group. This approach proposed HASM protocol that efficiently ensures secured mobile multicasting in WMNs. Fig. 5 shows a gateway at the higher level of hierarchy in which multicast tree is rooted which connect MRs that serve as multicast agents (MA). Each multicast group member, i.e., Mobile Host (MH), is being registered with and serviced by an MA.

The objective of HASM is to minimize the overall network communication overhead incurred for security, group membership maintenance, and mobility management tasks. This method guarantees both backward and forward secrecy properties but lacks addressing the security attacks that arise within the group.

2.3 Bandwidth-Efficient Key Distribution for Secure Multicast in Dynamic Wireless Mesh Networks [7]

Data confidentiality in group communication is achieved by encrypting the message with a group key that is known to all the group members. To ensure secure group communication, the group key must be updated when there is a change in the membership of the multicast group. This situation is termed as rekeying. Dipping the BW utilization of rekeying is a central problem to guarantee enough BW for reliable data delivery when multicast-based services are provided over wireless networks. This approach defines the metric that represents the expected BW consumption of rekeying for given key tree. The adaptive and bandwidth-reducing (ABR) tree is a BW-efficient key tree management approach designed for

WMN when the group membership is dynamically changed. When a new member joins the group, this scheme assigns to the new member the proper KEKs to keep the expected BW consumption of the key tree as low as possible. Using this approach, ABR tree effectively reduces the actual BW consumption used for re-keying compared to traditional key tree management schemes. The demerit of this approach is that the deletion event is not optimized to a minimum cost level.

2.4 Design of Certification Authority Using Secret Redistribution and Multicast Routing in Wireless Mesh Networks [10]

In common, public key infrastructure (PKI) has a certification authority (CA), which is trusted by all the nodes in a network. But there is no trusted third party (TTP) in self-organizing networks such as WMNs. As a result, CA functions should be distributed over MR. MRs in WMNs are with enough power and capacity, so they are all able to participate in CA function distribution, and actively participating MRs can be changed from time to time. In order to achieve secret sharing and redistribution, the fast verifiable share redistribution (FVSR) scheme works for threshold cryptography and minimizes the possibility of secret disclosure when some shareholders are compromised by adversaries. This method adopts multicasting based on Ruiz tree that optimally reduces the operation overhead. It can update, revoke, and verify certificates of WMN nodes in a secure and well-organized manner. The demerit of this approach incurs additional overhead and cost due to Transferring MeCA functions.

2.5 Secure Group Overlay Multicast [5]

This approach provides data confidentiality such that only valid group members are allowed to access to the data sent to the group and to secure the primary protocols and effective designing of key management schemes. This approach uses ODMRP as a multicast routing protocol. Every authenticated client has a pair of public/private keys and a client certificate that maps its public key to its user ID. The CA is responsible for authorizing clients by issuing them a group member certificate. This member certificate binds the client to the group ID (group IP address) that provides proof of the client's membership. The goal is to ensure group secrecy property such that it is technically infeasible for outside adversaries to discover the group data. In addition to it, it also ensures forward and backward secrecy properties. The advantages of this approach is that it incurs less computational overhead, communication overhead, and latency without compromising the security. The limitation of this approach is that it lacks consideration of attacks against the multicast protocol itself.

2.6 An Improved Security Mechanism for High-Throughput Multicast Routing in Wireless Mesh Network Against Sybil Attack [11]

The objective of this approach is to detect Sybil attacks against high-throughput multicast protocols in WMN. In Sybil attack, a node maliciously claims multiple identities. Each node identity in the multicast group is validated using random key pre-distribution (RKP) technique, in which nodes create secure links to neighboring nodes. Using RKP, a random group of keys to each node is being assigned in such a way that each node can compute common keys that it share with its neighbors. These are called secret session keys that are used to ensure node-to-node secrecy. The limitation of this technique is not scalable when the attacker increases to high level.

3 Performance Analysis

This section analyzes the performance of existing approaches of secure multicasting in WMN against different parameters, viz. multicast protocol, routing metric, performance metrics, security issues addressed, merits and demerits. Table 1 shows the comparative study of existing approaches in secure multicasting protocols of WMN.

In S-ODMRP [8], the high-throughput metric (ETX, SPP) leads to increase in the ratio of attacker, but the defense mechanism is very effective against drop-only, LMM and GMM attack with the PDR of 95 %. The overhead of S-ODMRP is due to the periodic flooding of authenticated query packets that is common in all scenarios. In HASM [6], the total communication cost is much smaller than SPT due to the hybrid hierarchical multicast structure. The ABR tree [7] is effective for reducing the BW consumption of rekeying, and it achieves around 80 % reduction in total BW consumption compared to conventional tree approach. MeCA [10] minimizes the secret key discloser and improves the efficiency by incorporating multicasting but incurs high control overhead due to transferring MeCA functions over several MRs. The SeGrOM [5] offers high delivery ratio with minor encryption overhead, but the security attack against the multicast protocol is not addressed. Finally, RKP [11] tries to overcome the drawbacks of S-ODMRP, but it fails to scale for many attackers.

The analyses of multicasting protocols are performed using multicast routing metrics and performance metrics. The description of the multicast routing and performance metrics is as follows:

Table1 Comparison of existing approaches

Secure protocols	Multicast protocol	Routing metric	Security issues addressed	Performance metrics	Advantages	Drawbacks
SODMRP [8]	ODMRP	ETX, SPP	Resource consumption, mesh structure, data forwarding attacks	PDR-95 %, BW overhead-0.95 kbps, data-transmission efficiency-2–3 pkts	High throughput, security	Single accusation at time, detection is based on threshold value
HASM [6]	HASM	Hop count	Forward and backward secrecy	Avg. total communication cost/sec-low (10–20 times)	Minimizes the overall communication cost	Security attacks from within group not addressed
			User mobility		Ensure forward and backward secrecy	Reliability, QOS requirements not addressed
BW-efficient key distribution [7]	ABR tree	Expected BW consumption for rekeying	Data confidentiality	Total BW consumption-80 % reduction	Reduces network BW consumption of rekeying	Deletion event needs to be optimized
Design of CA using secret redistribution (MeCA) [3]	MCT	Hop count	Exposure attack, compromise attack	Control overhead-high, data overhead-small	Strengthens security, improves efficiency	Transferring MeCA functions incurs much overhead
SeGrOM [5]	ODMRP	Hop count	Data confidentiality	Delivery ratio is very high. computation, BW, latency overhead of join and leave events—increases linearly with data rate	Higher performance, smaller overhead	Attacks against multicast protocol are not considered
RKP [11]	S-ODMRP	ETX	Sybil attacks	PDR-high, BW overhead-low	High throughput, high security	Not scalable

3.1 Multicast Routing Metrics

In multicasting, route selection is carried out based on the metric designed to improve throughput. The commonly used routing metrics for multicasting are defined below.

(a) Expected Transmission Count

ETX is the total number of transmissions needed to successfully deliver a packet over a link from a source to destination [12].

$$\mathrm{ETX} = 1/d_f \tag{1}$$

where d_f is the loss rate of the link in forward direction.

(b) Success Probability Product

SPP is used to provide the probability for the receiver to receive a packet over a link. SPP value for a path of j links between a source S and a receiver R is [8],

$$\mathrm{SPP}_{S \to R} = \prod_{i=1}^{j} \mathrm{SPP}_i \tag{2}$$

where $\mathrm{SPP}_i = d_f$ and d_f is defined in ETX.

(c) Expected Transmission Time

ETT is the product between ETX and the average time required to deliver a single data packet. Let S be the size of the probing packet and B be the measured BW of a link; then, the ETT of this link is defined as follows [12]:

$$\mathrm{ETT} = \mathrm{ETX} * S/B \tag{3}$$

(d) Hop Count

A hop-count metric is a metric that counts router hops.

The description of the performance metrics is as follows:

3.2 Performance Metrics

The following metrics are frequently used to evaluate the performance of a secure multicast protocol:

(a) *Packet delivery ratio*

The amount of packets that are received successfully at the receiver to the total number of packets that are sent by source.

(b) *End-to-end delay*

The average time taken for a packet to reach the destination after it leaves the source.

(c) *Routing overhead*

The amount of control messages that every multicast router sends on average per unit of time.

(d) *Avg. total communication costs*

It is the number of wireless transmissions needed per operation of multicast members. It consists of cost of mobility management, the cost for security key management and the cost for group membership management.

4 Conclusion

The research in secure multicast routing in WMN is still in its infancy. In summary, the major security requirements, routing metrics, performance metrics, multicast protocol and their merits and demerits for the efficient secure multicast routing protocols in wireless mesh network are analyzed. In addition to this, various security issues in the WMN are discussed. WMN is a technology suitable for next generation wireless networking stimulating the application setups to its rapid development. Nevertheless, to make stronger market penetration, more research is needed in the area of secure mobile multicasting which accomplishes QOS for different application services with least cost, less BW consumption, and high throughput on WMN.

References

1. Ian F. Akyildiz, Xudong Wang, Weilin Wang. "Wireless mesh networks: a survey", Elsevier transactions on computer networks, 2005.
2. Guokai Zeng, Bo Wang, Yong Ding, Li Xiao, Matt W. Mutka, "Efficient Multicast Algorithms for Multichannel Wireless Mesh Networks", IEEE Transactions on Parallel and Distributed Systems, vol 21, no 1 January 2010.
3. Uyen Trang Nguyen, "On multicast routing in wireless mesh networks", Elsevier Computer Communications, pp 1385-1399, January 2008.
4. Pedro M.Ruiz, Francisco J.Galera, "Efficient Muticast Routing in Wireless Mesh Networks connected to Internet", IEEE, 2006.

5. Jing Dong, Kurt Erik Ackermann, Cristina Nita-Rotaru, "Secure Group Communication in Wireless Mesh Networks", IEEE conference on World of Wireless, Mobile and Multimedia Networks, 2008, pp.1-7.
6. Yinan Li, Ing-Ray Chen, "Hierarchical Agent- Based Secure Multicast for Wireless Mesh Networks", proc.IEEE Communication Society, Proceedings of IEEE International Conference on Communications, 2011.
7. Seungjae Shin, Junbeom Hur, Hanjin Lee, Hyunsoo Yoon, "Bandwidth Efficient Key Distribution for Secure Multicast in Dynamic Wireless Mesh Networks", proc IEEE Communication Society, 2009
8. Jing Dong, Reza Curtmola, Cristina Nita-Rotaru, "Secure High Throughput Multicast Routing in Wireless Mesh Networks", IEEE Transactions on Mobile Computing, vol 10 no 5, pp 653-667, May 2011.
9. Yinan Li, Ing-Ray Chen, "Hierarchical Agent- Based Secure Multicast for Wireless Mesh Networks", proc.IEEE Communication Society, Proceedings of IEEE International Conference on Communications, 2011.
10. Jong Talc Kim, Saewoong Bahk, "Design of certification authority using secret redistribution and multicast routing in wireless mesh networks".
11. P.Anitha, G.N.Pavithra, P.S.Periasamy, "An Improved Security Mechanism for High Throughput Multicast Routing in Wireless Mesh Network Against Sybil Attack", proc International Conference on Pattern Recognition, Informatics and Medical Engineering, pp 125-130, March 2012.
12. Sabyasachi Roy, Dimitrios Koutsonikolas, Saumitra Das, and Y. Charlie Hu, "High-throughput Multicast Routing Metrics in Wireless Mesh Networks", IEEE, 2006.
13. Weichao Wang,Bharat Bhargava,"Key Distribution and Update for Secure Inter-group Multicast Communication".
14. Adarsh. R, Ganesh Kumar.R, Jitendranath Mungara, "Secure Data Transition over Multicast Routing In Wireless Mesh network", International Journal of Innovative Technology and Exploring Engineering (IJITEE), vol 1 Issue 3, pp 98-103, August 2012.
15. Adrian Perrig, Dawn Song J.D Tygar, "ELK, a New Protocol for Efficient Large-Group Key Distribution.
16. Jing Dong, Kurt Erik Ackermann, Cristina Nita-Rotaru, "Secure Group Communication in Wireless Mesh Networks", IEEE conference on World of Wireless, Mobile and Multimedia Networks, 2008, pp. 1-7.
17. S.Sasikala Devi, Dr.Antony Selvadoss Danamani, "A Survey on Multicast rekeying for secure group communication", International Journal of Computer Tech Application, vol 2(3) pp 385-391.
18. Jing Dong, Cristina Nita Rotaru, "Enabling Confidentiality for Group Communication in Wireless Mesh Networks".
19. Shafiullah Khan, Kok-Keong Loo, Noor Mast, Tahir Naeem, "SRPM: Secure Routing Protocol for IEEE 802.11 Infrastructure Based Wireless Mesh Networks"
20. M.Iqbal, X.Wang, S.Li, T.Ellis, "Qos scheme for multimedia multicast communications over wireless mesh networks", IEt Commun, vol 4, Issue 11, pp 1312-1324, 2010
21. Guokai Zeng, Bo Wang, Yong Ding, Li Xiao, Matt W. Mutka, "Efficient Multicast Algorithms for Multichannel Wireless Mesh Networks", IEEE Transactions on Parallel and Distributed Systems, vol 21, no 1 January 2010
22. Sung-Hwa Lim, Young-Bae Ko, Cheolgi Kim, Nitin H.Vaidya, "Design and Implementation of multicast for multi-channel multi-interface wireless mesh networks", Springer, pp 955-972, February 2011.
23. Qin Xin, Fredrik Manne, Yan Zhang, Xin Wang, "Almost optimal distributed M2 M multicasting in wireless mesh networks", Elsevier, Theoretical Computer Science, pp69-82, March 2012.
24. Goukai Zeng, Bo Wang, Matt Mutka, Li Xiao, Eric Torng, "Efficient Multicast for Link-Heterogeneous Wireless Mesh Networks".

25. Uyen Trang Nguyen, "On multicast routing in wireless mesh networks", Elsevier Computer Communications, pp 1385-1399, January 2008.
26. Sung-Ju Lee, Mario Gerla, Ching-Chuan Chiang, "On-Demand Multicast Routing Protocol."
27. Ashish Raniwala, Tzi-cker Chiueh, "Architecture and Algorithms for an IEEE 802.11-Based Multi-Channel Wireless Mesh Network"
28. Huan-Wen Tsai, Hsu-Cheng Lin, Chih-Lun Chou, Sheng-Tzong Cheng, "Multicast-Tree Construction and Streaming Mechanism for Intra 802.16 Mesh Networks", International Conference on Networking and Distributed Computing".
29. Uyen Trang Nguyen, Jin Xu, "Multicast Routing in Wireless Mesh Networks: Minimum Cost Trees or Shortest Path Trees?"
30. Kyoung Jin Oh, Chae Y.Lee, "Multicast Routing Protocol with Low Transmission Delay in Multirate, Multi-radio Wireless mesh Networks", proc IEEE Computer Society, 2010.
31. unaid Qadir, Chun Tung Chou, Archan Misra, "Exploiting Rate Diversity for Multicasting in Multi-Radio Wireless Mesh Networks".
32. Brian Keegan, Karol Kowalik and Mark Davis, "Optimisation of Multicast Routing in Wireless Mesh Networks".
33. Mingquan Wu, Hayder Radha, "Distributed network embedded FEC real-time multicast applications in multi-hop wireless networks", Springer, pp 1447-1458, October 2009 .
34. Zheng Liu, Min Yang, Heng Dai, Jufeng Dai, "Concurrent Transmission Scheduling for Multi-hop Multicast in Wireless Mesh Networks".
35. Vineet Khisty, "Link Selection for Point-to-Point 60Ghz Networks".

Part IV
Computational Models

A New Reversible SMG Gate and Its Application for Designing Two's Complement Adder/Subtractor with Overflow Detection Logic for Quantum Computer-Based Systems

S. Manjula Gandhi, J. Devishree and S. Sathish Mohan

Abstract Reversible computation plays an important role in the synthesis of circuits having application in quantum computing-, low-power CMOS design-, bioinformatics-, and nanotechnology-based systems. Conventional logical circuits are not reversible. A reversible circuit maps each input vector, into a unique output vector and vice versa. A new 4 × 4 reversible full-adder gate called as SMG gate is suggested in this paper. Three approaches to design reversible two's complement adder/subtractor with overflow detection logic are also proposed. The first approach is based on Toffoli and Feynman gates, second approach is based on Peres gate, and third approach is based on the new SMG gate. The proposed reversible circuits are evaluated in terms of number of quantum cost.

Keywords Nanotechnology · Quantum computer · Quantum cost · Quantum gates · Qubits · Reversible gates · Two's complement adder/subtractor

1 Introduction

Many scientists say that "single-atom transistor is the end of Moore's Law" [1], which states that "within a period of 18 months, the number of transistors in an IC doubles all along with its efficiency." He also predicted that the addition of extra

S. M. Gandhi (✉)
Department of MCA, Coimbatore Institute of Technology, Coimbatore, Tamil Nadu, India
e-mail: manjulagandhi@cit.edu.in

J. Devishree
Department of EEE, Coimbatore Institute of Technology, Coimbatore, Tamil Nadu, India
e-mail: devi_567@yahoo.com

S. Sathish Mohan
Enterprise Information Management, Cognizant Technology Solutions, Coimbatore, Tamil Nadu, India
e-mail: sathishmohan.selvaraj@cognizant.com

G. S. S. Krishnan et al. (eds.), *Computational Intelligence, Cyber Security and Computational Models*, Advances in Intelligent Systems and Computing 246,
DOI: 10.1007/978-81-322-1680-3_28,

number of transistors will not be possible in next few years and even if so, it may cause power consumption problems. Also, according to Landauer's principle, loss of 1 bit of information dissipates at least $kTln_2$ joule of energy, where $k = 1.3806505 \times 10^{-23}\ m^2\ kg\ s^{-2}\ k^{-1}$ (joule/Kelvin) is the Boltzman constant and T is the operating temperature [2]. Bennett showed that $kTln_2$ energy dissipation would not occur if a computation is carried out in a reversible way [3]. The design that does not result in information loss is reversible.

A new 4 × 4 reversible full-adder gate called as SMG gate is proposed in this paper. The conventional circuit considered in this work is four-bit two's complement adder/subtractor with overflow detection logic [4]. The paper is organized as follows: The various full-adder gates and the proposed full-adder gate are discussed in Sect. 2. Section 3 contains the literature survey. The conventional four-bit two's complement adder/subtractor with overflow detection logic is discussed in Sect. 4. In Sect. 5, the proposed designs of reversible four-bit two's complement adder/subtractor with overflow detection logic using reversible gates have been performed. Result and analysis of the proposed designs are given in Sect. 6. Conclusions are contained in Sect. 7.

2 Proposed Full-Adder Gate

A full adder [6] is a device that takes two input bits with a carry-in (C_{in}) bit and produces the output with the sum of the bits and the carry-out (C_{out}).

2.1 Full Adder Using Toffoli and Feynman Gates

Full adder constructed using Toffoli [5] and Feynman gates [6] is given in Fig. 1. The circuit consists of two Feynman gates and two Toffoli gates; hence, its quantum cost is 12. The inputs *A*, *B*, and C_{in} are given in qubit1, qubit2, and qubit3, respectively. The outputs sum and C_{out} are measured in qubit3 and qubit4, respectively. Hence, when the input is |1>|1>|0>|0>, the output is |1>|1>|0>|1>.

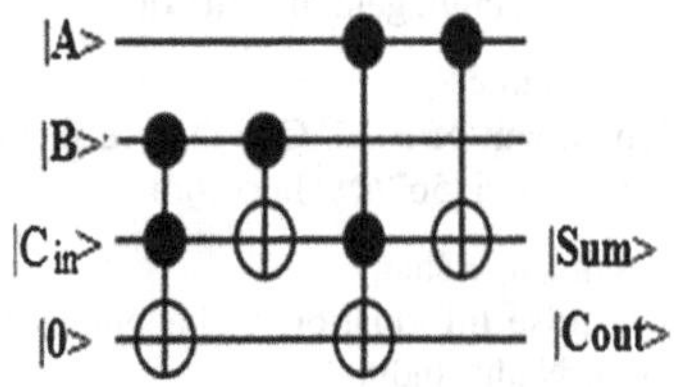

Fig. 1 Full adder using Toffoli and Feynman gates

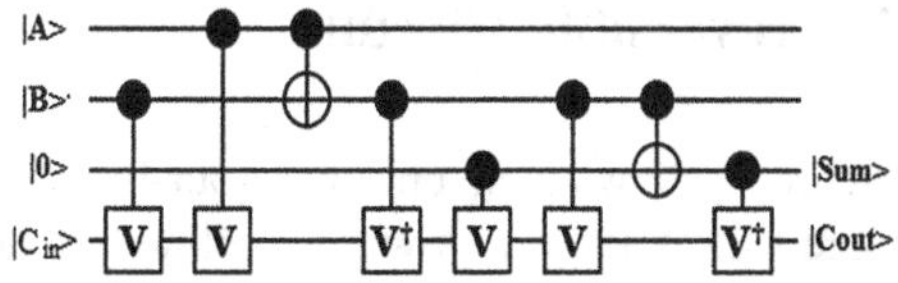

Fig. 2 Quantum realization of full adder using two Peres gates

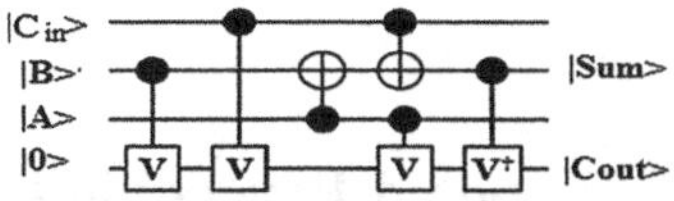

Fig. 3 SMG gate

2.2 Full Adder Using Peres Gate

The quantum realization of a full adder realized using Peres gates is given in Fig. 2. Since two Peres gates are used, its quantum cost is 8.

2.3 Full Adder Using SMG Gate

In this paper, a new full-adder gate called as SMG gate is proposed. The SMG gate is a 4 × 4 reversible gate shown in Fig. 3. The quantum cost of SMG gate is 6. The inputs C_{in}, B, and A are given in qubit1, qubit2, and qubit3, respectively. The outputs sum and C_{out} are measured in qubit2 and qubit4, respectively. Hence, when the input is |0>|1>|1>|0>, the output is |0>|0>|1>|1>.

3 Literature Survey

Himanshu Thapliyal et al. [7] proposed a 3 × 3 reversible TKS gate with two of its outputs working as 2:1 multiplexer. The gate is used to design a reversible half adder and multiplexer. Rangaraju et al. [8] discussed various adder/subtractor circuit designs and proposed a reversible eight-bit parallel binary adder/subtractor with three designs. Md. Saiful Islam et al. [9], designed full-adder circuit using IG gates and MIG gates, and designed fault tolerant reversible carry look-ahead and carry-skip adders. Bhagyalakshmi et al. [10] proposed a reversible design of a multiplier using DPG and BVF gate. Vandana Shukla et al. [11] proposed a new reversible logic module to design a 4-bit binary two's complement circuit. A new 3 × 3 reversible two's complement gate is suggested in [12]. Two quantum models are offered for two's complement gate. Majid Mohammadi et al. [13] proposed a modular synthesis method to realize a reversible BCD-full-adder and subtractor circuit. In the paper, three approaches are designed to optimize all parts of a BCD-FA circuit using genetic algorithm and don't-care concept.

4 Irreversible Circuit

Figure 4 shows conventional four-bit adder/subtractor with overflow detection logic [4]. When S is low, the full adders produce the sum of A and B. When S is high, the full adders produce the difference of A and B. In Fig. 4, $V = 1$ indicates overflow condition when adding/subtracting signed two's complement numbers.

5 Proposed Reversible Designs

In this paper, we propose three designs for reversible two's complement adder/subtractor with overflow detection logic to achieve the optimized reversible circuit. Proposed design approaches have been presented in proceeding subsections.

5.1 *Design I*

In Design I, Toffoli and Feynman reversible gates have been used. To add/subtract a four-bit binary number, the input is $|A_0>|B_0>|S>|S>|0>|A_1>|B_1>|S>|0>|A_2>|B_2>|S>|0>|A_3>|B_3>|S>|0>$ and the outputs are S_0, S_1, S_2, and S_3 measured in qubit4, qubit8, qubit12, and qubit16, respectively. V is measured in qubit13, $V = 1$ states that an overflow has occurred. The reversible circuit of this design is given in Fig. 5.

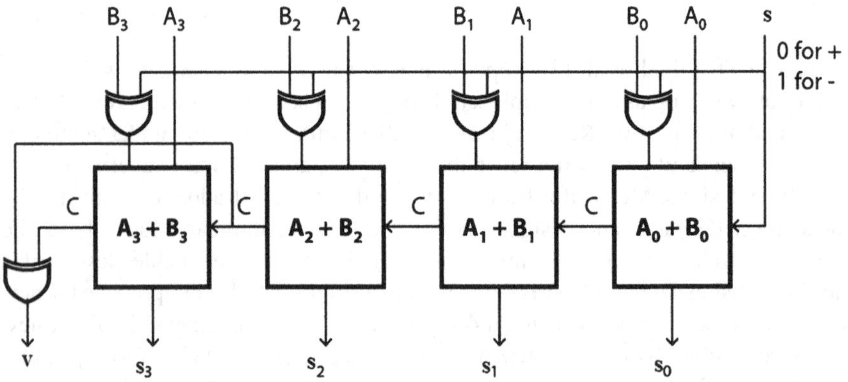

Fig. 4 Conventional four-bit adder/subtractor with overflow detection logic

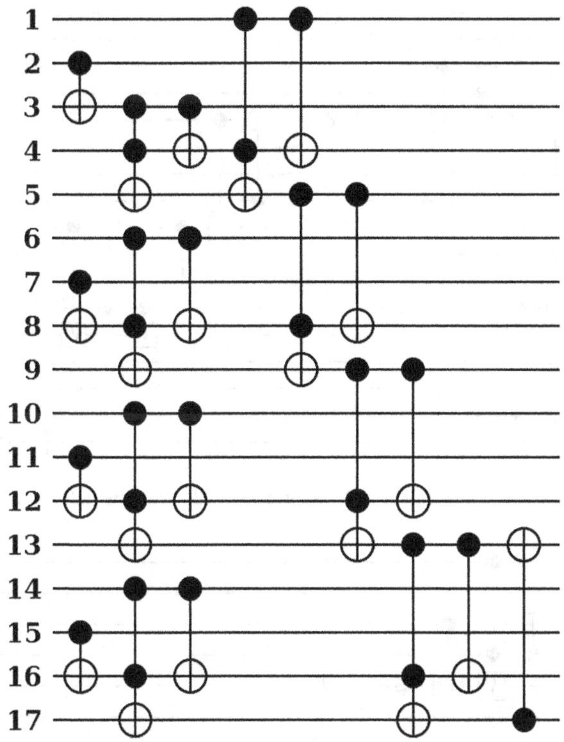

Fig. 5 Reversible two's complement adder/subtractor with overflow detection logic—Design I

5.2 Design II

In Design II, Peres gates have been used. To add/subtract a four-bit binary number, the input is $|A_0\rangle|B_0\rangle|S\rangle|S\rangle|0\rangle|A_1\rangle|B_1\rangle|S\rangle|0\rangle|A_2\rangle|B_2\rangle|S\rangle|0\rangle|A_3\rangle|B_3\rangle|S\rangle|0\rangle$ and the outputs are S_0, S_1, S_2, and S_3 are measured in qubit4, qubit5, qubit9, and qubit11, respectively. V is measured in qubit13, $V = 1$ states that an overflow has occurred. The reversible circuit of this design is given in Fig. 6.

5.3 Design III

In Design III, SMG gate have been used. To add/subtract a four-bit binary number, the input is $|S\rangle|B_0\rangle|S\rangle|A_0\rangle|0\rangle|B_1\rangle|S\rangle|A_1\rangle|0\rangle|B_2\rangle|S\rangle|A_2\rangle|0\rangle|B_3\rangle|S\rangle|A_3\rangle|0\rangle$ and the outputs are S_0, S_1, S_2, and S_3 measured in qubit3, qubit7, qubit11, and qubit15, respectively. V is measured in qubit13, $V = 1$ states that an overflow has occurred. The reversible circuit of this design is given in Fig. 7.

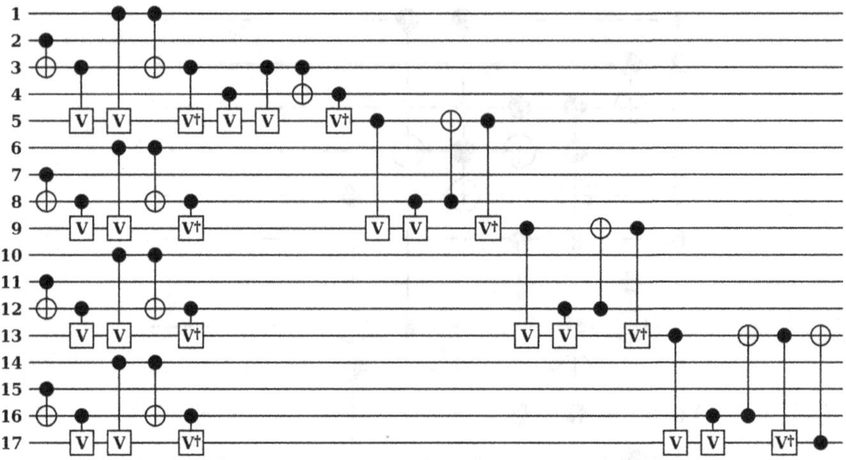

Fig. 6 Reversible two's complement adder/subtractor with overflow detection logic—Design II

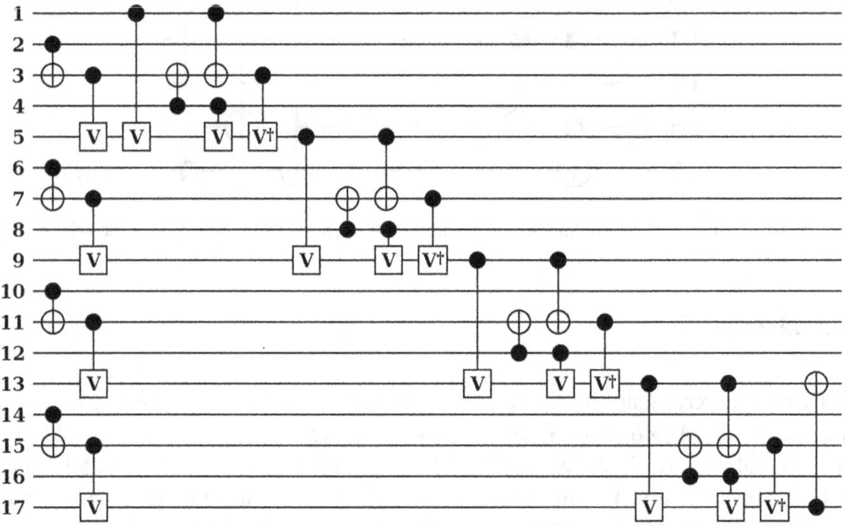

Fig. 7 Reversible two's complement adder/subtractor with overflow detection logic—Design III

6 Results

In this paper, three design types, viz. Design I, Design II, Design III of reversible four-bit two's complement adder/subtractor, are proposed. SMG gate is used to construct Design III. The comparison of reversible four-bit two's complement adder/subtractor in the terms of quantum cost is shown in Table 1. The performance of Design III is better in terms of quantum cost in comparison with Design II and Design I.

Table 1 Reversible four-bit two's complement adder/subtractor with overflow detection logic

	Quantum cost
Design I	53
Design II	37
Design III	29

7 Conclusions

In this paper, a new 4 $\times$ 4 reversible full-adder gate is presented. Three designs for reversible four-bit two's complement adder/subtractor with overflow detection logic has been proposed for the first time. Table 1 shows the results of the three proposed designs. According to the obtained result, Design III using SMG gate is more optimal than Design I and Design II in the terms of quantum cost. Finally, the proposed circuits have been simulated using QCViewer [14] and obtained results of simulation shows the correct operation of circuits.

References

1. M. A. Nielsen, I. L. Chuang, "*Quantum Computation and Quantum Information*", Cambridge Univ. Press, Cambridge, 2000
2. R. Landauer, "Irreversibility and Heat Generation in the Computational Process", *IBM Journal of Research and Development*, 5, pp. 183 – 191, 1961
3. C.H. Bennett, "Logical Reversibility of Computation", *IBM Journal of Research and Development*, volume 17, pp. 525-532, 1973
4. Donald P Leach, Albert Paul Malvino, Goutam Saha, "*Digital Principles and Applications*", Tata McGraw-Hill, New Delhi, 2006
5. T. Toffoli, "Reversible Computing", *Tech. Memo MIT/LCS/TM-151*, MIT Lab for Computer Science, 1980
6. R. Feynman, "Quantum mechanical computers", *Optics News* vol. 11, issue 2, pp. 11–20, 1985
7. Himanshu Thapliyal, M. B. Srinivas and Mandalika B, "Novel design and reversible logic synthesis of multiplexer based full adder and multipliers", *Forty Eighth Midwest Symposium on Circuits and Systems*, Vol. 2, pp. 1593-1596, 2005
8. Rangaraju H G, Venugopal U, Muralidhara K N, Raja K B, "Low Power Reversible Parallel Binary Adder/Subtractor", *International Journal of VLSI design & Communication Systems,* Vol. 1, No. 3, pp. 23-34, 2010
9. Md. Saiful Islam, Muhammad Mahbubur Rahman, Zerina begum, and Mohd. Zulfiquar Hafiz, "Efficient Approaches for Designing Fault Tolerant Reversible Carry Look-Ahead and Carry Skip Adders", *MASAUM Journal of Basic and Applied Sciences*, Vol. 1, No. 3, 2009
10. Bhagyalakshmi H R, Venkatesha M K, "An Improved Design of a Multiplier using Reversible Logic Gates", *International Journal of Engineering Science and Technology*, Volume 2, Issue 8, pp 3838-3845, 2010
11. Vandana Shukla, O P Singh, G R Mishra, R K Tiwari, "Design of a 4-bit 2's Complement Reversible Circuit for Arithmetic Logic Unit Applications", *The International Conference on Communication, Computing and Information Technology, 2012*

12. Soudebeh Boroumand, "A Novel Nanometric Reversible Four-bit Signed-magnitude Adder/ Subtractor", *Life Science Journal*, volume 9 issue 3, 2012
13. Majid Mohammadi, Majid Haghparast, Mohammad Eshghi, Keivan Navi, "Minimization and Optimization of reversible BCD-Full adder/subtractor using Genetic Algorithm and Don't care concept", *International Journal of Quantum Information*, Volume 7, Number 5, pp. 969-989, 2009
14. Alex Parent and Jacob Parker, "*QCviewer0.8*", May 2012, http://qcirc.iqc.uwaterloo.ca

An Algorithm for Constructing Graceful Tree from an Arbitrary Tree

G. Sethuraman and P. Ragukumar

Abstract A function f is called graceful labeling of a graph G with m edges if f is an injective function from $V(G)$ to $\{0, 1, 2, \ldots, m\}$ such that if every edge uv is assigned the edge label $|f(u) - f(v)|$, then the resulting edge labels are distinct. A graph that admits graceful labeling is called a graceful graph. The popular graceful tree conjecture states that every tree is graceful. The graceful tree conjecture remains open over four decades. In this paper, we introduce a new method of constructing graceful trees from a given arbitrary tree by designing an exclusive algorithm.

Keywords Graceful tree · Graceful tree conjecture · Graceful tree embedding · Graceful labeling · Graph labeling

1 Introduction

All the graphs considered in this paper are finite and simple graphs. The terms that are not defined here can be referred from [11]. In 1963, Ringel posed his celebrated conjecture, called Ringel's conjecture [7], which states that, K_{2n+1}, the complete graph on $2n + 1$ vertices can be decomposed into $2n + 1$ isomorphic copies of the Ringel's conjecture that the complete graph K_{2n+1} can be cyclically decomposed into $2n + 1$ copies of a given tree with n edges. In an attempt to solve both Ringel's conjecture and Kotzig's conjecture, in 1967, Rosa in his classical paper [8] introduced hierarchical series of labelings called ρ, σ, β and α labelings as a tool to attach both Ringel's conjecture and Kotzig's conjecture. Later, β-labeling was called graceful labeling by Golomb, and now, this term is being widely used. A function

G. Sethuraman (✉) · P. Ragukumar
Department of Mathematics, Anna University, Chennai 600025, India
e-mail: sethu@annauniv.edu

G. S. S. Krishnan et al. (eds.), *Computational Intelligence, Cyber Security and Computational Models*, Advances in Intelligent Systems and Computing 246,
DOI: 10.1007/978-81-322-1680-3_29, © Springer India 2014

f is called graceful labeling of a graph G with m edges, if f is an injective function from $V(G)$ to $\{0, 1, 2, \ldots, m\}$ such that if every edge uv is assigned the edge label $|f(u) - f(v)|$, then the resulting edge labels are distinct. A graph that admits graceful labeling is called a graceful graph. The popular graceful tree conjecture by Ringel, Kotzig, and Rosa states that all trees are graceful. In spite of number of interesting and significant results [1, 2, 4, 5, 6, 9, 10] proved on graceful tree conjecture, it remains open over four decades. For an exhaustive survey on graceful tree conjecture and other related results, refer the excellent dynamic survey by Gallian [3]. In this paper, we introduce a new method of constructing graceful trees from a given arbitrary tree by designing an exclusive algorithm.

2 Main Result

In this section, first we present labeling algorithm, which generates distinct labels on vertices and edges of given arbitrary input tree T with m edges. Using the labeling algorithm, we present a graceful tree embedding algorithm that will generate graceful tree T^* containing the given input arbitrary tree T as its subtree.

Labeling Algorithm

Input: Arbitrary tree T with m edges

Step 1: Initialization

Identify a longest path P of T. Let d be the length of the longest path P. Let u_0 be the origin of P. Consider the tree T as a rooted tree having u_0 as its root. Then, T has $d + 1$ levels. Arrange the vertices in every level of the rooted tree T in such a way that each vertex of the longest path P always appears as the leftmost vertex of the respective level of the vertex. For each r, $1 \le r \le d$, describe the vertices of the rth level as $u_{r,1}, u_{r,2}, \ldots, u_{r,\gamma_r}$ from left to right such that $u_{r,1}$ is the vertex of the longest path P in the rth level, where γ_r denotes the number of vertices of the rth level.

Step 2: Labeling Vertices

Step 2.1: Labeling vertices in each even level

$$l(u_{0,1}) = l(u_{0,\gamma_0}) = l(u_0) = 0$$

For each r, $1 \le r \le \lfloor \frac{d}{2} \rfloor$, define the labels of the vertices of the 2rth even level,

$$l(u_{2r,1}) = l\left(u_{2(r-1),\gamma_{2(r-1)}}\right) + 1$$
$$l(u_{2r,i}) = l(u_{2r,i-1}) + 1, \; 2 \le i \le \gamma_{2r}$$

Step 2.2: Labeling vertices in each odd level

Step 2.2.1: Labeling vertices in the last odd level

Find $w = \max\{\gamma_{2r} : 0 \le r \le \lfloor \frac{d}{2} \rfloor\}$ (maximum over the number of vertices on every even level of T).

If d is even, then define the labels of the vertices of the last odd level, $(d-1)$th level,

$$l\left(v_{d-1,\gamma_{d-1}}\right) = l\left(u_{d,\gamma_d}\right) + 2w,$$
$$l\left(v_{d-1,\gamma_{d-1}-i}\right) = l\left(v_{d-1,\gamma_{d-1}-(i+1)}\right) + 2w, \text{ for } 1 \le i \le \gamma_{d-1} - 1$$

If d is odd, then define the labels of the vertices of the last odd level, dth level,

$$l\left(v_{d,\gamma_d}\right) = l\left(u_{d-1,\gamma_{d-1}}\right) + 2w,$$
$$l\left(v_{d,\gamma_d-i}\right) = l\left(v_{d,\gamma_d-(i-1)}\right) + 2w, \text{ for } 1 \le i \le \gamma_d - 1$$

Step 2.2.2: Labeling vertices in the remaining odd levels

For each r, $0 \le r \le \lceil \frac{d}{2} \rceil - 2$, define the labels of the vertices of the $(2r+1)$th odd level,

$$l\left(v_{2r+1,\gamma_{2r+1}}\right) = l\left(v_{2r-1,\gamma_{2r-1}}\right) + 2w,$$
$$l\left(v_{2r+1,\gamma_{2r+1}-i}\right) = l\left(v_{2r+1,\gamma_{2r+1}-(i-1)}\right) + 2w, \text{ for } 1 \le i \le \gamma_{2r+1} - 1$$

Step 3: Edge labels of the edges of *T*

For every edge uv of T, define the edge label $l'(uv) = |l(u) - l(v)|$.

Observation 1: Vertex labels defined in labeling algorithm are distinct.

From the labeling algorithm, observe that the vertex labels of the vertices in the even levels from left to right where the order of the even levels increased in the top to bottom followed by the vertex labels of the vertices in the odd levels from right to left where the order of the odd levels decreased in the bottom to top form a monotonically increasing sequence. Thus, the vertex labels of the vertices of the input tree T are distinct.

Observation 2: The edge labels of all the edges of *T* that is defined in the labeling algorithm are distinct.

The edge labels of the edges incident at the vertices of each of the odd levels from right to left where the order of the odd levels decreased in the bottom to top form a monotonically increasing sequence. Thus, the edge labels of the edges of the input tree T are distinct.

Note 1: For a given arbitrary tree T with m edges, we run the labeling algorithm and obtain the labeled output tree. This labeled output tree is referred as T'. For the convenience, hereafter the vertices of output tree T' are referred by their vertex labels and the edges by their edge labels. Thus, for the output tree T', we consider $V(T') = \{0, 1, 2, \ldots, p-1, p, \alpha_1, \alpha_2, \ldots, M\}$ and
$E(T') = \{l'(e_1), l'(e_2), \ldots, l'(e_m)\}$.

Graceful Tree Embedding Algorithm

Input: Any arbitrary tree *T* with *m* edges

Step 1:

Step 1.1:

Run labeling algorithm on input tree T and get the output tree T'.

Step 1.2:

For the tree T', define

Vertex label set $V = V(T') = \{0, 1, 2, \ldots, p-1, p, \alpha_1, \alpha_2, \ldots, M\}$, where the elements of V are the vertex labels of the vertices of the input tree T that is defined in the labeling algorithm.

Edge label set $E = E(T') = \{l'(e_1), l'(e_2), \ldots, l'(e_m)\}$, where $l'(e_i)$is the edge label of the edge e_i, for $1 \leq i \leq m$ of T that is defined in the labeling algorithm.

All label set $X = \{0, 1, 2, \ldots, M\}$,

Common label set $I = V \cap E$,

Exclusive vertex label set $\hat{V} = (V - \{0\}) - I$,

Exclusive edge label set $\hat{E} = E - I$ and

Missing vertex label set $\hat{X} = X - V$.

Initiate $T^* \leftarrow T'$, $V(T^*) \leftarrow V(T')$, $E(T^*) \leftarrow E(T')$.

Step 2:

While $\hat{X} \neq \emptyset$, find min $\hat{X} = a$.

Step 3:

If $a \notin \hat{E}$, then consider a new vertex with label a and add a new edge between the vertex with label 0 and the new vertex with label a to T^*.

Update $T^* \leftarrow T^* + (0, a)$, $V(T^*) \leftarrow V(T^*) \cup \{a\}$, $E(T^*) \leftarrow E(T^*) \cup \{(0, a)\}$.

Delete a from $\hat{X}$ and go to Step 2.

Step 4:

If $a \in \hat{E}$, then find min $\hat{V} = b$ and $\beta = a - b$. Consider a new vertex with label a and add a new edge between the vertex labeled β and the new vertex with label a to T^*.

Update $T^* \leftarrow T^* + (\beta, a)$, $V(T^*) \leftarrow V(T^*) \cup \{a\}$, $E(T^*) \leftarrow E(T^*) \cup \{(\beta, a)\}$.

Delete a from $\hat{X}$ and delete b from $\hat{V}$and go to Step 2.

Theorem 1 *The output tree T^*generated by graceful tree embedding algorithm for an input arbitrary tree T is graceful and contains the input arbitrary tree T as its subtree.*

Proof For an input arbitrary tree T, obtain tree T^*generated by the graceful tree embedding algorithm. Consider the sets V, E, I, X, $\hat{V}, \hat{E}, \hat{X}$ that are defined in Step 1.2 of embedding algorithm. By Step 1.2 of graceful tree embedding algorithm, we have $\hat{X} = X - V$. Then, we have $X = \hat{X} \cup V$. Since $\hat{E} \subset \hat{X}$, we can write $X = \hat{X} \cup V = \left(\left(\hat{X} - \hat{E}\right) \cup \hat{E}\right) \cup V$. Observe that by definition of $\hat{E}$, $\hat{E} \cap V = \emptyset, \hat{E} \cap \left(\hat{X} - \hat{E}\right) = \emptyset$, and $V \cap \left(\hat{X} - \hat{E}\right) = \emptyset$. That is, the sets $\left(\hat{X} - \hat{E}\right), \hat{E}$ and V are mutually disjoint. Note that V consists of all the vertex labels of T. $\hat{E}$ consists of the edge labels of T that are not vertex labels of T. $\hat{X} - \hat{E}$ consists of the members of X, which are neither the vertex labels of T nor the edge labels of T. Consider $a =$ min $\hat{X}$ obtained by an execution of Step 2 of graceful tree embedding algorithm. If

$a \notin \hat{E}$, then by Step 3 of graceful tree embedding algorithm, the vertex label a is obtained in the updated tree T^* by adding the new edge (0, a) to the current tree T^*. Also, a is removed from $\hat{X}$. Since a was removed from $\hat{X}$, the vertex label a will never be obtained again.

If $a \in \hat{E}$, then by Step 4 of graceful tree embedding algorithm, the vertex label a is obtained in the updated tree T^* by adding the new edge (β, a) in the current tree T^*, where $\beta = a - b$ and $b = \min \hat{V}$. Since a is removed from $\hat{X}$, the vertex label a will never be obtained again. Thus, after executing Step 3 of graceful tree embedding algorithm $|\hat{X} - \hat{E}|$ times and Step 4 of graceful tree embedding algorithm $|\hat{E}|$ times, T^* contains all the vertex labels 0, 1, 2, …, M. Observe that all the vertex labels obtained from embedding algorithm are distinct and belong to $X - V$. By Observation 1, all the vertex labels of T are also distinct. Thus, vertex labels of all the vertices of T^* are distinct and the final updated tree T^* has $M + 1$ vertices with vertex set $V(T^*) = \{0,1,2,\ldots, M\}$ (where a vertex of T^* is referred by its corresponding label).

We can write the set $X - \{0\} = \hat{X} \cup (V - \{0\}) = (\hat{X} - \hat{E}) \cup E \cup \hat{V}$. Observe that the sets $(\hat{X} - \hat{E}), \hat{E}, \hat{V}$, and I are mutually disjoint. Elements in $\hat{E}$ and I are already existing as edge labels in T. Consider $\min\hat{X} = a$, obtained by executing Step 2 of the embedding algorithm. If $a \notin \hat{E}$, then by Step 3 of graceful tree embedding algorithm, the edge label a is obtained in the updated tree T^* by adding the new edge (0, a) to the current tree T^* and a is removed from $\hat{X}$. Since a was removed from $\hat{X}$, the edge label a will never be obtained again. If $a \in \hat{E}$, then a unique $b \in \hat{V}$, where $b = \min\hat{V}$ is found by executing Step 4 of graceful tree embedding algorithm, and the edge label b is obtained in the updated tree T^* from the new edge (β, a), which was added to the current tree T^*, where $\beta = a - b$ and b is removed from $\hat{V}$. Also, a is removed from $\hat{X}$. Since a is removed from $\hat{X}$ and b is removed from $\hat{V}$, the edge label b will never be obtained again. Since $|\hat{V}| = |\hat{E}|$, whenever $a \in \hat{E}$, corresponding unique $b \in \hat{V}$ is found. We see that every element of $\hat{V}$ is obtained as edge label in the final updated tree T^*. Thus, after executing Step 3 of graceful tree embedding algorithm $|\hat{X} - \hat{E}|$ times and Step 4 of graceful tree embedding algorithm $|\hat{V}| (= |\hat{E}|)$ times in the final updated tree T^*, the edge labels belonging to $(X - \{0\}) - E$ are all obtained as distinct edge labels. As T^* was initiated with m edges having distinct edge labels belonging to the set E, the final updated tree T^* has M edges with distinct edge labels 1,2,3,…,M. Thus, the final updated tree T^* is graceful.

3 Illustration

Figures 1 and 2.

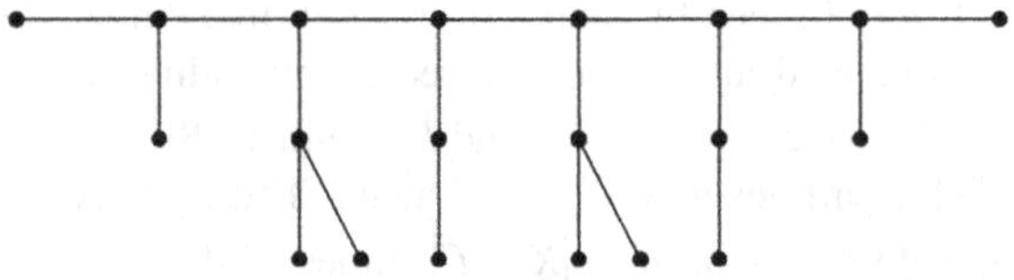

Fig. 1 Input tree *T* with 19 edges

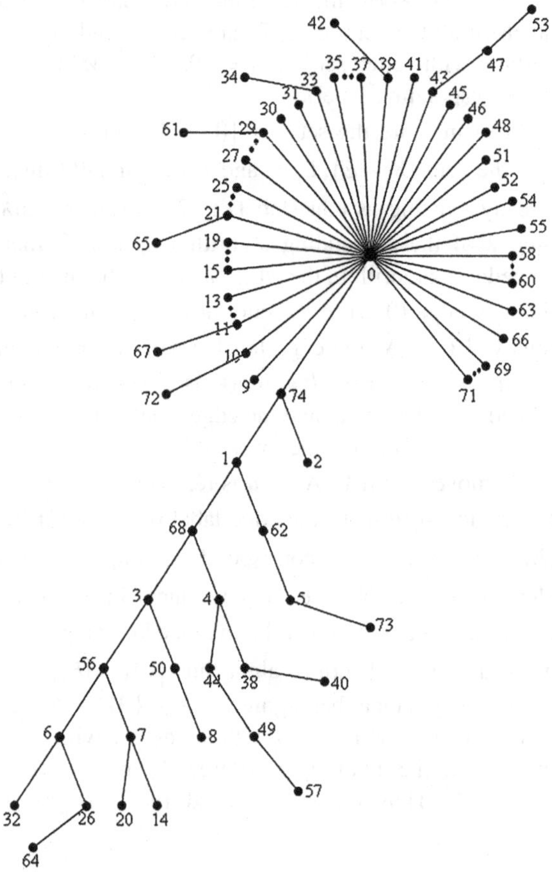

Fig. 2 Graceful tree *T** with 74 edges having the arbitrary tree *T* as its subtree

Acknowledgments The second author gratefully acknowledges Centre for Research, Anna University, Chennai, under the Grant Ref: CR/ACRF/Jan.2011/31.

References

1. B.D. Acharya, S.B. Rao, S. Arumugam, *Embedding and NP-Complete problems fro Graceful Graphs, Labelings of Discrete Structures and Applications*, B.D. Acharya, S. Arumugam, Alexander Rosa, eds., (2008), 57–62, Narosa Publishing House, New Delhi.
2. Bloom G.S, *A chronology of the Ringel-Kotzig conjecture and the continuing quest to call all trees graceful*, Ann. N.Y. Acad. Sci., 326, (1979), 35–51.
3. J.A. Gallian, *A Dynamic Survey of Graph Labeling*, The Electronic Journal of Combinatorics, 19, (2012), #DS6.
4. Golomb S.W, *How to number a graph*, Graph Theory and Computing R.C. Read, ed., Academic Press, New York, 1972, 23–37.
5. Jeba Jesintha J and Sethuraman G, *All arbitrary fixed generalized banana trees are graceful*, Math. Comput. Sci, 5, (2011), 1, 51–62.
6. Kotzig A, *Decompositions of a complete graph into 4 k-gons* (in Russian), Matematicky Casopis, 15, (1965), 229–233.
7. Ringel G, *Problem 25, in Theory of Graphs and its Applications*, Proc. Symposium Smolenice, Prague, (1963) page-162.
8. Rosa A, *On certain valuations of the vertices of a graph*, Theory of graphs, (International Symposium, Rome, July 1966), Gordon and Breach, N.Y. and Dunod Paris, (1967), 349–355.
9. Sethuraman G. and Venkatesh S, Decomposition of complete graphs and complete bipartite graphs into α-labeled trees, Ars Combinatoria, 93, (2009), 371–385.
10. Van Bussel F., *Relaxed graceful labelings of trees*, The Electronic Journal of Combinatorics, 9, (2002), #R4.
11. West D.B., *Introduction to Graph Theory*, Prentice Hall of India, 2nd Edition, 2001.

References

Characterization of Semi-open Subcomplexes in Abstract Cellular Complex

N. Vijaya and G. Sai Sundara Krishnan

Abstract The concept of abstract cellular complexes was introduced by Kovalevsky (Computer Vision, Graphics and Image Processing, 46:141–161, 1989) and established that the topology of cellular complex is the only possible topology of finite sets to describe the structure of images. Further, the topological notions of connectedness and continuity in abstract cellular complexes were introduced while using the notions of an open subcomplex, closed subcomplex, and boundary of a subcomplex, etc. In this paper, the notion of semi-open subcomplex in abstract cellular complex is introduced and some of its basic properties are studied by defining the notions of semi-closure, semi-frontier, and semi-interior. Further, a homogeneously n-dimensional complex is characterized while using the notion of semi-open subcomplexes. Introduced is also the concept of a quasi-solid in subcomplex. Finally, a new algorithm for tracing the semi-frontier of an image is presented.

Keywords Semi-open subcomplex · Semi-closed subcomplex · Semi-frontier · Semi-interior · Semi-closure · Quasi-solid · Semi-region · Semi-frontier tracing algorithm

N. Vijaya (✉)
Adhiyamaan College of Engineering, Hosur, India
e-mail: viji_prathiksha@yahoo.co.in

G. S. S. Krishnan
PSG College of Technology, Coimbatore, India
e-mail: g_ssk@yahoo.com

G. S. S. Krishnan et al. (eds.), *Computational Intelligence, Cyber Security and Computational Models*, Advances in Intelligent Systems and Computing 246,
DOI: 10.1007/978-81-322-1680-3_30,

1 Introduction

Digital image processing is a rapidly growing discipline with a broad range of applications in medicine, environmental sciences, and in many other fields. The field of digital image processing refers to processing two-dimensional pictures by a digital computer.

Rosenfeld [2, 3] represented a digital image by a graph whose nodes are pixels and whose edges are linking adjacent pixels to each other. He named the resultant graph as the neighborhood graph. But this representation contains two paradoxes, namely connectivity and boundary paradoxes. Kovalevsky [4] introduced the notion of abstract cellular complexes to study the structure of digital images and introduced axiomatic digital topology [8], which has no paradoxes. Moreover, he showed that every finite topological space with separation property is isomorphic to an abstract cellular complex. Further, Kovalevsky [5, 6] introduced the notions of a half plane, a digital line segment, etc., while using the notions of open sets, closed sets, closure, and interior.

The concepts of semi-open set and semi-continuity were introduced by Levine [10]. The half-open intervals $(a, b]$ and $[a, b)$ are characterized as semi-open subsets of the real line. Though the collections of semi-open sets do not form a classical topology on R, it satisfies the condition of a basis in R, and hence, both half-open intervals of the form $[a, b)$ and $(a, b]$ generate two special topologies, namely lower limit topology and upper limit topology, respectively, on R. This motivates us to study the notion of semi-open complex in abstract cellular complex. In this paper, we introduced the basic concepts of semi-open subcomplex in abstract cellular complex and studied some of their basic properties, which enable us to study the structure of digital images through semi-open subcomplexes.

In this paper, the concept of a semi-open subcomplex is introduced and some of its properties in abstract cellular complexes are studied. Further, the semi-open subcomplex is characterized by introducing the notions of semi-frontier, semi-interior, and semi-closure. Further, the relationship between a semi-open subcomplex and a homogeneously n-dimensional subcomplex is studied, and the notions of quasi-solid, semi-region are introduced. Finally, the algorithm to tracing the semi-frontier of an image using Kovalevsky's chain code is presented.

2 Preliminaries

In this section, some basic definitions are recalled.

Definition 2.1 [4] An abstract cellular complex (ACC) $C = (E, B, \dim)$ is a set E of abstract elements provided with an antisymmetric, irreflexive, and transitive binary relation $B \subset E \times E$ called the bounding relation, and with a dimension function dim: $E \to I$ from E into the set I of non-negative integers such that $\dim(e') < \dim(e'')$ for all pairs $(e', e'') \in B$.

Definition 2.2 [4] A subcomplex $S = (E', B')$ of a given K-complex $C = (E, B)$ is a k-complex whose set E' is the subset of E and the relation B' is an intersection of B with $E' \times E'$.

Definition 2.3 [4] A subcomplex S of C is called open in C if for every element e′ of S all elements of C which are bounded by e′ are also in S.

Definition 2.4 [4] The smallest subset of a set S which contains a given cell $c \in S$ and is open in S is called smallest neighborhood of c relative to S and is denoted by SON(c, S).

Definition 2.5 [4] The smallest subset of a set S which contains a given cell $c \in S$ and is closed in S is called the closure of c relative to S and is denoted by Cl(c, S).

Definition 2.6 [4] The frontier of a subcomplex S of an abstract cellular complex C relative to C is the subcomplex Fr(S, C) containing of all cells c of C such that the SON(c) contains cells both of S and of its complement C–S.

Definition 2.7 [9] Let t and T be subsets of the space S such that $t \subseteq T \subseteq S$. The set t-Fr (t, T) is called the interior of t in T, and it is denoted by Int(t, T).

3 Semi-open Subcomplexes in Abstract Cellular Complex

Definition 3.1 A subcomplex S in an abstract cellular complex C is called semi-open subcomplex if there exist an open subcomplex O such that $O \subseteq S \subseteq \mathrm{Cl}(O)$, where Cl denotes the closure operator in C (Fig. 1).

Remark 3.1 It follows from the Definition 3.1 that an example of a semi-open subcomplex can be a pixel with at least one element of its frontier.

Theorem 3.1 *A subcomplex S of an abstract cellular complex C is semi-open subcomplex if and only if* $S \subseteq \mathrm{Cl}(\mathrm{Int}(S))$.

Proof Suppose $S \not\subset \mathrm{Cl}(\mathrm{Int}(S))$, then there exist a cell $x \in S$ such that $x \notin \mathrm{Cl}(\mathrm{Int}(S))$. This implies that $x \notin \mathrm{Int}(S)$ and $x \notin \mathrm{Fr}(\mathrm{Int}(S))$. $X \notin \mathrm{Int}(S)$ implies that there exist no open subcomplex O containing x such that $O \subseteq S$, and $x \notin \mathrm{Fr}(\mathrm{Int}(S))$ implies that there exists no open subcomplex O contained in S such

Fig. 1 Semi-open subcomplex in 2D

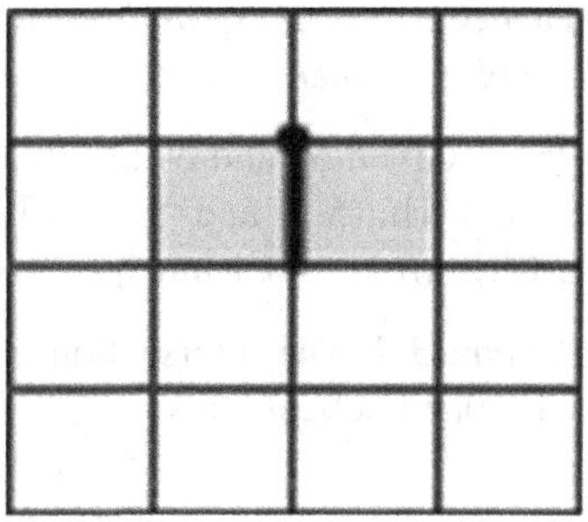

that $x \in \mathrm{Fr}(O)$. Hence, there exists no open subcomplex O such that $O \subseteq S \subseteq \mathrm{Int}(O) \cup \mathrm{Fr}(O) = \mathrm{Cl}(O)$. This is a contradiction to the assumption that S is a semi-open subcomplex in C. Converse part is obvious from the definition of $\mathrm{Int}(S)$, while $\mathrm{Int}(S)$ is an open subcomplex contained in S.

Theorem 3.2 *Every open subcomplex is a semi-open subcomplex.*

Proof Proof follows directly from Theorem 3.1.

Remark 3.2 The converse of the above Theorem 3.2 need not be true.

Lemma 3.1 *If S is a subcomplex of an abstract cellular complex C then the following equations hold*:

i $C–\mathrm{Int}(S) = \mathrm{Cl}(C–S)$
ii $C–\mathrm{Cl}(S) = \mathrm{Int}(C–S)$

Definition 3.2 A subcomplex S of an abstract cellular complex C is called semi-closed if $C–S$ is semi-open.

Theorem 3.3 *A subcomplex S of an abstract cellular complex C is semi-closed if and only if* $\mathrm{Int}(\mathrm{Cl}(S)) \subseteq S$.

Proof Proof follows directly from the Theorem 3.1 and Lemma 3.1.

Lemma 3.2 *If S_1 and S_2 are any two subcomplexes of an abstract cellular complex C and if $S_1 \subseteq S_2$, then*

i $\mathrm{Int}(S_1) \subseteq \mathrm{Int}(S_2)$
ii $\mathrm{Fr}(S_1) \subseteq \mathrm{Cl}(S_2)$
iii $\mathrm{Cl}(S_1) \subseteq \mathrm{Cl}(S_2)$

Proof
i Proof follows directly from the Definition 2.7
ii Let $x \in \mathrm{Fr}(S_1)$. This implies that $\mathrm{SON}(x)$ intersects with both S_1 and $C–S_1$. If $\mathrm{SON}(x) \subseteq S_2$, then $x \in \mathrm{Int}(S_2)$ and if $\mathrm{SON}(x) \not\subseteq S_2$, then $x \in \mathrm{Fr}(S_2)$. Hence, $x \in \mathrm{Int}(S_2) \cup \mathrm{Fr}(S_2) = \mathrm{Cl}(S_2)$.
iii Proof follows directly from (i) and (ii)

Theorem 3.4 *If S_1 and S_2 are any two subcomplexes of an abstract cellular complex C, then $S_1 \cup S_2$ is also semi-open subcomplex.*

Proof Given S_1 and S_2 are two semi-open subcomplexes of C. This implies that $S_1 \subseteq \mathrm{Cl}(\mathrm{Int}(S_1))$ and $S_2 \subseteq \mathrm{Cl}(\mathrm{Int}(S_2))$. This implies that $S_1 \cup S_2 \subseteq \mathrm{Cl}(\mathrm{Int}(S_1)) \cup \mathrm{Cl}(\mathrm{Int}(S_2)) \subseteq \mathrm{Cl}(\mathrm{Int}(S_1 \cup S_2))$. Hence, $S_1 \cup S_2$ is a semi-open subcomplex in C.

Remark 3.3 The intersection of any two semi-open subcomplexes need not be a semi-open subcomplex.

Theorem 3.5 *Let* S_1 *be a semi-open subcomplex of an abstract cellular complex C and* $S_1 \subseteq S_2 \subseteq \mathrm{Cl}(S_1)$. *Then* S_2 *is semi-open.*

Proof Given S_1 is semi-open. This implies that there exists an open subcomplex O such that $O \subseteq S_1 \subseteq \mathrm{Cl}(O)$. Hence, $O \subseteq S_2$. By the Lemma 3.2 (iii), we get $\mathrm{Cl}(S_1) \subseteq \mathrm{Cl}(O)$.Therefore, S_2 is also semi-open.

Theorem 3.6 *If S is homogeneously n-dimensional subcomplex of an n-dimensional complex C, then* $\mathrm{Fr}(S) = \mathrm{Fr}(\mathrm{Int}(S))$.

Proof Suppose $\mathrm{Fr}(S) \neq \mathrm{Fr}(\mathrm{Int}(S))$, it implies that there is at least one lower-dimensional cell k of Fr(S) does not belong to Fr(Int(S)). This implies that the cell k does not bound any n-cell of S. This contradicts the fact that S is homogeneously n-dimensional.

Theorem 3.7 *If S is homogeneously n-dimensional subcomplex of an n-dimensional complex C, then it is semi-open.*

Proof Given S is homogeneously n-dimensional subcomplex. By definition of interior, all the principal cells of S belong to Int(S). Suppose $S \not\subseteq \mathrm{Cl}(\mathrm{Int}(S))$, then there exists a lower-dimensional cell $c \in S$ such that $c \notin \mathrm{Cl}(\mathrm{Int}(S))$. This implies that the cell c does not bound any n-cell of Int(S). This contradicts the fact that S is homogeneously n-dimensional.

Theorem 3.8 *If S is strongly connected homogeneously n-dimensional subcomplex of an n-dimensional complex C, then it is semi-open.*

Proof Proof follows directly from the Theorem 3.6 and the definition of semi-open.

Theorem 3.9 *If a subcomplex S of an n-dimensional complex C is solid, then it is semi-open.*

Proof Proof follows directly from the definition of solid and semi-open.

Definition 3.3 Let S be a non-empty subcomplex of an abstract cellular complex C. Then, the semi-frontier of S is the set of all elements k of C–S, such that each neighborhood of k contains elements of both S and its complement C–S. It is denoted by SFr.

Lemma 3.3 *If S is a subcomplex of an abstract cellular complex C, then* $S\mathrm{Fr}(S) \subseteq \mathrm{Fr}(S)$.

Lemma 3.4 *If S is a subcomplex of an abstract cellular complex C, then* $S\mathrm{Fr}(S) \cup S\mathrm{Fr}(C\text{–}S) = \mathrm{Fr}(S)$.

Proof Proof follows directly from the definition of Fr and SFr

Definition 3.4 Let S be a subcomplex of an abstract cellular complex C. The subcomplex S–SFr(S) is called the semi-interior of S, and it is denoted by SInt.

Lemma 3.5 *If S is a subcomplex of an abstract cellular complex C, then* Int(S) $\subseteq$ SInt(S).

Lemma 3.6 *If S is a subcomplex of an abstract cellular complex C, then* SInt(S) *is semi-open.*

Theorem 3.10 *A subcomplex S of an abstract cellular complex C is semi-open if and only if* SInt(S) $=$ S.

Proof Proof follows directly from the definition of semi-frontier and semi-interior.

Definition 3.5 Let S be a subcomplex of an abstract cellular complex C. The subcomplex $S \cup S$Fr(S) is called the semi-closure of S. It is denoted by SCl.

Definition 3.6 A subcomplex S of an abstract cellular complex C is called semi-open connected if S is connected and semi-open.

Remark 3.4 If a subcomplex S of an abstract cellular complex C is open connected, then it is semi-open connected. But the converse is not true.

Definition 3.7 A subcomplex S of an abstract cellular complex C is called semi-region if S is semi-open, connected, and solid.

Theorem 3.11 *Every semi-open connected subcomplex S of an n-dimensional complex C is homogeneously n-dimensional.*

Proof Proof follows directly from the definition of semi-open connected and homogeneously n-dimensional.

Definition 3.8 A subcomplex S of an n-dimensional complex C is called quasi-solid if and only if it is homogeneously n-dimensional and is contained in Cl(Int(S, C),C).

Theorem 3.12 *If a subcomplex* S^n *of an n-dimensional complex is quasi-solid, then it is semi-open.*

Proof Proof follows directly from the definition of quasi-solid and semi-open.

Theorem 3.13 *Every solid subcomplex is quasi-solid.*

Proof Proof follows directly from the definition of solid and by the Theorem 3.9.

Remark 3.5 The converse of Theorem 3.13 need not be true.

4 Algorithm on Tracing the Semi-Frontier of an Image

This algorithm is defined to tracing the semi-frontier of an image. The proposed algorithm is more efficient for the image, which contains less number of components. Conceptually, the algorithm is divided into two major steps.

Step 1: Before starting the tracing, the membership of the lower-dimensional cells (semi-open subcomplex) must be defined by the user whether it belongs to foreground or background of the image. The user can make the decision on the ground of some knowledge about the image.

Step 2: Then, the image must be scanned row by row to find the starting point of each component. After finding the starting point, make the step along the boundary crack to the next boundary point using Kovalevsky's chain code [7]. The crack and end point of the crack both belong to semi-frontier if it does not belong to foreground. The process stops when the starting point is reached again.

During each pass, the already visited cracks must be labeled to avoid multiple tracing.

4.1 Algorithm

The following is the formal description of the algorithm:

Input: Given a digital pattern as a two-dimensional abstract cellular complex **Image** containing points, cracks, and pixels.
Output: A sequence **SF** of semi-frontier cracks and points.
Let p denote the current semi-frontier point.
Let c denote the current semi-frontier crack.
Begin

- Set **Label** to be empty
- Set **SF** to be empty
- Scan the **Image** row by row until two subsequent pixels of different colors are found
- Set the upper end point of the crack c lying between the pixels of different colors as starting point s
- Insert s, c in **SF** if it does not belong to foreground of the image
- Fix the direction as 1
- Move to next boundary point p along boundary crack **c**
- Do
  - Insert p, c in **SF** if it does not belong to foreground of the image
  - To recognize the next boundary crack test **left** and **right** pixels lying ahead of actual crack
  - If **Image** [**left**] is foreground

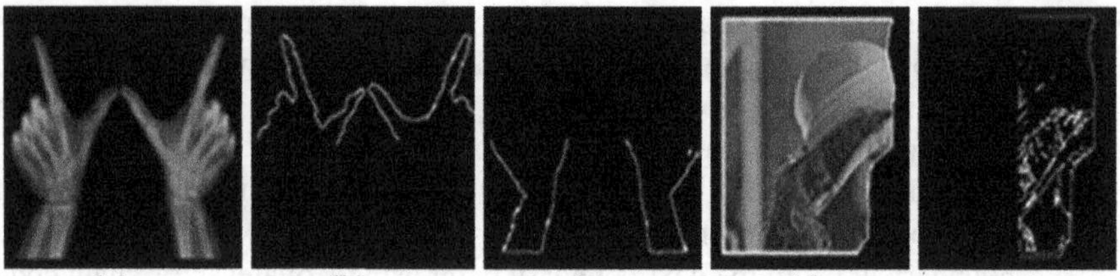

Fig. 2 **a** X-ray image of hands. **b** and **c** Semi-frontier of (**a**). **d** Image of torn photograph. **e** Semi-frontier of (**d**)

- Change the direction into (direction +1) %4
- If **Image [right]** is background
- Change the direction into (direction +3) %4
- Insert c in **Label**, if the direction is 1
- Move to the next boundary point p

End while if p is equal to s

End

4.2 Results and Discussion

Our proposed algorithm that extracts semi-frontier finds potential applications in pattern recognition. With the simple preprocessing steps, the algorithm directly traces the semi-frontier of an image. It is thus computationally effective in extracting semi-frontier of an image. The semi-frontier elements are generally a small subset of the total number of elements that represent a boundary. Therefore, the allocation of memory space is highly reduced and also the amount of computation is reduced when the images are processed by means of certain semi-frontier features. The time complexity of an algorithm is $O(n^4)$. The proposed algorithm is implemented in MATLAB (Fig. 2).

References

1. P.Alexandroff and H.Hopf, Topologie I, Springer, 1935.
2. Azriel Rosenfled, Digital topology, The American Mathematical Monthly, 8, pp 621–630, 1979.
3. Azriel Rosenfled, and A.C. Kak, Digital picture processing, Academic Press, 1976.
4. V.Kovalevsky, “Finite topology as applied to image analysis”, Computer Vision, Graphics and Image processing, 46, pp 141–161, 1989.
5. V.Kovalevsky, Digital geometry based on the topology of abstract cellular complexes, in proceedings of the Third International Colloquium “discrete Geometry for computer Imagery”, University of Strasbourg, pp 259–284, 1993.

6. V.Kovalevsky, “Algorithms and data structures for computer topology”, in G.Bertrand et al.(Eds), LNCS 2243, Springer, pp 37–58, 2001.
7. V.Kovalevsky, Algorithms in digital geometry based on cellular topology In R.Klette. and J.Zunic(Eds.), LNCS 3322, Springer, pp 366–393, 2004.
8. V.Kovalevsky, Axiomatic Digital Topology, Springer Math image vis 26, pp 41–58, 2006.
9. V.Kovalevsky, Geometry of Locally Finite spaces, Publishing House Dr.Baerbel Kovalevski, Berlin, 2008.
10. N.Levine, Semi-open sets and semi-continuity in topological spaces, American Mathematical Monthly, 70, pp 36–41, 1963.

Fluid Queue Driven by an $M/M/1$ Queue Subject to Catastrophes

K. V. Vijayashree and A. Anjuka

Abstract In this paper, we present the stationary analysis of a fluid queueing model modulated by an $M/M/1$ queue subject to catastrophes. The explicit expressions for the joint probability of the state of the system and the content of the buffer under steady state are obtained in terms of modified Bessel function of first kind using continued fraction methodology.

Keywords Buffer content distribution · Continued fractions · Laplace transform · Modified Bessel function of first kind

1 Introduction

In recent years, fluid queues have been widely accepted as appropriate models for modern telecommunication [4] and manufacturing systems [7]. This modelling approach ignores the discrete nature of the real information flow and treats it as a continuous stream. In addition, fluid models are often useful as approximate models for certain queueing and inventory systems where the flow consists of discrete entities, but the behaviour of individuals is not important to identify the performance analysis [3].

Steady-state behaviour of Markov-driven fluid queues has been extensively studied in the literature. Parthasarathy et al. [8] present an explicit expression for the buffer content distribution in terms of modified Bessel function of first kind using Laplace transforms and continued fractions. Silver Soares and Latouche [11]

K. V. Vijayashree (✉) · A. Anjuka
Department of Mathematics, Anna University, Chennai, India
e-mail: vkviji@annauniv.edu

A. Anjuka
e-mail: anjukaatlimuthu@gmail.com

G. S. S. Krishnan et al. (eds.), *Computational Intelligence, Cyber Security and Computational Models*, Advances in Intelligent Systems and Computing 246,
DOI: 10.1007/978-81-322-1680-3_31,

expressed the stationary distribution of a fluid queue with finite buffer as a linear combination of matrix exponential terms using matrix analytic methods. Besides, fluid queues also have successful applications in the field of congestion control [14] and risk processes [10]. Fluid models driven by an $M/M/1/N$ queue with single and multiple exponential vacations were recently studied by Mao et al. [5, 6] using spectral method.

In this paper, we analyse fluid queues driven by an $M/M/1$ queue subject to catastrophes. The effect of catastrophes in queueing models induces the system to be instantly reset to zero at some random times. Such models find a wide range of applications in diverse fields [12]. More specifically, birth–death stochastic models subject to catastrophes are popularly used in the study of population dynamics [1, 13], biological process [2], etc. With the arrival of a negative customer into the system (*referred to as catastrophe*), it induces the positive customers, if any, to immediately leave the system. For example, in computer systems which are not supported by power backup, the voltage fluctuations will lead to the system shut down momentarily and in that process all the work in progress and the jobs waiting to be completed will be lost and the system begins afresh. Such situations can be modelled as a single server queueing model with catastrophes. Under steady-state conditions, explicit analytical expressions for the joint system size probabilities of the modulating process and the buffer content distribution are obtained using continued fraction methodology.

2 Model Description

Let $X(t)$ denote a number of customer in the background queueing model at time t wherein customers arrive according to a Poisson process at an average rate λ and the service times are exponentially distribution with parameter μ. Further, the catastrophes are assumed to occur according to a Poisson process with rate γ. For such a model, the steady-state probabilities p_j's are given by

$$p_0 = 1 - \rho \quad \text{and} \quad p_j = (1-\rho)\rho^j, \quad j = 1, 2, 3\ldots,$$

where

$$\rho = \frac{\lambda + \mu + \gamma - \sqrt{\lambda^2 + \mu^2 + \gamma^2 + 2\lambda\gamma + 2\mu\gamma - 2\lambda\mu}}{2\mu}$$

See Nelson [9].

Let $\{C(t),\ t \geq 0\}$ represent the buffer content process where $C(t)$ denotes the content of the buffer at time t. During the busy period of the server, the fluid accumulates in an infinite capacity buffer at a constant rate $r > 0$. The buffer depletes the fluid during the idle periods of the server at a constant rate $r_0 < 0$ as long as the buffer is nonempty. Hence, the dynamics of the buffer content process is given by

$$\frac{\mathrm{d}C(t)}{\mathrm{d}t} = \begin{cases} 0, & \text{if } C(t) = 0, \quad X(t) = 0 \\ r_0, & \text{if } C(t) > 0, \quad X(t) = 0 \\ r, & \text{if } C(t) > 0, \quad X(t) > 0 \end{cases}$$

Clearly, the two-dimensional process $\{(X(t), C(t)), t \geq 0\}$ constitutes a Markov process, and it possesses a unique stationary distribution under a suitable stability condition. To ensure the stability of the process $\{(X(t), C(t)), t \geq 0\}$, we assume the mean aggregate input rate to be negative, that is, $r_0 p_0 + r\sum_{i=1}^{\infty} p_i < 0$.

3 Solution Methodology

Letting

$$F_j(t,x) = \Pr\{X(t) = j;\ C(t) \leq x\},\ t, x \geq 0,\ j = 0, 1, 2, \ldots,$$

the Kolmogorov forward equations for the Markov process $\{X(t), C(t)\}$ are given by

$$\frac{\partial F_0(t,x)}{\partial t} + r_0 \frac{\partial F_0(t,x)}{\partial x} = -\lambda F_0(t,x) + \mu F_1(t,x) + \gamma \sum_{k=1}^{\infty} F_k(t,x)$$

and for $j = 1, 2, 3, \ldots,$

$$\frac{\partial F_j(t,x)}{\partial t} + r\frac{\partial F_j(t,x)}{\partial x} = \lambda F_{j-1}(t,x) - (\lambda + \mu + \gamma)F_j(t,x) + \mu F_{j+1}(t,x).$$

Assume that the process is in equilibrium so that $\frac{\partial F_j(t,x)}{\partial t} \equiv 0$ and $\lim_{t \to \infty} F_j(t,x) \equiv F_j(x)$. The above system then reduces to a system of ordinary differential equations given by

$$r_0 \frac{\mathrm{d}F_0(x)}{\mathrm{d}x} = -\lambda F_0(x) + \mu F_1(x) + \gamma \sum_{k=1}^{\infty} F_k(x), \text{ and} \tag{1}$$

$$r\frac{\mathrm{d}F_j(x)}{\mathrm{d}x} = \lambda F_{j-1}(x) - (\lambda + \mu + \gamma)F_j(x) + \mu F_{j+1}(x) \quad j = 1, 2, 3, \ldots. \tag{2}$$

When the net input rate of fluid flow into the buffer is positive, the buffer content increases and the buffer cannot stay empty. Hence, it follows that the solution to Eqs. (1) and (2) must satisfy the boundary conditions,

$$\begin{aligned} F_j(0) &= 0, \quad j = 1, 2, \ldots, \text{ and} \\ F_0(0) &= a, \quad \text{for some constant } 0 < a < 1. \end{aligned}$$

The condition $F_0(0) = a$ suggests that with some positive probability, say a, the buffer content remains empty when the server in the background queuing model is idle.

Taking Laplace transform of Eq. (1) leads to

$$r_0\left[s\hat{F}_0(s) - F_0(0)\right] = -\lambda\hat{F}_0(s) + \mu\hat{F}_1(s) + \gamma\sum_{k=1}^{\infty}\hat{F}_k(s) \tag{3}$$

which upon simplification yields

$$\hat{F}_0(s) = \frac{a}{s + \frac{\lambda}{r_0} - g(s)}, \tag{4}$$

where

$$g(s) = \frac{\mu\hat{F}_1(s)}{r_0\hat{F}_0(s)} + \frac{\gamma\sum_{k=1}^{\infty}\hat{F}_k(s)}{r_0\hat{F}_0(s)}.$$

Again, Laplace transform of Eq. (2) gives

$$(rs + \lambda + \mu + \gamma)\hat{F}_j(s) - \mu\hat{F}_{j+1}(s) = \lambda\hat{F}_{j-1}(s), \quad j = 1, 2, 3, \ldots$$

which leads to the continued fraction representation as

$$\frac{\hat{F}_j(s)}{\hat{F}_{j-1}(s)} = \frac{\lambda\mu}{\mu(rs + \lambda + \mu + \gamma)-} \quad \frac{\lambda\mu}{rs + \lambda + \mu + \gamma-} \quad \frac{\lambda\mu}{rs + \lambda + \mu + \gamma-}\cdots.$$

Assume

$$f(s) = \frac{\lambda\mu}{rs + \lambda + \mu + \gamma-} \quad \frac{\lambda\mu}{rs + \lambda + \mu + \gamma-}\cdots.$$

Then, $\frac{\hat{F}_j(s)}{\hat{F}_{j-1}(s)} = \frac{1}{\mu}f(s)$ and hence,

$$\hat{F}_j(s) = \frac{f(s)}{\mu}\hat{F}_{j-1}(s) = \left(\frac{f(s)}{\mu}\right)^j\hat{F}_0(s). \tag{5}$$

Also $f(s)$ can be rewritten as,

$$f(s) = \frac{\lambda\mu}{rs + \lambda + \mu + \gamma - f(s)} = \frac{\frac{\lambda\mu}{r}}{s + \frac{\lambda+\mu+\gamma}{r} - \frac{f(s)}{r}}$$

which leads to the quadratic equation given by

$$\frac{f(s)^2}{r} - \left(s + \frac{\lambda + \mu + \gamma}{r}\right)f(s) + \frac{\lambda\mu}{r} = 0.$$

Upon solving the above equation, we get

$$f(s) = \frac{p - \sqrt{p^2 - \alpha^2}}{\frac{2}{r}} \tag{6}$$

where $p = s + \frac{\lambda+\mu+\gamma}{r}$ and $\alpha = \frac{2\sqrt{\lambda\mu}}{r}$. Now, substituting for $\hat{F}_k(s)$ in $g(s)$ from Eq. (5), we get

$$g(s) = \frac{f(s)}{r_0}\left[1 + \frac{\gamma}{\mu - f(s)}\right].$$

Therefore, from Eq. (4), we get

$$\hat{F}_0(s) = \frac{a}{s + \frac{\lambda}{r_0} - g(s)} = \frac{a}{s + \frac{\lambda}{r_0}} + a\sum_{k=1}^{\infty}\frac{g(s)^k}{\left(s + \frac{\lambda}{r_0}\right)^{k+1}}.$$

which on inversion leads to

$$F_0(x) = a\exp^{\frac{-\lambda x}{r_0}} + a\sum_{k=1}^{\infty}\left(\frac{x^k}{k!}\exp^{-\frac{\lambda}{r_0}x}\right) * h(x) \tag{7}$$

where

$$h(x) = \frac{1}{r_0^k}\sum_{n=0}^{k}\binom{k}{n}\left(\frac{\gamma}{\mu}\right)^n\sum_{j=0}^{\infty}\binom{n+j-1}{j}\left(\frac{r}{2}\right)^{j+k}\frac{1}{\mu^j}\left(\frac{(j+k)I_{j+k}(\alpha x)\alpha^{j+k}}{x}\right)\exp^{-\left(\frac{\lambda+\mu+\gamma}{r}\right)x}.$$

The other steady-state probabilities are computed from Eq. (5) as follows

$$\begin{aligned}\hat{F}_j(s) &= \left(\frac{f(s)}{\mu}\right)^j \hat{F}_0(s)\\ &= \left(\frac{p - \sqrt{p^2 - \alpha^2}}{\frac{2\mu}{r}}\right)^j \hat{F}_0(s)\\ &= \left(\frac{r}{2\mu}\right)^j\left[p - \sqrt{p^2 - \alpha^2}\right]^j \hat{F}_0(s)\end{aligned}$$

which on the inversion yields,

$$F_j(x) = \left(\frac{r}{2\mu}\right)^j\left(\frac{jI_j(\alpha x)\alpha^j}{x}\right)\exp^{-\left(\frac{\lambda+\mu+\gamma}{r}\right)x} * F_0(x) \tag{8}$$

where $F_0(x)$ is explicitly given by Eq. (7). It still remains to determine the constant a which represents $F_0(0)$. Towards this end, adding the Eqs. (1) and (2) yields

$$r_0\frac{\mathrm{d}F_0(x)}{\mathrm{d}x} + \sum_{j=1}^{\infty} r\frac{\mathrm{d}F_j(x)}{\mathrm{d}x} = 0. \tag{9}$$

Integrating (9) from zero to infinity gives

$$r_0(F_0(0) - F_0(\infty)) + \sum_{j=0}^{\infty} r(F_j(0) - F_j(\infty)) = 0.$$

Note that $F_j(\infty) = \lim_{t\to\infty} \Pr(X(t) = j, C(t) < \infty) = \lim_{t\to\infty} \Pr(X(t) = j) = p_j$, where p_j's are the steady-state probabilities of the background queueing model. Using the boundary conditions, we obtain

$$F_0(0) = \frac{r_0 p_0 + \sum_{j=1}^{\infty} r p_j}{r_0} = \frac{(r_0 - r)p_0 + r}{r_0}.$$

Therefore, the constant a is explicitly given by

$$F_0(0) = a = \frac{(r_0 - r)(1 - \rho) + r}{r_0} \tag{10}$$

Remark 1 When the catastrophe parameter $\gamma = 0$, after some direct calculations, it can be shown that $\hat{F}_0(s)$ and $\hat{F}_j(s)$ becomes

$$\hat{F}_0(s) = \frac{a}{s + \frac{\lambda}{r_0}} + a \sum_{k=1}^{\infty} \left(\frac{r}{2r_0}\right)^k \frac{\left(p - \sqrt{p^2 - \alpha^2}\right)^k}{\left(s + \frac{\lambda}{r_0}\right)^{k+1}}$$

and

$$\hat{F}_j(s) = a\left(\frac{r}{2\mu}\right)^j \sum_{k=0}^{\infty} \left(\frac{r}{2r_0}\right)^k \frac{\left(p - \sqrt{p^2 - \alpha^2}\right)^{k+j}}{\left(s + \frac{\lambda}{r_0}\right)^{k+1}} \quad j = 1, 2, \ldots$$

which is evidently the same as Eqs. (15) and (19) in Parthasarathy et al. [8].

Theorem 1 *The stationary buffer content distribution of the fluid model under consideration is given by*

$$\begin{aligned} F(x) = \Pr\{C < x\} &= \sum_{j=0}^{\infty} F_j(x) \\ &= \int_0^x exp^{-\left(\frac{\lambda+\mu+\gamma}{r}\right)u} \sum_{j=0}^{\infty} \left(\frac{r}{2\mu}\right)^j \frac{jI_j(\alpha u)\alpha^j}{u} F_0(x - u)\, du, \end{aligned}$$

where $F_0(x)$ is given by Eq. (7).

4 Conclusion

We provide explicit analytical expressions for the joint probability of the number of the customer in the background queueing model and the content of the buffer under steady state in terms of modified Bessel function of first kind. Also, the stationary buffer content distribution is obtained in closed form and is shown to coincide with the result of [8] as a special case when the catastrophe parameter γ equals to zero. To the best of our knowledge, the present paper is the first of its kind to analyse fluid models driven by queues subject to disasters, although earlier several authors have studied fluid queues modulated by vacation queueing models. Further extension to the present work can possibly include to analyse the model in time-dependent regime and also to study the process by considering the repair time which arise due to catastrophe.

References

1. X. Chao, Y. Zheng, "Transient analysis of immigration-birth-death process with total catastrophes", *Probability in the Engineering and Informational Sciences*, Vol. 17, pp. 83-106, 2003.
2. A. Di Crescenzo, V. Giorno, A.G. Nobile, L.M. Ricciardi, "On the M/M/1 queue with catastrophes and its continuous approximation", *Queueing Systems*, Vol. 43, pp. 329-347, 2003.
3. V.G. Kulkarni, "Fluid models for single buffer systems", In. J.H. Dshalalow (ed.) *Frontiers in Queueing*, CRC Press, Boca Raton, pp. 321-338, 1997.
4. G. Latouche and P.G. Taylor, "A stochastic fluid model for an ad hoc mobile network", *Queueing Systems*, Vol. 63, pp. 109-129, 2009.
5. B. Mao, F. Wang and N. Tian, "Fluid model driven by an *M*/*M*/1/*N* queue with single exponential vacation", *International Journal of Information and Management Science*, Vol. 21, pp. 29-40, 2010.
6. B. Mao, F. Wang and N. Tian, "Fluid model driven by an *M*/*M*/1/*N* queue with multiple exponential vacations", *Journal of Computational Information Systems*, Vol. 6, pp. 1809-1816, 2010.
7. D. Mitra, "Stochastic theory of a fluid model of producers and consumers couple by a buffer" *Advance in Applied Probability*, Vol. 20, pp. 646-676, 1988.
8. P.R. Parthasarathy, K.V. Vijayashree and R.B. Lenin, "An *M*/*M*/1 driven fluid queue - continued fraction approach", *Queueing Systems*, Vol. 42, pp. 189-199, 2002.
9. Randolph Nelson, "Probability, Stochastic Processes and Queueing Theory, The Mathematics of computer performance modeling", *Springer Verlag*, 1995.
10. J. Ren, "Perturbed risk processes analyzed as fluid flows", *Stochastic Models*, Vol. 25, pp. 522-544, 2009.
11. A. Silver Soares and G. Latouche, "Matrix analytic methods for fluid queues with finite buffers", *Performance Evaluation*, Vol. 63, pp. 295-314, 2006.
12. D. Stirzaker, "Processes with catastrophes", *Mathematical Scientist*, Vol. 31, pp. 107-118, 2006.
13. R.J. Swift, "Transient probabilities for a simple birth–death–immigration process under the influence of total catastrophes", *International Journal of Mathematics and Mathematical Sciences*, Vol. 25, pp. 689-692, 2001.
14. M. Veatch and J. Senning, "Fluid analysis of an input control problem", *Queueing Systems*, Vol. 61, pp. 87-112, 2009.

Fuzzy VEISV Epidemic Propagation Modeling for Network Worm Attack

Muthukrishnan Senthil Kumar and C. Veeramani

Abstract An epidemic vulnerable—exposed—infectious—secured—vulnerable (VEISV) model for the fuzzy propagation of worms in computer network is formulated. In this paper, the comparison between classical basic reproduction number and fuzzy basic reproduction number is analyzed. Epidemic control strategies of worms in the computer network—low, medium, and high—are analyzed. Numerical illustration is provided to simulate and solve the set of equations.

Keywords Epidemic threshold · Reproduction number · Fuzzy logic · Worm propagation

1 Introduction

Recent myriad research contributions in Internet technology are considered to secure the worm attacks and to analyze the dynamics of worms in network. To study the dynamics of worms in network, a mathematical model is to be developed to analyze the propagation of worms. Worms behave like infectious diseases and are epidemic in nature. The propagation of worms throughout a network can be studied by using epidemiological models for disease propagation [1–4]. Using Kermack and McKendrick SIR classical epidemic model [5–7], dynamical models for malicious object propagation were proposed, providing estimations for temporal evolution of nodes depending on network parameters considering topological

M. Senthil Kumar (✉) · C. Veeramani
Department of Applied Mathematics and Computational Sciences, PSG College of Technology, Coimbatore, Tamil Nadu 641004, India
e-mail: ms_kumar_in@yahoo.com

C. Veeramani
e-mail: veerasworld@yahoo.com

G. S. S. Krishnan et al. (eds.), *Computational Intelligence, Cyber Security and Computational Models*, Advances in Intelligent Systems and Computing 246,
DOI: 10.1007/978-81-322-1680-3_32,

aspects of the network [1–4]. The similarity between the spread of a biological virus and malicious worm propagation motivates the researchers to adopt an epidemic model to the network environment [8]. Recent research in epidemic models such as SIR [8–10], SIS [8], SEIR [1, 11–14], SIRS [15, 16], SEIQV [17], and vulnerable—exposed—infectious—secured—vulnerable (VEISV) [18] is proposed to study the worm propagation by developing different transaction states based on the behavior of the virus or the worm.

In particular, epidemic systems in computer networks have strong nonlinearity and should be treated in a different way. The nonlinearity is due to the force of epidemic of an infectious agent. This intrinsically includes the fuzzy logic analysis. Fuzzy epidemic modeling for human infectious diseases has been studied in many research contributions [19–22]. Recently, Mishra and Pandey [23] proposed fuzzy epidemic model for the transmission of worms in computer network. This motivated us to consider the fuzzy epidemic model for VEISV propagation for network worm attack. Our model generalizes Mishra and Pandey [23] work.

2 Classical VEISV Epidemic Model

Toutonji et al. [18] proposed VEISV epidemic model for security countermeasures that have been used to prevent and defend against worm attacks. Thus, they used the state name *vulnerable* → *exposed* → *infectious* → *secured* → *vulnerable*. The parameters and notations used in this model are given with explanation in Table 1. The schematic representation of this model is shown in Fig. 1. The vulnerable state includes all hosts which are vulnerable to worm attack. Exposed state includes all hosts which are exposed to attack but not infectious due to the latent time requirement. Infectious state includes all hosts which were attacked and actively scanning and targeting new victims. Secured state includes all hosts which gained one or more security countermeasures, providing the host with a temporary or permanent immunity against the malicious worm. The following assumptions are considered:

1. The total number of hosts N is fixed and defined by

$$N = V(t) + E(t) + I(t) + S(t). \tag{1}$$

2. Initially, all hosts are vulnerable to attack. The total number of quarantined hosts, without considering the quarantine time, will move to the secure state after installing the required security patches or updates.
3. The number of replaced hosts is equal to the number of dysfunctional hosts, and the model is closed network defined as

Table 1 Notation and parameters for VEISV model

Notation	Explanation
$V(t)$	Number of vulnerable hosts at time t
$E(t)$	Number of exposed hosts at time t
$I(t)$	Number of infectious hosts at time t
$S(t)$	Number of secured hosts at time t
β	Contact rate
α	State transition rate from E to I
ψ_1	State transition rate from V to S
ψ_2	State transition rate from E to S
γ	State transition rate from I to S
ϕ	State transition rate from S to V
N	Total number of hosts
θ	Dysfunctional rate
μ_1	Replacement rate

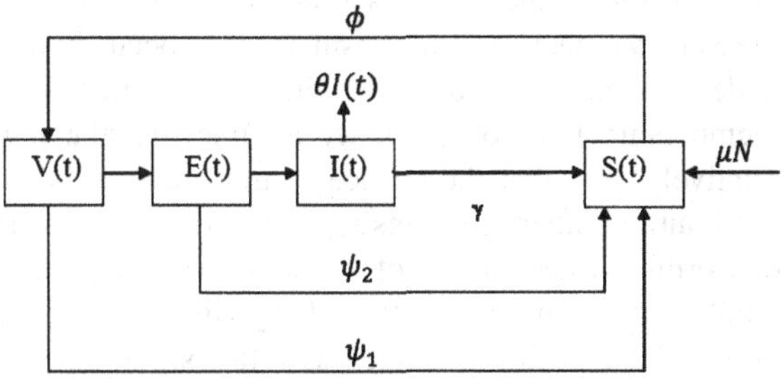

Fig. 1 VEISV model

$$\Gamma = \{(V, E, I, S) \in R_+^4 | V + E + I + S = N\}. \tag{2}$$

Since the number of hosts is large, we defined the incident of infection as $\beta \frac{V(t)}{N} I(t)$. β represents the number of incidents occurring in a unit of time. The transition of hosts from V state to E state in terms of Δt is

$$\Delta VE = \beta \frac{V(t)}{N} I(t) \Delta t. \tag{3}$$

Since $\alpha E(t)$ is the number of transitioning vulnerable hosts from time t to $(t + \Delta t)$ by the following equation:

$$V(t + \Delta t) - V(t) = -fE(t)V(t)\Delta t - \psi_1 V(t)\Delta t + \phi S(t). \tag{4}$$

We followed the mathematical approach [18] to derive the theoretical part. This set of differential equations governs the VEISV model:

$$
\begin{aligned}
\frac{dV}{dt} &= -fEV - \psi_1 V + \phi S, \\
\frac{dE}{dt} &= fEV - (\alpha + \psi_2)E, \\
\frac{dI}{dt} &= \alpha E - (\gamma + \theta)I, \\
\frac{dS}{dt} &= \mu_1 N + \psi_1 V + \psi_2 E + \gamma I - \phi S.
\end{aligned} \tag{5}
$$

3 Fuzzy VEISV Epidemic Model

During a worm attack, dysfunction occurred in infectious state; thereby, the hosts are taken over by a worm and are not capable of performing properly. The higher the worm load, the higher will be the chance of worm transmission. Let $\beta = \beta(x)$ measure the chance of a transmission to occur in a meeting between a vulnerable node and an exposed node with a large number of worms x. To obtain the membership function $\beta(x)$, we assume that the number of worms in a node is relatively low, that the chance of transmission is negligible, and that there are a minimum number of worms $x_{\min}$ needed to cause transmission. For certain number of worms x_M, the chance of transmission is maximum and equal to 1. Further, the number of worms in a node is always limited to $x_{\max}$. So the membership function of β (refer Fig. 2a) is defined as follows:

$$
\beta(x) = \begin{bmatrix} 0 & \text{if } x < x_{\min} \\ \frac{x - x_{\min}}{x_M - x_{\min}} & \text{if } x_{\min} < x < x_M \\ 1 & \text{if } x_M < x < x_{\max}. \end{bmatrix} \tag{6}
$$

To obtain the membership function of α, in latent period we assume that the number of worms exposed in a node is relatively low, that the chance of transmission is negligible, and that there are a minimum number of worms $x_{\min}$ needed

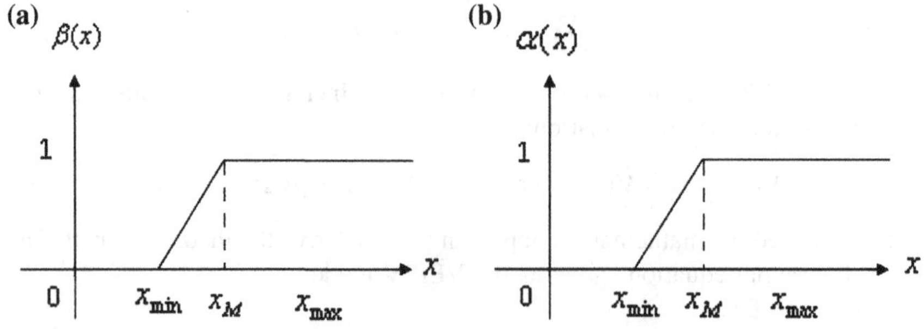

Fig. 2 **a** Membership function of β and **b** membership function of α

to cause transmission. For certain number of worms x_M, the chance of transmission is maximum and equal to 1. So the membership function of α (refer Fig. 2b) is defined as follows:

$$\alpha(x) = \begin{cases} 0 & \text{if } x < x_{\min} \\ \frac{x - x_{\min}}{x_M - x_{\min}} & \text{if } x_{\min} < x < x_M \\ 1 & \text{if } x_M < x < x_{\max}. \end{cases} \tag{7}$$

Now, the vulnerable node's recovery rate $\psi_1 = \psi_1(x)$ is also a function of worm load. The higher the worm load, the longer it will take to recover from infection; i.e., ψ_1 should be decreasing function of x. The membership function of ψ_1 (refer Fig. 3a) is as follows:

$$\psi_1(x) = \frac{\psi_{10} - 1}{x_{\max}} x + 1 \tag{8}$$

where ψ_{10} is the lowest recovery rate from vulnerable state to secured state. Also, the exposed node's recovery rate $\psi_2 = \psi_2(x)$ is also a function of worm load. The higher the worm load, the longer it will take to recover from infection; i.e., ψ_2 should be decreasing function of x. The membership function of ψ_2 (refer Fig. 3b) is as follows:

$$\psi_2(x) = \frac{\psi_{20} - 1}{x_{\max}} x + 1 \tag{9}$$

where ψ_{20} is the lowest recovery rate from exposed state to secured state. Furthermore, the infected node's recovery rate $\gamma = \gamma(x)$ is also a function of worm load. The higher the worm load, the longer it will take to recover from infection; i.e., γ should be decreasing function of x. The membership function of γ (refer Fig. 4a) is as follows:

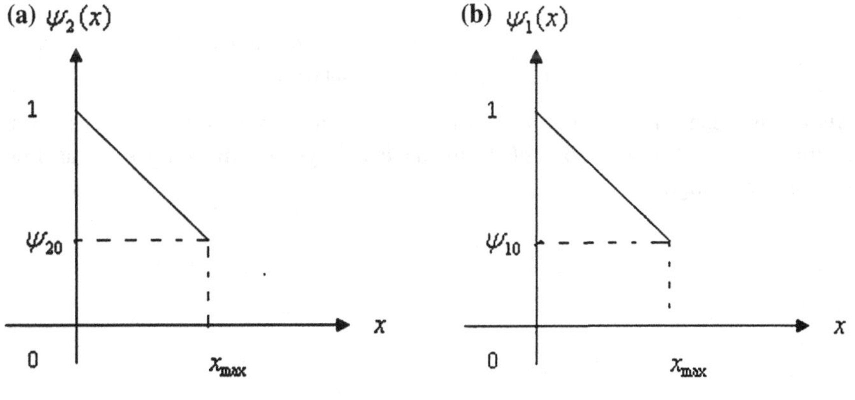

Fig. 3 **a** Membership function of ψ_1 and **b** membership function of ψ_2

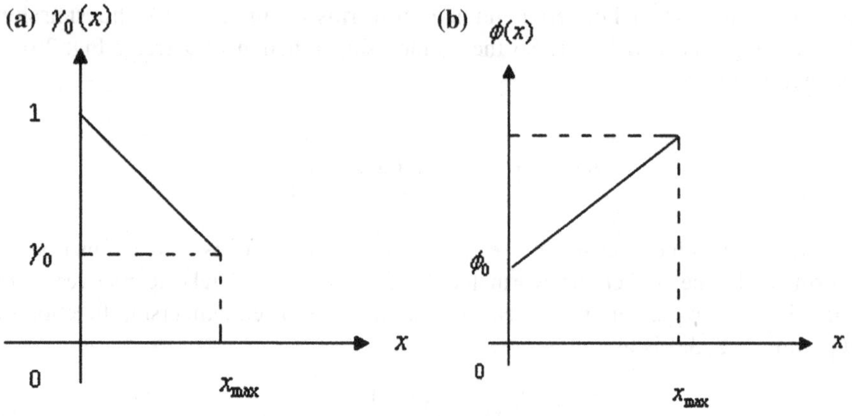

Fig. 4 **a** Membership function of γ and **b** membership function of ϕ

$$\gamma(x) = \frac{\gamma_0 - 1}{x_{\max}} x + 1 \tag{10}$$

where γ_0 is the lowest recovery rate from infected state to secured state. Here, ϕ is the rate of vulnerable after recovery, that is, the secured nodes may be vulnerable again. The higher we use the secondary devices and/or Internet services, the higher it will be vulnerable after recovery. The membership function of ϕ (refer Fig. 4b) will be increasing function of x. It is defined as follows:

$$\phi(x) = \frac{1 - \phi_0}{x_{\max}} x \tag{11}$$

where $\phi_0 > 0$ and (<1) is the lowest vulnerability after recovery. The number of worms differs in different nodes of the computer network. So we assume that x can be seen as fuzzy number, and hence, the membership function is defined as follows:

$$\rho(x) = \begin{bmatrix} 1 - \frac{|x - \bar{x}|}{\delta} & \text{if } x \in [\bar{x} - \delta, \bar{x} + \delta] \\ 0 & \text{otherwise} \end{bmatrix} \tag{12}$$

where $\bar{x}$ is a central value and δ gives the dispersion of each node of the fuzzy sets assumed by x. For a fixed $\bar{x}$,$\rho(x)$ can have a linguistic meaning such as low, medium, and high.

4 Stability Analysis of Fuzzy VEISV Model

Since $S(t) = N - V(t) - E(t) - I(t)$, we can use the reduction method by considering only the first three Eqs. of (5) to analyze the model

$$\begin{aligned} \frac{dV}{dt} &= \phi N - fEV - (\psi_1 + \phi)V - \phi(E + I), \\ \frac{dE}{dt} &= fEV - (\alpha + \psi_2)E, \\ \frac{dI}{dt} &= \alpha E - (\gamma + \theta)I. \end{aligned} \tag{13}$$

Then, for the equilibrium points, we take $\frac{dv}{dt} = 0, \frac{dE}{dt} = 0$, and $\frac{dI}{dt} = 0$. For $\frac{dE}{dt} = 0$, the equilibrium occurs at

$$E^* = 0 \text{ or } E^* > 0 \text{ and } v^* = \frac{\alpha + \psi_2}{\beta\alpha} N. \tag{14}$$

For $E^* = 0$, the worm-free equilibrium occurs at

$$\Pi_{wf} = (v_1^*, E_1^*, I_1^*) = \left(\frac{\phi}{\psi_1 + \phi} N, 0, 0 \right). \tag{15}$$

For $E^* > 0$, the worm-epidemic equilibrium is

$$\begin{aligned} \Pi_{we} &= (v_2^*, E_2^*, I_2^*) \\ &= \left(\frac{\alpha + \psi_2}{\beta\alpha} N, \frac{\phi - \frac{(\alpha+\psi_2)}{\beta\alpha}(\psi_1 - \phi)}{\alpha + \psi_2 + \phi\left(1 + \frac{\alpha}{\gamma+\theta}\right)} N, \frac{\alpha}{\gamma + \theta} E_2^* \right). \end{aligned} \tag{16}$$

Now, taking into account worm load, we have

$$\Pi_{we} = \left(\frac{\alpha(x) + \psi_2(x)}{\beta(x)\alpha(x)} N, \frac{\phi(x) - \frac{(\alpha(x)+\psi_2(x))}{\beta(x)\alpha(x)}(\psi_1(x) - \phi(x))}{\alpha(x) + \psi_2(x) + \phi(x)\left(1 + \frac{\alpha(x)}{\gamma(x)+\theta}\right)} N, \frac{\alpha(x)}{\gamma(x) + \theta} E_2^* \right). \tag{17}$$

As $\frac{f_1(x)N}{f_2(x)} < 1$, where $f_1(x) = \alpha(x)\phi(x) - \frac{\alpha(x)+\psi_2(x)}{\beta(x)}(\psi_1(x) - \phi(x))$ and $f_2(x) = (\gamma(x) + \theta)\left(\alpha(x) + \psi_2(x) + \phi(x)\left(1 + \frac{\alpha(x)}{\gamma(x)+\theta}\right)\right)$, so a value of bifurcation for x is x^*, the solution of the equation $f_1(x)N = f_2(x)$.

5 Basic Reproduction Number

The basic reproduction number (R_0) is obtained through the analysis of the stability of the trivial equilibrium point. For the classical VEISV propagation model, $R_0 = \frac{\alpha\beta\phi}{(\psi_1+\phi)(\alpha+\psi_2)}$. Based on the definition of R_0, the worm-free equilibrium is locally asymptotically stable when $R_0 \leq 1$ and unstable when $R_0 \geq 1$. As in this case, we taken $\alpha = \alpha(x)$, $\beta = \beta(x)$, $\phi = \phi(x)$, $\psi_1 = \psi_1(x)$, and $\psi_2 = \psi_2(x)$, then we write, $R_0(x) = \frac{\alpha(x)\beta(x)\phi(x)}{(\psi_1(x)+\phi(x))(\alpha(x)+\psi_2(x))}$.

We consider max$R_0(x) < 1$ to control the worm transmission. But we take an average value of $R_0(x)$ because it can be an extreme attitude. For this, we consider the distribution of the worm load as given by a triangular fuzzy number $\rho(x)$. Then, fuzzy basic reproduction number is defined as follows:

$$R_0^f = \frac{1}{\gamma_0}\text{FEV}[\gamma_0 R_0(x)] \tag{18}$$

where FEV is fuzzy expected value. Suppose that $R_0(x) > 1$, but $\gamma_0 R_0(x) \leq 1$, so that the value of R_0^f is well defined. This is defined as the average number of secondary cases by just one infected node introduced into entirely susceptible nodes.

To obtain FEV$[\gamma_0 R_0(x)]$, we define a fuzzy measure μ and use the possibility measure:

$$\mu(A) = \sup_{x\in A}\rho(x), \quad A \subset R.$$

This measure shows that the infectivity of a group is the one presented by the node belonging to the group with the maximal infectivity. R_0^f is estimated by assuming that the number of worms classified as low, medium, and high. The fuzzy set is given by the membership function $\rho(x)$ for different cases:

(i) low, if $\bar{x} + \delta < x_{\min}$,
(ii) medium, if $\bar{x} - \delta > x_{\min}$ and $\bar{x} + \delta < x_M$ and
(iii) high, if $\bar{x} - \delta > x_M$.

Case (i): It is noted that $R_0^f < 1$ if x is low.

Case (ii): Since $R_0(x) = \frac{\alpha(x)\beta(x)\phi(x)}{(\psi_1(x)+\phi(x))(\alpha(x)+\psi_2(x))}$ is an increasing function of x, $H(\lambda) = [x', x_{\max}] = \sup_{x' \leq x \leq x_{\max}}\rho(x)$. Here $(\lambda) = \mu\{I(x,t) \geq \lambda\}$, FEV$[I(x,t)]$ is the fixed point of $H(\lambda)$ and x' is the solution of the equation $\gamma_0 \frac{\alpha(x)\beta(x)\phi(x)}{(\psi_1(x)+\phi(x))(\alpha(x)+\psi_2(x))} = \lambda$. Since the fixed point of $H(\lambda)$ is same as FEV$[\gamma_0 R_0^f(x)]$.
Hence,

$$H(\lambda) = \begin{cases} 0 & \text{if } 0 \le \lambda \le \gamma_0 \frac{\alpha(\overline{x})\beta(\overline{x})\phi(\overline{x})}{(\psi_1(\overline{x})+\phi(\overline{x}))(\alpha(\overline{x})+\psi_2(\overline{x}))} \\ \rho(x') & \text{if } \gamma_0 \frac{\alpha(\overline{x})\beta(\overline{x})\phi(\overline{x})}{(\psi_1(\overline{x})+\phi(\overline{x}))(\alpha(\overline{x})+\psi_2(\overline{x}))} \le \lambda \\ & \le \gamma_0 \frac{\alpha(\overline{x}+\delta)\beta(\overline{x}+\delta)\phi(\overline{x}+\delta)}{(\psi_1(\overline{x}+\delta)+\phi(\overline{x}+\delta))(\alpha(\overline{x}+\delta)+\psi_2(\overline{x}+\delta))} \\ 0 & \text{if } \gamma_0 \frac{\alpha(\overline{x}+\delta)\beta(\overline{x}+\delta)\phi(\overline{x}+\delta)}{(\psi_1(\overline{x}+\delta)+\phi(\overline{x}+\delta))(\alpha(\overline{x}+\delta)+\psi_2(\overline{x}+\delta))} < \lambda \le 1. \end{cases} \quad (19)$$

For $\delta > 0$, H is a continuous and decreasing function, and in this case, $\text{FEV}[\gamma_0 R_0(x)]$ is equal to fixed point of H. By direct manipulation, we have

$$\frac{\alpha(\overline{x})\beta(\overline{x})\phi(\overline{x})}{(\psi_1(\overline{x})+\phi(\overline{x}))(\alpha(\overline{x})+\psi_2(\overline{x}))} < \frac{\text{FEV}[\gamma_0 R_0(x)]}{\gamma_0} < \frac{\alpha(\overline{x}+\delta)\beta(\overline{x}+\delta)\phi(\overline{x}+\delta)}{(\psi_1(\overline{x}+\delta)+\phi(\overline{x}+\delta))(\alpha(\overline{x}+\delta)+\psi_2(\overline{x}+\delta))}$$

So, $R_0(\overline{x}) < R_0^f < R_0(\overline{x}+\delta)$.

Case (iii): From the previous case, we have

$$\frac{1}{(\psi_1(\overline{x})+\phi(\overline{x}))(\alpha(\overline{x})+\psi_2(\overline{x}))} < R_0^f < \frac{1}{(\psi_1(\overline{x}+\delta)+\phi(\overline{x}+\delta))(\alpha(\overline{x}+\delta)+\psi_2(\overline{x}+\delta))}$$

It guarantees that the worms invade since $R_0^f > 1$.

6 Comparison Between R_0 and R_0^f

Here, we have analyzed the three cases discussed in the previous section related to the three classifications for the number of infections: low, medium, and high worm load. In any of the three cases, we have

$$\frac{\alpha(\overline{x})\beta(\overline{x})\phi(\overline{x})}{(\psi_1(\overline{x})+\phi(\overline{x}))(\alpha(\overline{x})+\psi_2(\overline{x}))} < \frac{\text{FEV}[\gamma_0 R_0(x)]}{\gamma_0} < \frac{\alpha(\overline{x}+\delta)\beta(\overline{x}+\delta)\phi(\overline{x}+\delta)}{(\psi_1(\overline{x}+\delta)+\phi(\overline{x}+\delta))(\alpha(\overline{x}+\delta)+\psi_2(\overline{x}+\delta))}$$

i.e., $R_0(\overline{x}) < R_0^f < R_0(\overline{x}+\delta)$. Since the function $R_0 = \frac{\alpha(\overline{x})\beta(\overline{x})\phi(\overline{x})}{(\psi_1(\overline{x})+\phi(\overline{x}))(\alpha(\overline{x})+\psi_2(\overline{x}))}$ is continuous, by intermediate mean value theorem we have $\tilde{x}$ such that $\overline{x} < \tilde{x} < \overline{x}+\delta$. So,

$$R_0^f = R_0(\tilde{x}) > R_0(\overline{(x)}) \quad (20)$$

It means that R_0^f (fuzzy) and R_0 (classical) coincide if the number of infection is $\tilde{x}$.

7 Concluding Remarks

In summary, this paper describes an epidemic VEISV model for the fuzzy propagation of worms in computer network. The comparison between R_0 and R_0^f notices the importance of fuzzy logic in worm propagation. In future, fuzzy-based analysis of epidemic processes on large networks paves the way to more realistic models.

References

1. Bimal Kumar Mishra, D.K. Saini, SEIRS epidemic model with delay for transmission of malicious objects in computer network, Appl. Math. Comput. 188 (2) (2007) 1476–1482.
2. Bimal Kumar Mishra, Dinesh Saini, Mathematical models on computer viruses, Appl. Math. Comput. 187 (2) (2007) 929–936.
3. Bimal Kumar Mishra, Navnit Jha, Fixed period of temporary immunity after run of anti-malicious software on computer nodes, Appl. Math. Comput.190 (2) (2007) 1207–1212.
4. E. Gelenbe, Dealing with software viruses: a biological paradigm, Inform. Secur. Tech. Rep. 12 (4) (2007) 242–250.
5. W.O. Kermack, A.G. McKendrick, Contributions of mathematical theory to epidemics, Proc. R. Soc. Lond. Ser. A 115 (1927) 700–721.
6. W.O. Kermack, A.G. McKendrick, Contributions of mathematical theory to epidemics, Proc. R. Soc. Lond. Ser. A 138 (1932) 55–83.
7. W.O. Kermack, A.G. McKendrick, Contributions of mathematical theory to epidemics, Proc. R. Soc. Lond. Ser. A 141 (1933) 94–122.
8. J. Kim, S. Radhakrishana, J. Jang, Cost optimization in SIS model of worm infection, ETRI J. 28 (5) (2006) 692–695.
9. H. Zhou, Y. Wen, H. Zhao, Modeling and analysis of active benign worms and hybrid benign worms containing the spread of worms, in: Proceedings of the IEEE International Conference on Networking (ICN07), 2007.
10. M.H.R. Khouzani, S. Sarkar, E. Altman, Maximum damage malware attack in mobile wireless networks, in: IEEE Proceedings, INFOCOM10, 1419 Mar.2010, pp. 1–9.
11. X.Z. Li, L.L. Zhou, Global stability of an SEIR epidemic model with vertical transmission and saturating contact rate, Chaos Soliton. Fract. 40 (2007) 874–884.
12. G. Li, J. Zhen, Global stability of an SEI epidemic model with general contact rate, Chaos Soliton. Fract. 23 (2004) 997–1004.
13. B.K. Mishra, N. Jha, SEIQRS model for the transmission of malicious objects in computer network, Appl. Math. Modell. 34 (2009) 1207–1212.
14. N. Yi, Q. Zhang, K. Mao, D. Yang, Q. Li, Analysis and control of an SEIR epidemic system with nonlinear transmission rate, Math. Comput. Modell. 50(2009) 1498–1513.
15. Y. Jin, W. Wang, S. Xiao, An SIRS model with a nonlinear incidence rate, Chaos Soliton. Fract. 34 (2007) 1482–1497.
16. Q. Liu, R. Xu, S. Wang, Modelling and analysis of an SIRS model for worm propagation, in: Proceedings of the International Conference Computational Intelligence and Security, CIS 09, vol. 2, 1114 Dec. 2009, pp. 361–365.
17. F. Wang, Y. Zhang, C. Wang, J. Ma, S. Moon, stability analysis of SEIQV epidemic model for rapid spreading worms, Comput. Secur. 29 (2010) 410–418.
18. Ossama A. Toutonji, Seong-Moo Yoo, Moongyu Park, Stability analysis of VEISV propagation modeling for network worm attack, Appl. Math. Model. 36 (2012) 2751–2761.

19. E. Massad, M.N. Burattini, N.R.S. Ortega, Fuzzy logic and measles vaccination: designing a control strategy, Int. J. Epidemiol. 28 (3) (1999) 550–557.
20. N.R.S. Ortega, P.C. Sallum, E. Massad, Fuzzy dynamical systems in epidemic modelling, Kybernetes 29 (12) (2000) 201–218.
21. E. Massad, et al., Fuzzy Logic in Action: Applications and Epidemiology and Beyond, in: STUDFUZZ, vol. 232, Springer-Verlag, Berlin, Heidelberg, 2008.
22. L.C. Barros, R.C. Bassanezi, M.B.F. Leite, The epidemiological models SI with a fuzzy transmission, Comput. Math. Appl. 45 (2003) 1619–1628.
23. B K Mishra and S K Pandey, Fuzzy epidemic model for the transmission of worms in computer network,Nonlinear Analysis: Real World Applications 11 (2010) 4335–4341.

Hexagonal Prusa Grammar Model for Context-Free Hexagonal Picture Languages

T. Kamaraj and D. G. Thomas

Abstract Prusa Grammar is a recently introduced rectangular picture languages generating model which exploits the parallel application of two-dimensional context-free rules. We introduce the hexagonal version of Prusa grammar and generate images. We compare this model with other hexagonal array generating devices for the description of its generative power.

Keywords Hexagonal array languages · Hexagonal tiling system · Regional hexagonal tile rewriting grammars · Prusa grammars

1 Introduction

Hexagonal arrays generated by grammars are found in the literature with the insight into computations in pattern recognition, picture processing, and scene analysis [4, 6]. Some of classical formalisms to generate hexagonal arrays are hexagonal kolam array grammars (HKAG) [6] and hexagonal array rewriting grammars (HAG) [7]. In HKAG model, hexagonal arrays on triangular grid were viewed as two-dimensional representation of three-dimensional blocks, and also several scene analytic transformations were discussed. Hexagonal array rewriting grammars (HAG) are the generalization for HKAG. Hexagonal Wang system (HWS) and hexagonal tiling system (HTS) were an equivalent pair of formal devices for recognizable class of hexagonal arrays (HREC) [1]. Hexagonal tile

T. Kamaraj (✉)
Department of Mathematics, Sathyabama University, Chennai, Tamil Nadu 600119, India
e-mail: kamaraj_mx@yahoo.co.in

D. G. Thomas
Department of Mathematics, Madras Christian College, Chennai, Tamil Nadu 600059, India
e-mail: dgthomasmcc@yahoo.com

G. S. S. Krishnan et al. (eds.), *Computational Intelligence, Cyber Security and Computational Models*, Advances in Intelligent Systems and Computing 246,
DOI: 10.1007/978-81-322-1680-3_33,

rewriting grammars (HTRG) [8] and regional hexagonal tile rewriting grammars (RHTRG) [2] are the recent tiling-based hexagonal array rewriting models, which have more generative capacity. In the generalization process of rectangular grammar models, Prusa [5] has recently defined a grammar device which extended the rectangular context-free kolam array grammar model, attaining some generative capacity. This model generates the rectangular pictures by parallel application of rules and thereby maintains a grid-like structure in each stage of derivation of picture. But hexagonal grammar models with parallel derivation rules are very rare in literature and that too have limited growth patterns. With this quest of generalization of hexagonal models, we propose a hexagonal version of Prusa grammar to generate context-free hexagonal picture languages and study some comparisons with the other existing models for their generative capacity.

The paper is organized in the following manner. In Sect. 2, we recall some basic notions of hexagonal pictures and languages. In Sect. 3, we introduce hexagonal Prusa grammar (HPG) and examples for illustration. In Sect. 4, we present comparison results of HPG with other hexagonal models with respect to the generating capacity.

2 Preliminaries

Let T be a finite alphabet of symbols. A hexagonal picture p over T is a hexagonal array of symbols of T.

With respect to a triad 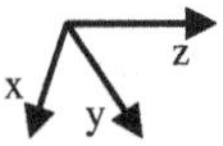

of triangular axes x, y, z, the coordinates of each element of hexagonal picture can be fixed [1]. The origin of reference is the upper left vertex, having co-ordinates (1, 1, 1).

The set of all hexagonal arrays over of the alphabet T is denoted by T^{**H} and set of all non-empty hexagonal arrays over T is denoted by T^{++H}. T^+ denotes set of all non-empty strings in the three directions parallel to the triangular axes. A non-empty hexagonal picture language L over T is a subset of T^{++H}.

For $p \in T^{++H}$, let $\hat{p}$ be the hexagonal array obtained by surrounding p with a special boundary symbol $\# \notin T$.

Given a picture $p \in T^{++H}$, if ℓ, m, n denote the number of elements in the borders of p, parallel to x-, y-, z-directions, respectively, then the triple (ℓ, m, n) is called the size of the picture p denoted by $|p| = (\ell, m, n)$. Let p_{ijk} denotes the symbol in p with coordinates (i, j, k) where $1 \leq i \leq \ell$, $1 \leq j \leq m$, $1 \leq k \leq n$. Let $T^{(\ell,\ m,\ n)H}$ be the set of hexagonal pictures of size (ℓ, m, n). A typical hexagonal array of size (ℓ, m, n) can be denoted by $[p_{ijk}]^{(\ell,m,n)H}$. A hexagonal picture of size (2, 2, 2) is called a hexagonal tile. We denote the set of all hexagonal tiles contained in a picture $\hat{p}$ by $[[\hat{p}]]$.

The notions of non-convex hexagonal arrays called arrowheads, and arrowhead catenations in six directions are adapted as in [1, 6].

3 Hexagonal Prusa Grammar

In this section, we give a formal definition of hexagonal Prusa grammar and some simple examples of languages generated by these grammars.

Definition 1 A Hexagonal Prusa Grammar (HPG) is a tuple $\langle N, T, P, S\rangle$ where N is a finite set of non-terminals, T is a finite set of terminals, $P \subseteq N \times [(N \cup T)^{++H} \cup (N \cup T)^{+}]$ and $S \in N$ is the start symbol.

Definition 2 Let $G = \langle N, T, P, S\rangle$ be a HPG. We define a hexagonal picture language $L(G, C)$ over T for every $C \in N$, by the following recursive rules.

1. Terminal rule: If $C \to X$ is in P, and $X \in (T^{++H} \cup T^{+})$, then $X \in L(G, C)$.
2. Mixed rule: Let $C \to X$ be a production in P with

$$X \in \cup\, (N \cup T)^{(\ell', m', n')H},\ 1 \le \ell' \le \ell,\ 1 \le m' \le m \text{ and } 1 \le n' \le n$$

and Q_{ijk} ($1 \le i \le \ell$, $1 \le j \le m$, $1 \le k \le n$) be the pictures such that

(a) if $X_{ijk} \in T$, then $Q_{ijk} = X_{ijk}$.
(b) if $X_{ijk} \in N$, then $Q_{ijk} \in L[G, X]$.

Then, if $Q = [Q_{ijk}]^{(\ell', m', n')H}$ is defined through string catenation (or) arrowhead catenation, then $Q \in L[G, C]$.

The set $L[G, C]$ contains all and only pictures that can be obtained by applying a finite sequence of rules (1) and (2). The hexagonal language $L[G]$ generated by the grammar G is defined to be the language $L[G, S]$. $\mathcal{L}(\mathrm{HPG})$ is the class of all languages generated by these grammars. Languages in $\mathcal{L}(\mathrm{HPG})$ are called HPG languages.

Example 1 The language

$$L_1 = \left\{ \begin{array}{ccccc} & a & & a & \\ a & & a & & a \\ & a & & a & \end{array},\ \begin{array}{ccccccc} & a & & a & & a & \\ a & & a & & a & & a \\ & a & & a & & a & \end{array},\ \begin{array}{ccccccccc} & a & & a & & a & & a & \\ a & & a & & a & & a & & a \\ & a & & a & & a & & a & \end{array}, \ldots \right\}$$

is generated by HPG $G_1 = \langle N, T, P, S\rangle$ where $N = \{H, S\}$, $T = \{a\}$ and

$$P = \left\{ S \to SH,\quad S \to \begin{array}{ccccc} & a & & a & \\ a & & a & & a \\ & a & & a & \end{array},\quad H \to \begin{array}{cc} a & \\ & a \\ a & \end{array} \right\}.$$ By applying terminal rules, $H \to \begin{array}{cc} a & \\ & a \\ a & \end{array}$, $S \to \begin{array}{ccccc} & a & & a & \\ a & & a & & a \\ & a & & a & \end{array}$ we get $\begin{array}{cc} a & \\ & a \\ a & \end{array} \in L[G, H]$ and $\begin{array}{ccccc} & a & & a & \\ a & & a & & a \\ & a & & a & \end{array} \in L[G, S]$.

Now, by applying mixed rule $S \rightarrow SH$, we have

$$\begin{array}{ccccccc} & a & & a & & a & \\ a & & a & & a & & a \\ & a & & a & & a & \end{array} \in L[G,S].$$

The repeated application of mixed rule $S \rightarrow SH$ generates all the members of L_1.

Example 2 The language

$$L_2 = \left\{ \begin{array}{ccccc} & 1 & & 1 & \\ 1 & & 2 & & 1 \\ & 1 & & 1 & \end{array}, \begin{array}{ccccccc} & 1 & & 1 & & 1 & \\ 1 & & 2 & & 2 & & 1 \\ & 1 & & 1 & & 1 & \end{array}, \begin{array}{ccccccccc} & 1 & & 1 & & 1 & & 1 & \\ 1 & & 2 & & 2 & & 2 & & 1 \\ & 1 & & 1 & & 1 & & 1 & \end{array}, \ldots \right\}$$

can be generated by HPG, $G_2 = \langle N, T, P, S \rangle$ where $N = \{S, A, B, H\}$, and $T = \{1, 2\}$.

$$P = \left\{ S \rightarrow AH \;/\; \begin{array}{ccccc} & 1 & & 1 & \\ 1 & & 2 & & 1 \\ & 1 & & 1 & \end{array},\; A \rightarrow AB \;/\; \begin{array}{ccccc} & 1 & & 1 & \\ 1 & & 2 & & 2 \\ & 1 & & 1 & \end{array},\; B \rightarrow \begin{array}{cc} & 1 \\ 2 & \\ & 1 \end{array},\; H \rightarrow \begin{array}{cc} & 1 \\ 1 & \\ & 1 \end{array} \right\}$$

From the terminal rules, we have $\begin{array}{ccccc} & 1 & & 1 & \\ 1 & & 2 & & 1 \\ & 1 & & 1 & \end{array} \in L[G,S]$, $\begin{array}{cc} & 1 \\ 2 & \\ & 1 \end{array} \in L[G, B]$.

Parallel application of mixed rules $A \rightarrow AB$, $S \rightarrow AH$ repeatedly produces the language L_2.

We now introduce non-terminal normal form for hexagonal Prusa grammars.

Definition 3 A hexagonal Prusa grammar $G = \langle N, T, P, S \rangle$ is in non-terminal normal form (NNF) iff every rule in P has the form either $C \rightarrow t$ or $C \rightarrow X$, where $C \in N$, $X \in (N^{++H} \cup N^{+})$ and $t \in T$.

4 Comparison Results

In this section, we present comparison results of HPG with other hexagonal models with respect to its generative power. We use $\mathcal{L}[X]$ to denote the family of languages generated or recognized by the device X.

Theorem 1

$$\mathcal{L}[\text{CFHAG}] \subset \mathcal{L}[\text{HPG}].$$

Proof Consider a context-free hexagonal array grammars (CFHAG) [7] G in Chomsky normal form. It may contain six types of rules:

$C \rightarrow A \;(\leftarrow)\; B$, $C \rightarrow A \;(\nearrow)\; B$, $C \rightarrow A \;(\searrow)\; B$ and its duals.

$C \rightarrow A \;(\rightarrow)\; B$, $C \rightarrow A \;(\swarrow)\; B$ and $C \rightarrow A \;(\nwarrow)\; B$.

Now, the rule $C \rightarrow A \;(\leftarrow)\; B$ is equivalent to a HPG rule of the form $C \rightarrow$ BA.

The rule $C \rightarrow A \; \textcircled{\nearrow} \; B$ is equivalent to a HPG rule of the form $C \rightarrow A^B$.

The rule $C \rightarrow A \; \textcircled{\searrow} \; B$ is equivalent to a HPG rule of the form $C \rightarrow A_B$.

Similarly, we can have equivalent rules in HPG for the corresponding dual rules in CFHAG. The terminal rules $C \rightarrow t$ are identical in both the grammars. This shows that $\mathcal{L}[\text{CFHAG}] \subseteq \mathcal{L}[\text{HPG}]$.

The inclusion is strict as the language L_2 in Example 2 cannot be generated by any CFHAG, which was proved in [2].

Theorem 2 *$\mathcal{L}[\text{HPG}]$ and $\mathcal{L}[\text{HTS}]$ are incomparable but not disjoint.*

Proof It is already proved in [2] that the language L_1 in Example 1 is a hexagonal tiling system [1] recognizable language. So $L_1 \in \mathcal{L}(\text{HPG}) \cap \mathcal{L}(\text{HTS})$.

Now, consider the language L_4, which consists of palindromic left arrowheads over $T = \{a, b\}$.

L_4 can be generated by the HPG, $G = \langle N, T, P, S \rangle$ with $N = \{S\}$, $T = \{a, b\}$,

$$P = \left\{ S \rightarrow \begin{array}{cc} & a \\ S & \\ & a \end{array}, \; S \rightarrow \begin{array}{cc} & b \\ S & \\ & b \end{array}, \; S \rightarrow \begin{array}{cc} & a \\ a & \\ & a \end{array} \Big/ \begin{array}{cc} & a \\ b & \\ & a \end{array} \Big/ \begin{array}{cc} & b \\ a & \\ & b \end{array} \Big/ \begin{array}{cc} & b \\ b & \\ & b \end{array} \right\}$$

But there is no local strategy available to remember the characters that appearing in the corresponding positions above and below the arrowhead vertex. So L_4 cannot be generated by any HTS.

In [3], the authors have given a hexagonal tiling system to generate a language $L_{\text{parallel}} \swarrow$ which consists of all hexagonal pictures over $\Sigma = \{a, b\}$ with elements in the corresponding positions of the borders parallel to x-direction are identical. Since the productions of a HPG do not have control in producing such patterns, $L_{\text{parallel}} \swarrow$ cannot be generated by any HPG.

Theorem 3

$$\mathcal{L}(\text{HPG}) \subset \mathcal{L}[\text{RHTRG}].$$

Proof Consider a HPG grammar G in NNF. First, without loss of generality, we assume that the non-terminals in the right part of any rule of G are all different.

Let us define a RHTRG [2] G' equivalent to G. The terminal rules are easy to handle. For a non-terminal rule of G, $C \rightarrow A^B$, the equivalent rule in G' is

$$C \rightarrow \left[\left[\begin{array}{ccccccccccc}
 & & & & \# & & \# & & \# & & \\
 & & & \# & & B & & B & & \# & \\
 & & \# & & B & & B & & B & & \# \\
 & \# & & A & & A & & B & & \# & \\
\# & & A & & A & & A & & \# & & \\
 & \# & & A & & A & & \# & & & \\
 & & \# & & \# & & \# & & & &
\end{array} \right]\right]$$

Similarly, for the rules $C \rightarrow A_B$ and $C \rightarrow$ BA in G, equivalent rules can be formed in G'. Therefore, $\mathcal{L}(\text{HPG}) \subseteq \mathcal{L}(\text{RHTRG})$

For strict inclusion, consider the language L_5 over the alphabet $T = \{a, b, c, x\}$, consists of pictures of size $(2, 2, n)$, $n \geq 4$, with misaligned palindromic borders in the Z-direction, with no $(2, 2, 2)$ hexagonal subpicture contains the symbol c in both the z-directional borders. An RHTRG can be easily defined to generate L_5 with one of the variable size rules as

$$S \rightarrow \left[\left[\begin{array}{ccccccccccccccccc} & & \# & & \# & & \# & & \# & & \# & & \# & & \# & & \\ & \# & & A_1 & & A_1 & & A_1 & & A_1 & & C_1 & & C_1 & & \# & \\ \# & & X & & X & & X & & X & & X & & X & & X & & \# \\ & \# & & C_2 & & C_2 & & A_2 & & A_2 & & A_2 & & A_2 & & \# & \\ & & \# & & \# & & \# & & \# & & \# & & \# & & \# & & \end{array}\right]\right]$$

This language cannot be generated by any HPG G', as a production of G' in the form of $S \rightarrow \begin{array}{ccccc} & A & & C & \\ X & & X & & X \\ & C & & A & \end{array}$ where A generates the z-direction palindromic string, and C, X generate z-direction strings of c's and x's, respectively, cannot restrict the occurrence of c's so that we may have a hexagonal subpicture of the form $\begin{array}{ccccc} & d & & c & \\ x & & x & & x \\ & c & & d & \end{array}$ where $d \in \{a, b\}$, but which is not a picture in L_5.

5 Conclusions

HPG is the simple type of hexagonal array rewriting models with parallel application of rewriting rules in each derivation, which gives a considerable generative capacity than CFHAG. But lack of control in the production rules make this model less general than that of RHTRG. Since hexagonal tiling patterns and kolam patterns are the applications of CFHAG, HPG can also produce these patterns. Other pattern recognition tasks by this model remain to be explored and which may further depends on the development of good parsing algorithms.

References

1. K.S. Dersanambika, K. Krithivasan, C. Martin-Vide and K.G. Subramanian, Local and recognizable hexagonal picture languages, International Journal of Pattern Recognition and Artificial Intelligence, 19 (2005), 853–871.
2. T. Kamaraj and D.G. Thomas, Regional hexagonal tile rewriting grammars, In : R.P. Barneva et al (eds.), IWCIA 2012, LNCS, Vol. 7655, 181–195, Springer, Heidelberg, 2012.

3. T. Kamaraj, D.G. Thomas and T. Kalyani, Hexagonal picture recognizability by HWA, In Proc. of ICMCM 2011, Narosa Publishing, 378–388, 2012.
4. L. Middleton and J. Sivaswamy, Hexagonal image processing: A practical approach, Advances in Computer Vision and Pattern Recognition Series, Springer, 2005.
5. D. Prusa, Two-dimensional Languages, Ph.D. Thesis, 2004.
6. G. Siromoney and R. Siromomey, Hexagonal arrays and rectangular blocks, Computer Graphics and Image Processing, 5 (1976), 353–381.
7. K.G. Subramanian, Hexagonal array grammars, Computer Graphics and Image Processing, 10 (1979), 388–394.
8. D.G. Thomas, F. Sweety and T. Kalyani, Results on Hexagonal Tile Rewriting Grammars, G. Bebis et al. (Eds.), International Symposium on Visual Computing, Part II, LNCS, 5359, Springer-Verlag, Berlin, Heidelberg (2008), 945–952.

Iso-Triangular Pictures and Recognizability

V. Devi Rajaselvi, T. Kalyani and D. G. Thomas

Abstract *P* systems generating two-dimensional languages have been studied. In this paper, the variants of *P* systems that generate hexagonal pictures have been extended to iso-triangular pictures drawn on triangular grid. Local and recognizable iso-triangular picture languages have been introduced, and an approach of recognizing them by iso-triangular tiling system and automata has also been discussed.

Keywords *P* system · Iso-triangular picture languages · Tiling systems · Online tessellation automata

1 Introduction

The study of syntactic methods of describing pictures considered as connected digitized finite arrays in a two-dimensional plane has been of great interest. Formal models for describing rectangular array patterns [2], two-dimensional picture languages [6], iso-picture languages [1], and hexagonal picture languages [3] have

V. D. Rajaselvi (✉)
Sathyabama University, Chennai 600119, India
e-mail: devijerson@gmail.com

T. Kalyani
Department of Mathematics, St. Joseph's Institute of Technology,
Chennai 600119, India
e-mail: kalphd02@yahoo.com

D. G. Thomas
Department of Mathematics, Madras Christian College,
Chennai 600059, India
e-mail: dgthomasmcc@yahoo.com

G. S. S. Krishnan et al. (eds.), *Computational Intelligence, Cyber Security and Computational Models*, Advances in Intelligent Systems and Computing 246,
DOI: 10.1007/978-81-322-1680-3_34, © Springer India 2014

been studied. Motivated by these studies in this paper, we discuss the recognition of iso-triangular picture languages in triangular grid.

In Sect. 1, we have given a definition for iso-triangular picture languages and explain about triangular grid. In Sect. 2, we review the notions of local and recognizable iso-triangular picture languages inspired by the corresponding studies in picture languages. In Sect. 3, recognition of iso-triangular picture languages by iso-triangular tiling systems and two-direction iso-triangular online tessellation automata are studied. The main result of the paper is that these models describe the same class of iso-triangular picture languages. In Sect. 4, we describe how iso-triangular pictures can be generated by tissue-like *P* system with active membranes.

Membrane computing (or *P* system) is an area of theoretical computer science aiming to abstract computing models from the structure (or) functioning of living cells. In recent years, generation of two-dimensional picture languages using *P* systems is extensively studied. *P* systems generating arrays and Indian folk design (kolam patterns) are considered in [7]. Tissue-like *P* systems with active membranes that generate local picture languages are considered in [2]. Ceterchi et al. [2] introduced a variant of tissue-like *P* system to generate rectangular picture languages on rectangular grid. Dersanambika et al. [3] considered a *P* system on the outlines of [2] to generate hexagonal picture languages on triangular grid. Also Annadurai et al. [1] considered a *P* system on the outlines of [2] to generate iso-picture languages on rectangular grid.

In Sect. 4, the approach of [2] is extended to iso-triangular picture languages and analogous *P* systems for their generation on triangular grid are considered. The study exhibits the power of the novel technique used in [2] in a triangular grid.

2 Iso-triangular Pictures

In this section, we define the notion of iso-triangular picture.

Definition 1 An iso-triangular picture p over the alphabet Σ is an isosceles triangular arrangement of symbols over Σ. The set of all iso-triangular pictures over the alphabet Σ is denoted by Σ_T^{**}. An iso-triangular picture language over Σ is a subset of Σ_T^{**}. With respect to a triad of triangular axes *X*, *Y*, *Z*, the coordinates of each element of a triangular array can be fixed. Given an iso-triangular picture p, the number of rows (counting from bottom to top) denoted by $r(p)$ is the size of an iso-triangular picture. The empty picture is denoted by Λ. Given an iso-triangular picture p of size i for $k \leq i$. We denote by $B_k(p)$, the set of all iso-triangular subpictures p of size k, and $B_2(p)$ in fact an iso-triangular tile.

Definition 2 If $p \in \Sigma_T^{**}$ then $\hat{p}$ is the iso-triangular picture obtained by surrounding p with a special boundary symbol $\# \notin \Sigma$.

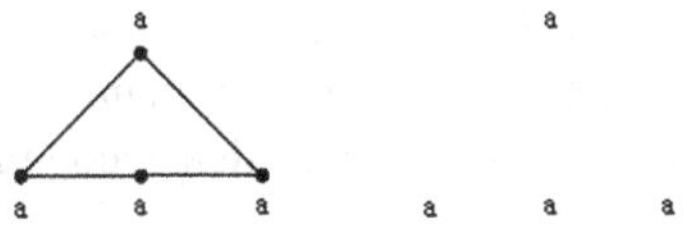

Fig. 1 Iso-triangular tile

3 Recognizability of Iso-Triangular Picture Languages

We first define an iso-triangular tile. An iso-triangular picture of the form shown in Fig. 1 is called an iso-triangular tile over the alphabet $\{a\}$.

Definition 3 Let Σ be a finite alphabet. An iso-triangular picture language $L \subseteq \Sigma_T^{**}$ is called local if there exists a finite set Δ of iso-triangular tiles over $\Sigma \cup \{\#\}$ such that $L = \left\{p \in \Sigma_T^{**} | B_2(\hat{p}) \subseteq \Delta\right\}$.

The family of local iso-triangular picture languages will be denoted by ITLOC.

Definition 4 An iso-triangular tiling system T is a 4-tuple $(\Sigma, \Gamma, \pi, \theta)$ where Σ and Γ are finite set of symbols π: $\Gamma \rightarrow \Sigma$ is a projection and θ is a set of iso-triangular tiles over the alphabet $\Gamma \cup \{\#\}$.

The iso-triangular picture language $L \subseteq \Sigma_T^{**}$ is tiling recognizable if there exists a tiling system $T = (\Sigma, \Gamma, \pi, \theta)$ such that $L = \pi(L(\theta))$. It is denoted by L(IT). The family of iso-triangular picture languages recognizable by iso-triangular tiling system is denoted by $\mathcal{L}$(ITTS). Here, we define a two-direction iso-triangular online tessellation automata referred as 2DIOTA to accept languages of iso-triangular picture languages.

Definition 5 A non-deterministic (deterministic) two-direction iso-triangular online tessellation automaton is defined by $A = (\Sigma, Q, q_0, F, \delta)$ where Σ is the input alphabet, Q is the finite set of states, $q_0 \in Q$ is the initial state, $F \subseteq Q$ is the set of final states, and δ: $Q \times Q \times \Sigma \rightarrow 2^Q$ (δ: $Q \times Q \times \Sigma \rightarrow Q$) is the transition function.

A 2DIOTA accepts the input iso-triangular picture p if there exists a run on $\hat{p}$ such that the state associated to the position $c(s, t, u)$ is a final state where $c(s, t, u)$ is the last symbol of the last row. The set of all iso-triangular pictures recognized by A is denoted by $L(A)$. Let $\mathcal{L}$(2DIOTA) be the set of iso-triangular picture languages recognized by 2DIOTA.

Theorem 1 $\mathcal{L}(\text{2DIOTA}) = \mathcal{L}(\text{ITTS})$.

To prove this theorem, we prove the following lemma.

Lemma 1 *If an iso-triangular picture language is recognized by a* 2DIOTA, *then it is recognized by a finite iso-triangular tiling system.*
i.e., $\mathcal{L}(\text{2DIOTA}) \subseteq \mathcal{L}(\text{ITTS})$.

Proof Let $L \subseteq \Sigma_T^{**}$ be a language recognized by a two-direction iso-triangular online tessellation automaton $A = (\Sigma, Q, q_0, F, \delta)$. We have to show that there exists tiling system that recognizes L. Consider the following set of tiles.

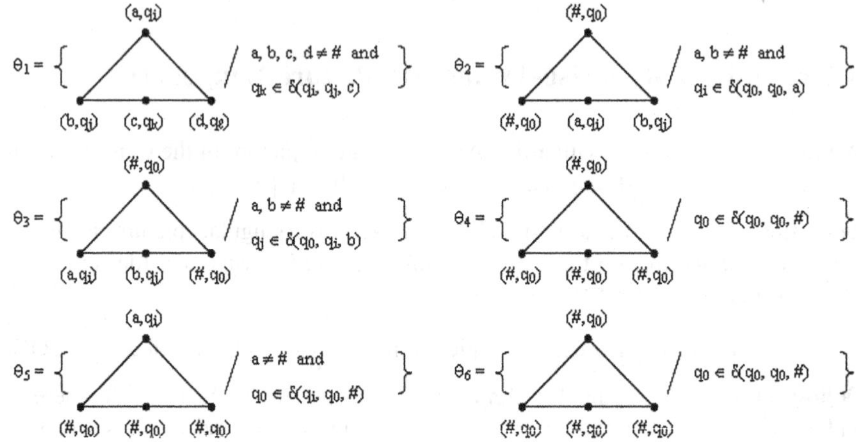

Let $T = (\Sigma, \Gamma, \pi, \theta)$ be a tiling system such that $\Gamma = \Sigma \cup \{\#\} \times Q$, $\theta = \theta_1 \cup \theta_2 \cup \theta_3 \cup \theta_4 \cup \theta_5 \cup \theta_6$, $\pi = \Sigma \cup \{\#\} \times Q \rightarrow \Sigma$.

The set θ is defined in such a way that a picture $\hat{p}$ of the underlying local iso-triangular picture language of L(IT) describes exactly a run of the 2DIOTA A on $p = \pi(\hat{p})$. It can be verified that the language recognized by the 2DIOTA is recognized by the iso-triangular tiling system defined above.

Lemma 2 *If a language is recognizable by a finite iso-triangular tiling system then it is recognizable by a two dimensional iso-triangular online tessellation automaton.* i.e., $\mathcal{L}(\text{ITTS}) \subseteq \mathcal{L}(\text{2DIOTA})$

Proof Let $L \subseteq \Sigma_T^{**}$ be a language recognized by the iso-triangular tiling system $T = (\Sigma, \Gamma, \pi, \theta)$ and L' the underlying local language represented by the set of tiles θ, i.e., $\pi(L') = L$. It is enough to prove that there exists a 2DIOTA recognizing $L' \subseteq \Gamma_T^{**}$.

4 *P* Systems for Iso-Triangular Pictures

In [1], the authors develop a class of tissue-like *P* systems with active membranes as a generative device for iso-picture languages. For generating two-dimensional pictures, tissue-like *P* systems with active membranes were proposed in [2], based on the features of tissue-like *P* systems [4] and *P* systems with active membranes [5] but with certain important differences.

We apply the formalism of P systems to generate iso-triangular pictures over a given alphabet $V \cup \{\#\}$ where # is the special symbol surrounding the borders of triangular pictures.

Our P system generates only pictures of an iso-triangular shape; we distinguish seven types of nodes of a triangular grid: upper most node, left-down corner, right-down corner, left margin, right margin, down margin nodes, and inner nodes.

We codify these grid positions in our P system by allowing seven symbols $(P_t)_{t \in T}$, such that at any moment, in any given membrane at least one of these symbols is present.

A triangular-grid-like P system with active membranes is a tissue-like P system with active membranes $A = (O, V, M, (\text{cont}_{m \in M}), R)$ such that

- $O = \{V \cup \{\#\} \cup P_t / t \in T\} \cup C \cup \{s, c, k\}$. The symbols $(\underline{P_t})_{t \in T}$ indicate the type of membrane with respect to the position into the triangular grid. The symbols s, c, and k are the output start symbol, the checking start symbol and the killer, respectively.
- $V \subseteq O$, is the output alphabet.
- $M \subseteq (x + y + z)^*$. A membrane label is considered as a multiset over x, y, and z. For $m + x$ means, the multiset obtained from m by following the co-ordinate system as discussed in Sect. 2, in the positive direction of x. A similar method is followed for $m + y$ and $m + z$ (respectively for the negative direction).
- cont_m is a multiset, which represents the content (of symbol objects) of the membrane labeled by m, for any $m \in M$.
- The set of rules R is divided into four groups as described below.

(i) The subset of creation rules Cre_A:
These rules are creating the grid of membranes and checking for the integrity (iso-triangular form) of the grid.

(1) $[P_{00}s]_m \to P_{00}\ \#\ c\ [P_{i0}s]_{m+z}\ [P_{i2}s]_{m-y},\ i \in \{1, 2\}$
(2) $[P_{10}s]_m \to P_{10}\ \#\ c\ [P_{\ell 0}\}]_{m-z}\ [P_{i0}s]_{m+z}\ [P_{ij}s]_{m-y},\ \ell \in \{0, 1\}$
(3) $[P_{12}]_m \to P_{12}\ \#\ c\ [P_{ij}]_{m+y}\ [P_{11}]_{m-x}\ [P_{\ell 1}s]_{m+z}\ [P_{\ell 2}s]_{m-y}$,
$i \in \{0, 1\}, j \in \{0, 1, 2\}, \ell \in \{1, 2\}$
(4) (a) When the inner node has four neighbors,
$[P_{11}s]_m \to P_{11}\ \#\ \mathrm{c}\ [P_{11}s]_{m+z}\ [P_{11}]_{m-y}\ [P_{12}]_{m+x}\ [P_{10}]_{m-x},\ j \in \{0, 1, 2\}$
(b) When the inner node has five neighbors,
$[P_{11}s]_m \to P_{11}\ \#\ c\ [P_{i1}s]_{m-z}\ [P_{i1}]_{m-y}\ [\mathrm{P}_{1j}]_{m+x}\ [P_{1\ell}]_{m-x}\ [P_{i1}]_{m+z}$,
$i \in \{1, 2\}, j \in \{1, 2\}, \ell \in \{0, 1\}$
(c) When the inner node has six neighbors,
$[P_{11}s]_m \to P_{11} \times \mathrm{c}\ [P_{i1}s]_{m+z}\ [P_{i1}s]_{m-y}\ [P_{1j}]_{m-z}\ [P_{1j}]_{m-z}\ [P_{1\ell}]_{m+y}\ [P_{ij}]_{m-x}$,
$i, j \in \{1, 2\}, \ell \in \{0, 1\}$
(5) $[P_{20}s]_m \to P_{20}\ \#\ \mathrm{c}\ [P_{ij}]_{m-z}\ [P_{ij}]_{m+x}\ i \in \{1, 2\}, j \in \{0, 1\}$
(6) $[P_{21}s]_m \to P_{21}\ \#\ \mathrm{c}\ [P_{ij}]_{m+x}\ [P_{ij}]_{m-z}\ [P_{ij}]_{m+y}\ [P_{ij}]_{m-x}\ i \in \{1, 2\}, j \in \{0, 1\}$
(7) $[P_{22}s]_m \to P_{22}\ \#\ \mathrm{c}\ [P_{ij}]_{m-x}\ [P_{ij}]_{m+y}\ i, j \in \{1, 2\}$

(ii) The subset of checking rules: If there are n checking rules, then the system generates all iso-triangular pictures on V.

(iii) The subset of contamination rules: These rules are contaminating the P system with the killer k that dissolves the membrane

(8) $[P_{ij}P_{\ell h}]_m \rightarrow k$ with $i, j, \ell, h \in \{0, 1, 2\}$ and $(i, j) \neq (\ell, h)$.

Two different markers in the same membrane produce a killer. Some contamination rules may also appear as an effect of the checking failure.

(iv) The subset of destruction rules: By these rules, the killer spreads from a contaminated membrane to all over the P system, while dissolving the membranes in which it appears.

(9) $[P_{00}k]_m \rightarrow [k]_{m+z}\ [k]_{m-y}$

The killer, in the presence of the marker P_{00} is sent forward down and backward down and dissolves the membrane m. Similarly, we have the following destruction rules.

(10) $[P_{10}k]_m \rightarrow [k]_{m+z}\ [k]_{m+x}\ [k]_{m-y}\ [k]_{m-z}$
(11) (a) $[P_{11}k]_m \rightarrow [k]_{m+z}\ [k]_{m-y}\ [k]_{m+x}\ [k]_{m-x}$
(b) $[P_{11}k]_m \rightarrow [k]_{m-z}\ [k]_{m-y}\ [k]_{m+x}\ [k]_{m-x}\ [k]_{m+z}$
(c) $[P_{11}k]_m \rightarrow [k]_{m+z}\ [k]_{m-y}\ [k]_{m+z}\ [k]_{m-z}\ [k]_{m+y}\ [k]_{m-x}$
(12) $[P_{12}k]_m \rightarrow [k]_{m+y}\ [k]_{m-z}\ [k]_{m+z}\ [k]_{m-y}$
(13) $[P_{20}k]_m \rightarrow [k]_{m-z}\ [k]_{m+x}$
(14) $[P_{21}k]_m \rightarrow [k]_{m-z}\ [k]_{m+x}\ [k]_{m+y}\ [k]_{m-x}$
(15) $[P_{22}k]_m \rightarrow [k]_{m-x}\ [k]_{m+y}$

A triangular-grid-like P system with active membranes $A = (O, V, M, \text{cont}_{m\in M}, R)$ is called initial over V if and only if $M = \{\lambda\}$ and $\text{cont}_m = \{P_{00}s\}$. With any stable alive grid-like P system with active membranes $A = (O, V, M, \text{cont}_{m\in M})$ we may associate an iso-triangular picture over the alphabet in the following way.

First, we define three natural numbers s, t, and u by

$$s = \max\{i/\text{there are } j \text{ and } k \text{ such that } x^i y^j z^k \in M\} - 2$$
$$t = \max\{j/\text{there are } i \text{ and } k \text{ such that } x^i y^j z^k \in M\} - 2$$
$$u = \max\{k/\text{there are } i \text{ and } j \text{ such that } x^i y^j z^k \in M\} - 2$$

If s, t, and u are all equal and greater than 1, then we consider the triangular picture $(p(i,j,k))$, with $(p(i,j,k)) = h$ if and only if $h \in \text{cont}_{x^i y^j z^k} \cap$ V. Otherwise, we consider the empty picture. We define iso-triangular picture languages generated by an initial grid-like P system with active membrane A, as the set of iso-triangular pictures associated with all grid-like P systems with active membranes from the stable universe of A. Thus, we obtain the following result.

Theorem 2 *An initial grid like P system with active membranes, over a given alphabet V, without any checking rules generates the language of all iso-triangular pictures over V.*

5 Conclusion

The notions of local and recognizable iso-triangular picture languages are introduced by extending the corresponding notion of triangular arrays. We find that the family of iso-triangular picture languages recognized by tiling systems and two-dimensional online tessellation automata are same. *P* systems for the generation of iso-triangular picture languages in triangular grid are presented, exhibiting the power of the system to generate two-dimensional picture languages.

References

1. S. Annadurai, T. Kalyani, V.R. Dare and D.G. Thomas, P systems generating iso-picture languages, Progress in Natural Science, 18 (2008), 617–622.
2. R. Ceterchi, R. Gramatovichi, N. Jonaska and K.G. Subramanian, Tissue like P systems with active membranes for picture generation, Fundamenta Informaticae (2003), 311–328.
3. K.S. Dersanambika, K. Krithivasan and K.G. Subramanian, P systems generating hexagonal picture languages. Workshop on Membrane Computing, (2003), 168–180.
4. K. Inoue and A. Nakamura, Some properties of two dimensional online tessellation acceptors, Information Sciences, 13 (1977), 95–121.
5. Gh. Păun, Computing with membranes (P Systems): Twenty Six research topics, Auckland University, CDMTCS Report NO. 119 (2000) (www.cs.auckland.ac.nz/CDMTCS).
6. D. Giammarresi and A. Restivo, Two dimensional languages. In A. Salomaa and G. Rozenberg, editors, Handbook of Formal Languages, Vol. 3, Beyond words, pages 215–267, Springer-Verlag, Berlin 1997.
7. K.G. Subramanian, R. Saravanan, T. Robinson, P system for array generation and application of Kolam patterns, Forma, 22 (2007), 47–54.

Job Block Scheduling with Dual Criteria and Sequence-Dependent Setup Time Involving Transportation Times

Deepak Gupta, Kewal Krishan Nailwal and Sameer Sharma

Abstract A two-machine flowshop scheduling with sequence-dependent setup time (SDST), job block, and transportation time is considered with the objective of minimizing makespan and the rental cost of machines taken on rent under a specified rental policy. The processing time of attributes on these machines is associated with probabilities. To find the optimal or near-optimal solution for these objectives, a heuristic algorithm is developed. To test the efficiency of the proposed heuristic algorithm, a numerical illustration is given.

Keywords Scheduling · Makespan · Job block · Sequence-dependent setup time · Processing time · Transportation time · Rental cost · Utilization time

1 Introduction

Scheduling is a decision-making process that is used on a regular basis in many manufacturing and services industries. It deals with the allocation of resources to tasks over a given time periods, and its goal is to optimize one or more objectives [19]. Many of the heuristics with objective as makespan given by [5, 14, 17, 18], assumes that the setup times are negligibly small or are included in the processing time, but in some applications, setup time has a major impact on the performance

D. Gupta
M. M. University, Mullana, Ambala, Haryana, India
e-mail: guptadeepak2003@yahoo.co.in

K. K. Nailwal (✉)
A. P. J. College of Fine Arts, Jalandhar City, India
e-mail: kk_nailwal@yahoo.co.in

S. Sharma
D. A. V. College, Jalandhar City, Punjab, India
e-mail: samsharma31@yahoo.com

G. S. S. Krishnan et al. (eds.), *Computational Intelligence, Cyber Security and Computational Models*, Advances in Intelligent Systems and Computing 246,
DOI: 10.1007/978-81-322-1680-3_35, © Springer India 2014

measure considered for scheduling problem. The term "sequence dependent" implies that the setup time depends on the sequence in which the jobs are processed on the machines, i.e., setup times depend on the type of job just completed as well as on the job to be processed [4]. A typical example is the manufacturing of different colors of paint, in which a cleaning operation time is needed, and is related to sequence of colors to be processed. Also, a setup changeover from black to white in painting industry takes longer duration than from white to black or dark gray. Scheduling with sequence-dependent setup time (SDST) has received significant attention in recent years. Corwin and Esogbue [6] minimized makespan by considering SDST. Gupta [12] proposed a branch-and-bound algorithm to minimize setup cost in n jobs and m machines flowshop with SDST. Rajendran and Ziegler [21] gave a heuristic for scheduling to minimize the sum of weighted flowtime of jobs in a flowshop with SDST of jobs. Rios and Bard [22] solved permutation flowshops with sequence-dependent setup times using branch and bound. Rabadi et al. [20] extended branch-and-bound algorithm for the early/tardy machine scheduling problem with a common due date and SDST. Tan and Narasimhan [23] minimized tardiness on a single processor with SDST using simulated annealing approach. A review of literature on scheduling research with various setup considerations has been presented by Allahverdi et al. [1]. Also, Allahverdi et al. [2] extended this survey up to the year 2006. Gajpal et al. [8] described an ant colony algorithm for scheduling in flowshop with SDST of jobs. Gupta and Smith [13] proposed two heuristics: a greedy randomized adaptive search procedure (GRASP) and a problem space-based local search heuristic. They showed that the space-based local search heuristic performs equally well when compared to ACO of Gagne et al. [7] while taking much less computational time. Gupta and Smith [13] also showed that GRASP gives much better solutions than ACO, while it takes much more computation time than ACO. The bicriteria scheduling problems are motivated by the fact that they are more meaningful from practical point of view. Some of the noteworthy heuristic approaches are due to Bagga and Bambani [3], Gupta et al. [9–11], Van Wassenhove and Gelders [24]. The idea of transportation has a practical significance when the material from one place of processing is to be carried to another place for further processing. Maggu and Das [15] introduced the concept of job block in flowshop scheduling. With job block, the importance lies in the fact that how to create a balance between a cost of providing priority in service and cost of providing service with non-priority customers, i.e., how much is to be charged extra from the priority customer(s) as compared to non-priority customer(s). Renting of machines is an affordable and quick solution for an industrial setup to complete assignment when one does not have resources. Renting enables saving working capital, gives option for having the equipment, and allows up gradation to new technology. The two criteria of minimizing the maximum makespan, and the utilization of machines or rental cost is the concern in this paper with the technological constraints as posed with job block and transportation times.

2 Assumptions, Notations, and Rental Policy

The proposed algorithm is based on the assumptions as discussed for flow shop scheduling in [4]. The following notations will be used all the way through this present paper:

S: Sequence of jobs 1, 2, 3… n; S_p: Sequence obtained by applying Johnson's procedure; M_k: Machine k, $k = 1, 2\ldots$ $a_{i,k}$: Processing time of ith attribute on machine M_k; $p_{i,k}$: Probability associated with the processing time $a_{i,k}$; $A_{i,k}$: Expected processing time of ith attribute on machine M_k; J_i: ith job, $i = 1, 2, 3\ldots$ n; $S_{ij,k}$: Setup time if job i is processed immediately after job j on kth machine; $L_k(S_p)$: The latest time when machine M_k is taken on rent for sequence S_p; $t_{ij,k}(S_p)$: Completion time of ith job processed immediately after jth job for sequence S_p on machine M_k; $I_{i,k}(S_p)$: Idle time of machine M_k for job i in the sequence S_p; $R(S_p)$: Total rental cost for the sequence S_p of the machines; C_i : Rental cost of ith machine; $t'_{ij,k}(Sp)$: Completion time of ith job processed immediately after jth job for sequence S_p on machine M_k when machine M_k starts processing jobs at time $L_k(S_p)$; $U_k(S_p)$: Utilization time for which machine M_k is required, when M_k starts processing jobs at time $L_k(S_p)$; β: Equivalent jobs for job block; $T_{i,k\rightarrow l}$: Transportation time required for ith job from machine k to machine l.

The machines will be taken on rent as and when they are required and are returned back as and when they are no longer required, i.e., the first machine will be taken on rent in the starting of the processing the jobs, second machine will be taken on rent at time when first job is completed on first machine, transported to second machine, and is in ready mode for processing on second machine.

3 Problem Formulation

Completion time of ith job processed immediately after jth job for sequence S on machine M_k is defined as $t_{ij,k} = \max(t_{(i-1)j,k}, t_{ij,k-1}) + A_{i,k} + S_{ij,k} + T_{i,(k-1)\rightarrow k}$; $k \geq 2$

Also, completion time if ith job processed immediately after jth job on machine M_k at latest time L_k is defined as

$$t'_{ij,k} = L_k + \sum_{q=1}^{i} A_{q,k} + \sum_{r=1}^{i-1} S_{rj,k} = \sum_{q=1}^{i} I_{q,k} + \sum_{q=1}^{i} A_{q,k} + \sum_{r=1}^{i-1} S_{rj,k}$$
$$= \max\left(t_{(i-1)j,k}, t'_{ij,k-1}\right) + A_{i,k} + S_{ij,k},$$

Let some job J_i ($i = 1, 2\ldots n$) are to be processed on two machines M_k ($k = 1, 2$) under the specified rental policy. Let there are n attributes of jobs on machine M_1 and m attributes of jobs on machine M_2. The mathematical model of the problem in matrix form can be stated as (Table 1).

Table 1 Attributes of jobs

	Machine M_2						
	1	2	3	–	j	–	m
Machine M_1							
1	J_1	–	J_2		J_3	–	–
2	–	J_4	–	–	–	–	J_5
3	–	–	J_6	–	–	–	–
–	–	–	–	–	–	–	–
i	–	–	–	–	J_i	–	–
–	–	–	–	–	–	–	–
n	J_{n-1}	–	–	–	–	–	J_n

Each job is characterized by its first attribute (row) on the first machine and second attribute (column) on the second machine

The processing times for various attributes on machine M_1 and M_2 are shown in Table 2.

The setup times for various attributes on machine M_k ($k = 1, 2$) are shown in Table 3.

Mathematically, the problem can be stated as minimize $t_{nj,2}(S)$ and minimize $U_k(S)$ or $R(S_i) = t_{n,1} \times C_1 + U_k(S_i) \times C_2$, subject to constraint: Rental Policy (P).

Table 2 Processing time with probability

Attributes	Machine M_1		Machine M_2	
1	$a_{1,1}$	$p_{1,1}$	$a_{1,2}$	$p_{1,2}$
2	$a_{2,1}$	$p_{2,1}$	$a_{2,2}$	$p_{2,2}$
3	$a_{3,1}$	$p_{3,1}$	$a_{3,2}$	$p_{3,2}$
–	–	–	–	–
m	$a_{m,1}$	$p_{m,1}$	$a_{m,2}$	$p_{m,2}$
–	–	–	–	–
n	$a_{n,1}$	$p_{n,1}$	–	–

Table 3 Setup time on machine M_k

Attributes	1	2	3	–	j	–	n
1	–	$S_{12,k}$	$S_{13,k}$	–	$S_{1j,k}$	–	$S_{1n,k}$
2	$S_{21,k}$	–	$S_{23,k}$	–	$S_{2j,k}$	–	$S_{2n,k}$
3	$S_{31,k}$	$S_{32,k}$	–	–	$S_{3j,k}$	–	$S_{3n,k}$
–	–	–	–	–	–	–	–
i	$S_{i1,k}$	$S_{i2,k}$	$S_{i3,k}$	–	–	–	$S_{in,k}$
–	–	–	–	–	–	–	–
n	$S_{n1,k}$	$S_{n2,k}$	–	–	$S_{nj,k}$	–	–

The attribute in row i is processed immediately after the attribute in column j

4 Theorems

The following theorems support the finding of optimal sequence of jobs processing.

Theorem 4.1 *Consider a flowshop problem consisting of two machines* M_1 *and* M_2 *and a set of n-jobs to be processed on these machines. Let* $A_{i,1}$ *and* $A_{i,2}$ *be the given processing time for each job i (* $1 \leq i \leq n$*) on machine* M_1 *and* M_2*, respectively. Each machine can handle at most one job at a time and the processing of each job must be finished on machine* M_1 *before it can be processed on machine* M_2*. It has been assumed that the order of treatments in the process* M_1 *and* M_2 *are same. Let* $T_{i,1\to2}$ *denote the transportation time of job i from machine* M_1 *to* M_2*. In the transportation process, several jobs can be handled simultaneously. Let* β *be an equivalent job for a given ordered set of jobs, then processing times* $A_{\beta,1}$ *and* $A_{\beta,2}$ *on machines* M_1 *and* M_2 *are given by*

$$A_{\beta,1} = (A_{k,1} + T_{k,1\to2}) + (A_{k+1,1} + T_{k+1,1\to2}) - \min\{A_{k,2} + T_{k,1\to2}, A_{k+1,1} + T_{k+1,1\to2}\}$$
$$A_{\beta,2} = (A_{k,2} + T_{k,1\to2}) + (A_{k+1,2} + T_{k+1,1\to2}) - \min\{A_{k,2} + T_{k,1\to2}, A_{k+1,1} + T_{k+1,1\to2}\}$$

and the transportation time of β *from machines* M_1 *to* M_2 *is given by* $T_\beta = 0$ *as given by Maggu and Das* [16].

Theorem 4.2 *The processing of jobs on* M_2 *at time* $L_2 = \sum_{i=1}^{n} I_{i,2}$ *keeps* $t_{nj,2}$ *unaltered as given by* [10].

5 Algorithm

The following proposed algorithm of bicriteria here can be referred to as $F_2/S_{sd}/R(S), C_{\max}$.

Step 1: Calculate the expected processing times of the given attributes .

Step 2: Introduce two fictitious machines G and H with processing times G_i and H_i, respectively, defined as $G_i = A_{i,1} + T_{i,1\to2}$ and $H_i = A_{i,2} + T_{i,1\to2}$

Step 3: Take equivalent job $\beta(k_1, m_1)$ and calculate the processing time $A_{\beta,1}$ and $A_{\beta,2}$ on the guide lines of Maggu and Das [15].

Step 4: Define a new reduced problem with the processing times $A_{i,k}$ as defined in step 1 and jobs (k_1,m_1) are replaced by single equivalent job β with processing time $A_{\beta,k}(k = 1, 2)$as defined in step 3.

Step 5: Using Johnson's technique [14], obtain the sequences S_p having minimum total elapsed time. Let these sequences be S_1, S_2, ————.

Step 6: Compute total elapsed time $t_{n,2}(S_p)$, p = 1, 2, 3,— of second machine by preparing in–out tables for sequence S_p.

Step 7: Compute $L_2(S_p) = t_{n,2} - \sum_{i=1}^{n} A_{i,2} - \sum_{i=1}^{n-1} S_{ij,2}$ for each sequence S_p.

Step 8: Find utilization time of 2nd machine $U_2(S_p) = t_{n,2}(S_p) - L_2(S_p)$.
Step 9: Find minimum of $\{(U_2(S_p))\}$; $p = 1, 2, 3, \ldots$.

Let it be the sequence S_γ. Then, S_γ is the optimal sequence, and minimum rental cost for the sequence S_γ is $R(S_\gamma) = t_{n,1}(S_\gamma) \times C_1 + U_2(S_\gamma) \times C_2$.

6 Numerical Illustration

Consider a two-stage furniture production system where each stage represents a machine. Seven jobs are to be processed on each machine. At stage one, sheets as raw materials (having six attributes) are cut and subsequently painted in the second stage (having four attributes) according to the market demand. A setup changeover is needed in cutting department when the thickness of two successive jobs differs substantially. In the painting department, a setup is required when the color of two successive jobs changes. The setup times are sequence dependent. Further, the machines M_1 and M_2 are taken on rent under rental policy P. The attributes, processing times as well as setup times on the first and second machines are shown in Tables 4, 5, 6, and 7, respectively.

Let the jobs 3 and 5 are processed as a job block $\beta = (3, 5)$. The rental cost per unit for the machines M_1 and M_2 be 8 and 10 units, respectively. Our objective is to find the sequence of jobs processing with minimum possible rental cost, when the machines are taken on rent under the specified rental policy.

Solution: As per steps 1, the expected processing times of the jobs on two machines for the possible attributes with transportation time $T_{i,1\to 2}$ is given in Table 8.

Using Johnson's technique [14] as per the algorithm, the sequence S_γ having minimum total elapsed time is

$$S_\gamma = J_1 - J_\beta - J_2 - J_4 - J_6 - J_7 = J_1 - J_3 - J_5 - J_2 - J_4 - J_6 - J_7.$$

The in–out flow table of jobs for the sequence S_γ gives the total elapsed time $t_{n,2}(S_\gamma) = 37.5$ units. Hence, the utilization time of machine M_2 is $U_2(S_\gamma) = t_{n,2}(S_\gamma) - L_2(S_\gamma) = 37.5 - 2.8 = 34.7$ units.

Total Minimum Rental Cost $= R(S_\gamma) = t_{n,1}(S_\gamma) \times C_1 + U_2(S_\gamma) \times C_2 =$ 587.8 units.

Now, the latest time at which machine M_2 should be taken on rent

$$L_2(S_\gamma) = t_{n,2}(S_\gamma) - \sum_{q=1}^{n} A_{q,2}(S_\gamma) - \sum_{j=1}^{n-1} S_{ij,2}(S_\gamma) = 37.5 - 20 - 9 = 8.5 \text{ units.}$$

The bi-objective in–out flow table for the sequence S_γ is given in Table 9.

Table 4 Attributes of jobs

	Machine M_2			
	1	2	3	4
Machine M_1				
1	–	J_3	–	–
2	J_1	–	J_6	–
3	–	J_5	–	–
4	J_2	–	–	–
5	–	–	J_4	–
6	–	–	–	J_7

Table 5 Processing times of attributes with probability

Attributes	Machine M_1		Machine M_2	
1	15	0.2	11	0.2
2	9	0.2	8	0.4
3	20	0.1	10	0.3
4	7	0.2	24	0.1
5	13	0.2		
6	25	0.1		

Table 6 Setup time on M_1

	Attributes					
	1	2	3	4	5	6
Attributes						
1	–	3	1	1	1	2
2	2	–	2	3	3	6
3	3	4	–	2	1	2
4	2	2	3	–	2	6
5	3	1	2	1	–	4
6	7	2	8	6	5	–

The attribute in row i is processed immediately after the attribute in column j

Table 7 Setup time on M_2

	Attributes			
	1	2	3	4
Attributes				
1	–	1	2	1
2	2	–	1	2
3	4	2	–	3
4	3	4	2	–

The attribute in row i is processed immediately after the attribute in column j

Table 8 Expected processing time

Jobs	J_1	J_2	J_3	J_4	J_5	J_6	J_7
M_1	1.8	1.4	3.0	2.6	2.0	1.8	2.5
$T_{i,1\to 2}$	1	2	3	1	2	4	5
M_2	2.2	2.2	3.6	3.0	3.6	3.0	2.4

Table 9 Bi-objective in–out table

Jobs	Machine M_1 In–out	$T_{i,1\to 2}$	Machine M_2 In–out
J_1	0.0–1.8	1	8.5–10.7
J_3	4.8–7.8	3	12.7–16.3
J_5	10.8–12.8	2	16.3–19.9
J_2	15.8–17.2	2	20.9–23.1
J_4	18.2–20.8	1	27.1–30.1
J_6	23.8–25.6	4	30.1–33.1
J_7	27.6–30.1	5	35.1–37.5

Therefore, the utilization time of machine M_2 is
$U_2(S_\gamma) = t'_{n,2}(S_\gamma) - L_2(S_\gamma) = 37.5 - 8.5 = 29$ units.
Total Minimum Rental Cost = $R(S_\gamma) = t_{n,1}(S_\gamma) \times C_1 + U_2(S_\gamma) \times C_2$.
Hence, effective decrease in the rental cost of machines = 587.8 − 530.8 = 57.0 units.

7 Conclusion

If the machine M_2 is taken on rent when it is required and is returned as soon as it completes the last job, the starting of processing of jobs at time $L_2(S_r) = t_{n,2}(S_r) - \sum_{i=1}^{n} A_{i,2}(S_r) - \sum_{i=1}^{n-1} S_{ij,2}(S_r)$ on M_2 will reduce its utilization time. Therefore, total rental cost of M_2 will be minimum. Also, rental cost of M_1 will always be minimum as idle time of M_1 is minimum always due to our rental policy. The study may further be extending by introducing the concept of weightage of jobs, breakdown interval, etc.

References

1. A. Allahverdi, J.N.D. Gupta, T. Aldowaisan, "A review of scheduling research involving setup considerations", OMEGA, The International Journal of Management Sciences, 27, 1999, pp. 219–239.
2. A. Allahverdi, C.T. Ng, T.C.E. Cheng, M.Y. Kovalyov, "A survey of scheduling problems with setup times or costs", European Journal of Operational Research, 187, 2008, pp. 985–1032.
3. P.C. Bagga and A. Bhambani, "Bicriteria in flow shop scheduling problem", Journal of Combinatorics, Information and System Sciences, 22, 1999, pp. 63–83.

4. K.R. Baker, "Introduction to Sequencing and Scheduling", Wiley: New York, 1974.
5. H.G. Campbell, R.A. Dudek, and M.L. Smith, "A heuristic algorithm for the n-job, m-machine sequencing problem", Management Science, 16, 1970, pp. 630–637.
6. B.D. Crown and A.O. Esogbue, "Two machine flow shop scheduling problems with sequence dependent setup time: a dynamic programming approach", Naval Research Logistics Quarterly, 21, 1974, pp. 515–523.
7. C. Gagne, W.L. Price, M. Gravel, "Comparing an ACO algorithm with other heuristics for the single machine scheduling problem with sequence-dependent setup times", Journal of the Operational Research Society 53, 2002, pp. 895–906.
8. Y. Gajpal, C. Rajendran and H. Ziegler, "An ant colony algorithm for scheduling in flow shop with sequence dependent set times of jobs", The international Journal of Advanced Manufacturing Technology, 30(5-6), 2006, pp. 416–424.
9. D. Gupta, S. Sharma, Seema and Shefali, "nx2 bicriteria flowshop scheduling under specified rental policy, processing time and setup time each associated with probabilities including job block", European Journal of Business and Management, 3(3) (2012) pp. 268–286.
10. D. Gupta, S. Sharma, Seema and K.K. Nailwal, "A bicriteria two machine flowshop scheduling with sequence dependent setup time", International Journal of Mathematical Sciences, 11(3-4), 2012, pp. 183–196.
11. D. Gupta, S. Sharma, Seema and K.K. Nailwal, "Bicriteria job block scheduling with sequence dependent setup time", Proceedings of the International Conference on Optimization, Computing and Business Analytics, 2012, pp. 8-13.
12. J.N.D. Gupta and W.P. Darrow, "The two machine sequence dependent flowshop scheduling problem", European Journal of Operational Research, 24(3) (1974) 439–446.
13. S.R. Gupta, J.S. Smith, "Algorithms for single machine total tardiness scheduling with sequence dependent setups", European Journal of Operational Research, 175, 2006, pp. 722–739.
14. S.M. Johnson, "Optimal two and three stage production schedule with set up times included", Naval Research Logistics Quarterly. 1(1) (1954) 61–68.
15. P.L. Maggu and G. Das, "Equivalent jobs for job block in job scheduling", Opsearch, 14(4) (1977) 277–281.
16. P.L. Maggu P and G. Das, "Elements of Advanced Production Scheduling", United Publishers and Periodical Distributors, 1985.
17. M. Nawaz, Jr.E.E. Enscore and I. Ham, "A heuristic algorithm for the m-machine n-job flowshop sequencing problem", OMEGA, International Journal of Management Science, 11, 1983, pp. 91–95.
18. D.S. Palmer, "Sequencing jobs through a multi-stage process in the minimum total time - a quick method of obtaining a near-optimum", Operational Research Quarterly, 16(1) (1965) 101–107.
19. M.L. Pinedo, "Scheduling: Theory, Algorithms, and Systems", Third Edition, Springer, 2008.
20. G. Rabadi, M. Mollaghasemi, and G.C. Anagnostopoulos, "A branch-and-bound algorithm for the early/tardy machine scheduling problem with a common due-date and sequence dependent setup time", Computers and Operations Research 31, 2004, pp. 1727–1751.
21. C. Rajendran, H. Ziegler, "A heuristic for scheduling to minimize the sum of weighted flowtime of jobs in a flowshop with sequence-dependent setup times of jobs", Computers and Industrial Engineering 33, 1997, pp. 281–284.
22. R.Z. Rıos-Mercado, J.F. Bard, "A branch-and-bound algorithm for permutation flow shops with sequence-dependent setup times", IIE Transactions 31, 1999, pp. 721–731.
23. K.C. Tan, R. Narasimhan, "Minimizing tardiness on a single processor with sequence-dependent setup times: A simulated annealing approach", OMEGA, 25, 1997, pp. 619–634.
24. L.N. Van Wassenhove and L.F. Gelders, "Solving a bicriteria scheduling problem", AIIE Tran 15 s, 1980, pp. 84–88.

Modular Chromatic Number of $C_m \square C_n$

N. Paramaguru and R. Sampathkumar

Abstract A modular k-coloring, $k \geq 2$, of a graph G is a coloring of the vertices of G with the elements in Z_k having the property that for every two adjacent vertices of G, the sums of the colors of their neighbors are different in Z_k. The minimum k for which G has a modular k-coloring is the modular chromatic number of G. In this paper, except for some special cases, modular chromatic number of $C_m \square C_n$ is determined.

Keywords Modular coloring · Modular chromatic number · Cartesian product

1 Introduction

For a vertex v of a graph G, let $N_G(v) = N(v)$, the *neighborhood of* v denotes the set of vertices adjacent to v in G. For a graph G, let $c{:}V(G) \to Z_k$, $k \geq 2$, be a vertex coloring of G where adjacent vertices may be colored the same. The *color sum* $\sigma(v) = \Sigma_{u \in N(v)} c(u)$ of a vertex v of G is the sum of the colors of the vertices in $N(v)$. The coloring c is called a *modular k-coloring* of G if $\sigma(x) \neq \sigma(y)$ in Z_k for all pairs x, y of adjacent vertices in G. The modular chromatic number mc(G) of G is the minimum k for which G has a modular k-coloring. This concept was introduced by Okamoto et al. [2].

The Cartesian product $G \square H$ of two graphs G and H has $V(G \square H) = V(G) \times V(H)$, and two vertices (u_1,u_2) and (v_1,v_2) of $G \square H$ are adjacent if and only if either $u_1 = v_1$ and $u_2v_2 \in E(H)$ or $u_2 = v_2$ and $u_1v_1 \in E(G)$.

N. Paramaguru (✉) · R. Sampathkumar
Annamalai University, Annamalainagar, Chidambaram 608002, India
e-mail: npguru@gmail.com

R. Sampathkumar
e-mail: sampathmath@gmail.com

G. S. S. Krishnan et al. (eds.), *Computational Intelligence, Cyber Security and Computational Models*, Advances in Intelligent Systems and Computing 246,
DOI: 10.1007/978-81-322-1680-3_36,

Okamoto et al. proved in [2] that every nontrivial connected graph G has a modular k-coloring for some integer $k \geq 2$ and $\mathrm{mc}(G) \geq \chi(G)$, where $\chi(G)$ denotes the chromatic number of G; for the cycle C_n of length n, $\mathrm{mc}(C_n)$ is 2 if $n \equiv 0 \bmod 4$ and it is 3 otherwise; every nontrivial tree has modular chromatic number 2 or 3; for the complete multipartite graph G, $\mathrm{mc}(G) = \chi(G)$; for the Cartesian product $G = K_r \square K_2$, $\mathrm{mc}(G)$ is r if $r \equiv 2 \bmod 4$ and it is $r + 1$ otherwise; for the wheel $W_n = C_n \vee K_1$, $n \geq 3$, $\mathrm{mc}(W_n) = \chi(W_n)$, where $\vee$ denotes the join of two graphs; for $n \geq 3$, $\mathrm{mc}(C_n \vee K_2^C) = \chi(C_n \vee K_2^C)$, where G^C denotes the complement of G; and for $n \geq 2$, $\mathrm{mc}(P_n \vee K_2) = \chi(P_n \vee K_2)$, where P_n denotes the path of length $n - 1$; and in [3] that for $m, n \geq 2$, $\mathrm{mc}(P_m \square P_n) = 2$.

Let $m \geq 3$ and $n \geq 2$. We have proved in [4] that $\mathrm{mc}(C_3 \square P_2) = 4$; if neither $m = 3$ and $n \in \{2,14,26,38,\ldots,12r + 2,\ldots\} \cup \{16,28,40,\ldots,12r + 4,\ldots\} \cup \{8,20,32,\ldots,12r + 8,\ldots\} \cup \{22,34,46,\ldots,12r + 10,\ldots\}$ nor $m \equiv 2 \bmod 4$ and $n \equiv 1 \bmod 4$, then $\mathrm{mc}(C_m \square P_n) = \chi(C_m \square P_n)$; if $m \equiv 2 \bmod 4$ and $n \equiv 1 \bmod 4$, then $\mathrm{mc}(C_m \square P_n) \leq 3$; and if $n \equiv 1 \bmod 4$, then $\mathrm{mc}(C_6 \square P_n) = 3$. For the leftover cases, we have conjectured in [4] that (a) if $n \in \{14,26,38,\ldots,12r + 2,\ldots\} \cup \{16,28,40,\ldots,12r + 4,\ldots\} \cup \{8,20,32,\ldots,12r + 8,\ldots\} \cup \{22,34,46,\ldots,12r + 10,\ldots\}$, then $\mathrm{mc}(C_3 \square P_n) = 3$, and (b) if $m \equiv 2 \bmod 4$, $m \geq 10$, and $n \equiv 1 \bmod 4$, $n \geq 5$, then $\mathrm{mc}(C_m \square P_n) = 3$.

In Sect. 2, except for some special cases, modular chromatic number of $C_m \square C_n$ is determined.

For the cycle C_v on v vertices, let $V(C_v) = \{1,2,\ldots, v\}$ and $E(C_v) = \{\{i,i + 1\}: i \in \{1,2,\ldots, v - 1\}\} \cup \{\{v,1\}\}$. For $m \geq 3$ and $n \geq 3$, $\chi(C_m \square C_n)$ is 2 if both m and n are even and it is 3 otherwise.

2 Results

Theorem 2.1 *If* $m \geq 4$ and $n \geq 4$ *are even integers and at least one of* m, n *is congruent to* 0 mod 4, *then* $\mathrm{mc}(C_m \square C_n) = \chi(C_m \square C_n)$.

Proof Assume, by symmetry, that $m \equiv 0 \bmod 4$. Define $c{:}V(C_m \square C_n) \to Z_2$ as follows: $c((i,j)) = 1$ if $(i,j) \in (\{1,5,9,\ldots,m - 3\} \times \{1,3,5,\ldots,n - 1\}) \cup (\{2,6,10,\ldots,m - 2\} \times \{2,4,6,\ldots,n\})$; and $c((i,j)) = 0$ otherwise.

Then, $\sigma((i,j)) = 1$ if $(i,j) \in (\{1,3,5,\ldots,m - 1\} \times \{2,4,6,\ldots,n\}) \cup (\{2,4,6,\ldots,m\} \times \{1,3,5,\ldots,n - 1\})$; and $\sigma((i,j)) = 0$ otherwise.

This completes the proof. ■

Theorem 2.2 *If* $n \geq 3$ *is an integer, then* $\mathrm{mc}(C_3 \square C_n) = \chi(C_3 \square C_n)$.

Proof Define $c{:}V(C_3 \square C_n) \to Z_3$ as follows. We consider three cases.

Case 1. $n \equiv 0 \bmod 4$.

$c((i,j)) = 1$ if $(i,j) \in (\{2\} \times \{1,5,9,\ldots,n-3\}) \cup (\{3\} \times \{3,7,11,\ldots,n-1\})$; $c((i,j)) = 2$ if $(i,j) \in (\{1\} \times \{3,7,11,\ldots,n-1\}) \cup (\{3\} \times \{1,5,9,\ldots,n-3\})$; and $c((i,j)) = 0$ otherwise.

Then, $\sigma((i,j)) = 0$ if $(i,j) \in (\{1\} \times \{1,5,9,\ldots,n-3\}) \cup (\{2\} \times \{3,7,11,\ldots,n-1\}) \cup (\{3\} \times \{2,4,6,\ldots,n\})$; $\sigma((i,j)) = 1$ if $(i,j) \in (\{1\} \times \{3,7,11,\ldots,n-1\}) \cup (\{2\} \times \{2,4,6,\ldots,n\}) \cup (\{3\} \times \{1,5,9,\ldots,n-3\})$; $\sigma((i,j)) = 2$ if $(i,j) \in (\{1\} \times \{2,4,6,\ldots,n\}) \cup (\{2\} \times \{1,5,9,\ldots,n-3\}) \cup (\{3\} \times \{3,7,11,\ldots,n-1\})$.

Case 2. $n \equiv 2 \bmod 4$.

$c((i,j)) = 1$ if $(i,j) \in (\{1\} \times \{3,7,11,\ldots,n-3\}) \cup (\{2\} \times \{1,5,9,\ldots,n-5\}) \cup \{(3, n-1)\}$; $c((i,j)) = 2$ if $(i,j) \in \{(1, n-1)\} \cup (\{2\} \times \{3,7,11,\ldots, n-3\}) \cup (\{3\} \times \{1,5,9,\ldots, n-5\})$; and $c((i,j)) = 0$ otherwise.

Then, $\sigma((i,j)) = 0$ if $(i,j) \in (\{1\} \times \{1,5,9,\ldots,n-5\}) \cup \{(1,n-2)\} \cup (\{2\} \times \{2,4,6,\ldots,n-4\}) \cup \{(2,n-1)\} \cup (\{3\} \times \{3,7,11,\ldots,n-3\}) \cup \{(3,n)\}$; $\sigma((i,j)) = 1$ if $(i,j) \in (\{1\} \times \{2,4,6,\ldots,n-4\}) \cup \{(1, n-1)\} \cup (\{2\} \times \{3,7,11,\ldots,n-3\}) \cup \{(2,n)\} \cup (\{3\} \times \{1,5,9,\ldots,n-5\}) \cup \{(3,n-2)\}$; $\sigma((i,j)) = 2$ if $(i,j) \in (\{1\} \times \{3,7,11,\ldots,n-3\}) \cup \{(1,n)\} \cup (\{2\} \times \{1,5,9,\ldots,n-5\}) \cup \{(2, n-2)\} \cup (\{3\} \times \{2,4,6,\ldots,n-4\}) \cup \{(3, n-1)\}$.

Case 3. $n \equiv 1 \bmod 2$.

Subcase 3.1. $n \in \{3,5\}$.

See Tables 1 and 2 for colors of the vertices and color sums of the vertices for $C_3 \square C_3$ and $C_3 \square C_5$, respectively.

Subcase 3.2. $n \equiv 3 \bmod 6$ and $n \geq 9$.

$c((i,j)) = 1$ if $(i,j) \in (\{1\} \times \{2,5,8,\ldots,n-1\}) \cup (\{2\} \times \{1,4,7,\ldots,n-2\}) \cup (\{3\} \times \{3,6,9,\ldots,n\})$; $c((i,j)) = 2$ if $(i,j) \in (\{1\} \times \{3,6,9,\ldots,n\}) \cup (\{2\} \times \{2,5,8,\ldots,n-1\}) \cup (\{3\} \times \{1,4,7,\ldots,n-2\})$; and $c((i,j)) = 0$ otherwise.

Then, $\sigma((i,j)) = 0$ if $(i,j) \in (\{1\} \times \{1,4,7,\ldots,n-2\}) \cup (\{2\} \times \{3,6,9,\ldots,n\}) \cup (\{3\} \times \{2,5,8,\ldots,n-1\})$; $\sigma((i,j)) = 1$ if $(i,j) \in (\{1\} \times \{2,5,8,\ldots,n-1\}) \cup (\{2\} \times \{1,4,7,\ldots,n-2\}) \cup (\{3\} \times \{3,6,9,\ldots,n\})$; $\sigma((i,j)) = 2$ if $(i,j) \in (\{1\} \times \{3,6,9,\ldots,n\}) \cup (\{2\} \times \{2,5,8,\ldots,n-1\}) \cup (\{3\} \times \{1,4,7,\ldots,n-2\})$.

Subcase 3.3. $n \equiv 5 \bmod 6$ and $n \geq 11$.

$c((i,j)) = 1$ if $(i,j) \in (\{1\} \times \{2,5,8,\ldots,n-9\}) \cup (\{2\} \times \{1,4,7,\ldots,n-10\}) \cup (\{3\} \times \{3,6,9,\ldots,n-8\}) \cup (\{1\} \times \{n-7, n-5, n-3, n-1\}) \cup (\{2\} \times \{n-6, n-2\}) \cup (\{3\} \times \{n-4, n\})$; $c((i,j)) = 2$ if $(i,j) \in (\{1\} \times \{3,6,9,\ldots, n-8\}) \cup (\{2\} \times \{2,5,8,\ldots,n-9\}) \cup (\{3\} \times \{1,4,7,\ldots,n-10\}) \cup (\{1\} \times \{n-4, n\}) \cup (\{2\} \times \{n-7, n-5, n-3, n-1\}) \cup (\{3\} \times \{n-6, n-2\})$; and $c((i,j)) = 0$ otherwise.

Table 1 $C_3 \square C_3$

Color c	Color sum σ
0 1 2	0 1 2
1 2 0	1 2 0
2 0 1	2 0 1

Then, $\sigma((i,j)) = 0$ if $(i,j) \in (\{1\} \times \{1,4,7,\ldots,n-10\}) \cup (\{2\} \times \{3,6,9,\ldots,n-11\}) \cup (\{3\} \times \{2,5,8,\ldots,n-9\}) \cup (\{1\} \times \{n-8, n-4\}) \cup (\{2\} \times \{n-6, n-2, n\}) \cup (\{3\} \times \{n-7, n-5, n-3, n-1\})$; $\sigma((i,j)) = 1$ if $(i,j) \in (\{1\} \times \{2,5,8,\ldots, n-9\}) \cup (\{2\} \times \{1,4,7,\ldots,n-10\}) \cup (\{3\} \times \{3,6,9,\ldots,n-11\}) \cup (\{1\} \times \{n-7, n-5, n-3, n-1\}) \cup (\{2\} \times \{n-8, n-4\}) \cup (\{3\} \times \{n-6, n-2, n\})$; $\sigma((i,j)) = 2$ if $(i,j) \in (\{1\} \times \{3,6,9,\ldots,n-11\}) \cup (\{2\} \times \{2,5,8,\ldots, n-9\}) \cup (\{3\} \times \{1,4,7,\ldots,n-10\}) \cup (\{1\} \times \{n-6, n-2, n\}) \cup (\{2\} \times \{n-7, n-5, n-3, n-1\}) \cup (\{3\} \times \{n-8, n-4\})$.

Subcase 3.4. $n \equiv 1 \bmod 6$ and $n \geq 7$.

$c((i,j)) = 1$ if $(i,j) \in (\{1\} \times \{2,5,8,\ldots,n-5\}) \cup (\{2\} \times \{1,4,7,\ldots,n-6\}) \cup (\{3\} \times \{3,6,9,\ldots,n-4\}) \cup (\{1\} \times \{n-3, n-1\}) \cup \{(2, n-2)\} \cup \{(3, n)\}$; $c((i,j)) = 2$ if $(i,j) \in (\{1\} \times \{3,6,9,\ldots,n-4\}) \cup (\{2\} \times \{2,5,8,\ldots,n-5\}) \cup (\{3\} \times \{1,4,7,\ldots,n-6\}) \cup \{(1,n)\} \cup (\{2\} \times \{n-3, n-1\}) \cup \{(3, n-2)\}$; and $c((i,j)) = 0$ otherwise.

Then, $\sigma((i,j)) = 0$ if $(i,j) \in (\{1\} \times \{1,4,7,\ldots,n-6\}) \cup (\{2\} \times \{3,6,9,\ldots, n-7\}) \cup (\{3\} \times \{2,5,8,\ldots,n-5\}) \cup \{(1, n-4)\} \cup (\{2\} \times \{n-2, n\}) \cup (\{3\} \times \{n-5, n-3, n-1\})$; $\sigma((i,j)) = 1$ if $(i,j) \in \{1\} \times \{2,5,8,\ldots, n-5\}) \cup (\{2\} \times \{1,4,7,\ldots,n-6\}) \cup (\{3\} \times \{3,6,9,\ldots,n-7\}) \cup (\{1\} \times \{n-3,n-1\}) \cup \{(2, n-4)\} \cup (\{3\} \times \{n-2, n\})$; $\sigma((i,j)) = 2$ if $(i,j) \in (\{1\} \times \{3,6,9,\ldots,n-7\}) \cup (\{2\} \times \{2,5,8,\ldots,n-5\}) \cup (\{3\} \times \{1,4,7,\ldots,n-6\}) \cup (\{1\} \times \{n-2, n\}) \cup (\{2\} \times \{n-3, n-1\}) \cup \{(3, n-4)\}$.

This completes the proof. ∎

Theorem 2.3 *If at least one of m, n is congruent to* 1 mod 2, $m \geq 4$, $n \geq 4$, *and none of the following conditions hold*:

$m \equiv 1 \bmod 4$ and $n \equiv 1 \bmod 4$; $m \equiv 3 \bmod 4$, $m \geq 7$ and $n \equiv 3 \bmod 4$, $n \geq 7$;
$m = 7$ and $n \neq 6$, $n \equiv 2 \bmod 4$; $m \neq 6$, $m \equiv 2 \bmod 4$ and $n = 7$;
$m = 5$ and $n \equiv 2 \bmod 4$; $m \equiv 2 \bmod 4$ and $n = 5$;
$m \equiv 1 \bmod 4$ and $n = 6$; $m = 6$ and $n \equiv 1 \bmod 4$;
then, $\mathrm{mc}(C_m \square C_n) = \chi(C_m \square C_n)$.

Proof. Assume, by symmetry, that $m \equiv 1 \bmod 2$. Define $c: V(C_m \square C_n) \rightarrow Z_3$ as follows: We consider 5 cases.

Case 1. $m \equiv 1 \bmod 4$ and $n \equiv 0 \bmod 4$.

$c((i,j)) = 1$ if $(i,j) \in (\{1,5,9,\ldots,m-4\} \times \{1,5,9,\ldots,n-3\}) \cup (\{3,7,11,\ldots, m-2\} \times \{3,7,11,\ldots,n-1\})$; $c((i,j)) = 2$ if $(i,j) \in (\{2,6,10,\ldots,m-3\} \times \{3,7,11,\ldots,n-1\})$; $c((m,j)) = 2$ if $j \in (\{1,5,9,\ldots,n-3\}$; and $c((i,j)) = 0$ otherwise.

Then, $\sigma((i,j)) = 1$ if $(i,j) \in (\{1,3,5,\ldots,m-2\} \times \{2,4,6,\ldots,n\}) \cup (\{2,4,6,\ldots, m-3\} \times \{1,3,5,\ldots,n-1\}) \cup (\{m-1\} \times \{3,7,11,\ldots,n-1\}) \cup (\{m\} \times \{1,5, 9,\ldots,n-3\})$; $\sigma((i,j)) = 2$ if $(i,j) \in (\{2,6,10,\ldots,m-3\} \times \{2,4,6,\ldots,n\}) \cup (\{3,5,7,\ldots,m-2\} \times \{3,7,11,\ldots,n-1\}) \cup (\{m-1\} \times \{1,5,9,\ldots,n-3\}) \cup (\{m\} \times \{2,4,6,\ldots,n\}) \cup (\{1\} \times \{1,3,5,\ldots,n-1\})$; and $\sigma((i,j)) = 0$ otherwise.

Table 2 $C_3 \square C_5$

Color c	Color sum σ
0 1 2 1 1	2 1 0 2 0
1 2 0 2 2	0 2 1 0 1
2 0 1 0 0	1 0 2 1 2

Case 2. $m \equiv 3 \bmod 4$ and $n \equiv 0 \bmod 4$.

$c((i,j)) = 1$ if $(i,j) \in (\{1,5,9,\ldots,m-6\} \times \{1,5,9,\ldots,n-3\}) \cup (\{3,7,11,\ldots,m-4\} \times \{3,7,11,\ldots,n-1\})$; $c((i,j)) = 2$ if $(i,j) \in (\{m-2\} \times \{1,5,9,\ldots,n-3\}) \cup (\{m\} \times \{3,7,11,\ldots,n-1\})$; and $c((i,j)) = 0$ otherwise.

Then, $\sigma((i,j)) = 1$ if $(i,j) \in (\{1,3,5,\ldots,m-4\} \times \{2,4,6,\ldots,n\}) \cup (\{2,4,6,\ldots,m-5\} \times \{1,3,5,\ldots,n-1\}) \cup (\{m-3\} \times \{3,7,11,\ldots,n-1\}) \cup (\{m\} \times \{1,5,9,\ldots,n-3\})$; $\sigma((i,j)) = 2$ if $(i,j) \in (\{1\} \times \{3,7,11,\ldots,n-1\}) \cup (\{m-3\} \times \{1,5,9,\ldots,n-3\}) \cup (\{m-2\} \times \{2,4,6,\ldots,n\}) \cup (\{m-1\} \times \{1,3,5,\ldots,n-1\}) \cup (\{m\} \times \{2,4,6,\ldots,n\})$; and $\sigma((i,j)) = 0$ otherwise.

Case 3. $m \equiv 1 \bmod 4$, $n \equiv 3 \bmod 4$ and $n \geq 7$.

$c((i,j)) = 1$ if $(i,j) \in (\{3,7,11,\ldots,m-6\} \times \{n-2\}) \cup (\{1,5,9,\ldots,m-4\} \times \{n\}) \cup (\{m-2\} \times \{5,9,13,\ldots,n-2\}) \cup (\{m\} \times \{3,7,11,\ldots,n-4\})$; $c((i,j)) = 2$ if $(i,j) \in (\{1,5,9,\ldots,m-4\} \times \{3,7,11,\ldots,n-4\}) \cup (\{3,7,11,\ldots,m-6\} \times \{1,5,9,\ldots,n-6\}) \cup (\{m-1\} \times \{5,9,13,\ldots,n-2\}) \cup \{(m-2,1)\}$; and $c((i,j)) = 0$ otherwise.

Then, $\sigma((i,j)) = 1$ if $(i,j) \in (\{1\} \times \{3,7,11,\ldots,n-4\}) \cup (\{2,4,6,\ldots,m-3\} \times \{n-2,n\}) \cup (\{3,7,11,\ldots,m-2\} \times \{n-3, n-1\}) \cup (\{1,5,9,\ldots,m-4\} \times \{1, n-1\}) \cup (\{m-3\} \times \{5,9,13,\ldots,n-6\}) \cup (\{m-2\} \times \{4,6,8,\ldots,n-5\}) \cup (\{m-1\} \times \{3,5,7,\ldots,n-2\}) \cup (\{m\} \times \{2,4,6,\ldots,n-3\}) \cup \{(m,n)\}$; $\sigma((i,j)) = 2$ if $(i,j) \in (\{2,4,6,\ldots,m-5\} \times \{1,3,5,\ldots,n-4\}) \cup (\{3,7,11,\ldots,m-6\} \times \{2,4,6,\ldots,n-5\}) \cup (\{1,5,9,\ldots,m-4\} \times \{2,4,6,\ldots,n-3\}) \cup (\{m-3\} \times \{3,7,11,\ldots,n-4\}) \cup \{(m-3,1)\} \cup (\{m-2\} \times \{5,9,13,\ldots,n-2\}) \cup (\{m-2\} \times \{2,n\}) \cup (\{m-1\} \times \{4,6,8,\ldots,n-1\}) \cup \{(m-1,1)\} \cup (\{m\} \times \{3,5,7,\ldots,n-2\}) \cup (\{3,7,11,\ldots,m-6\} \times \{n\})$; and $\sigma((i,j)) = 0$ otherwise.

Case 4. $m \equiv 3 \bmod 4$ and $n \equiv 2 \bmod 4$.

Subcase 4.1. $m = 7$ and $n = 6$.

See Table 3 for colors of the vertices and color sums of the vertices for $C_7 \square C_6$.

Subcase 4.2. $m \geq 11$ and $n \geq 6$.

$c((i,j)) = 1$ if $(i,j) \in (\{1,5,9,\ldots,m-10\} \times \{3,7,11,\ldots,n-3\}) \cup (\{3,7,11,\ldots,m-8\} \times \{1,5,9,\ldots,n-1\}) \cup (\{5,9,13,\ldots,m-6\} \times \{n-1\}) \cup (\{m-6\} \times \{3,7,11,\ldots,n-7\}) \cup (\{m-4\} \times \{5,9,13,\ldots,n-5\}) \cup \{(m,n)\}$; $c((i,j)) = 2$ if $(i,j) \in \{(1, n-2),(m-6, n-3),(m-5, n-2),(m-4,1),(m-4, n-1),(m-2, n-1)\} \cup (\{m-2\} \times \{3,7,11,\ldots,n-3\}) \cup (\{m\} \times \{1,5,9,\ldots,n-5\})$; and $c((i,j)) = 0$ otherwise.

Table 3 $C_7 \square C_6$

Color c	Color sum σ
1 0 0 0 1 0	0 1 2 1 0 2
0 0 2 0 0 0	1 2 1 2 1 0
0 0 1 0 0 0	0 1 2 1 0 2
0 0 0 0 0 2	2 0 1 0 2 1
0 0 0 0 0 1	1 2 0 2 1 2
0 2 0 2 0 0	2 0 1 0 2 1
0 0 0 0 0 0	1 2 0 2 1 0

Then, $\sigma((i,j)) = 1$ if $(i,j) \in (\{1\} \times \{2,4,6,\ldots,n\}) \cup (\{5,9,13,\ldots,m-10\} \times \{2,4,6,\ldots,n-4\}) \cup (\{3,7,11,\ldots,m-8\} \times \{2,4,6,\ldots,n-2\}) \cup (\{2\} \times \{1,3,5,\ldots,n-1\}) \cup (\{4,6,8,\ldots,m-9\} \times \{1,3,5,\ldots,n-3\}) \cup (\{m-7\} \times \{1,3,5,\ldots,n-5\}) \cup (\{m-6\} \times \{2,4,6,\ldots,n-6\}) \cup (\{m-5\} \times \{3,5,7,\ldots,n-3\}) \cup (\{m-4\} \times \{4,6,8,\ldots,n\}) \cup (\{m-3\} \times \{5,9,13,\ldots,n-1\}) \cup \{(m-2, n-2)\} \cup (\{m\} \times \{3,7,11,\ldots,n-3\}) \cup (\{m\} \times \{1,n-1\}) \cup (\{5,9,13,\ldots,m-6\} \times \{n\}) \cup \{(m-1, n)\}$; $\sigma((i,j)) = 2$ if $(i,j) \in (\{1\} \times \{1,5,9,\ldots,n-1\}) \cup \{(1,n-3)\} \cup (\{5,9,13,\ldots,m-6\} \times \{n-2\}) \cup \{(2, n-2)\} \cup (\{4,6,8,\ldots,m-5\} \times \{n-1\}) \cup (\{3,7,11,\ldots,m-8\} \times \{n\}) \cup \{(m-7, n-3), (m-6,n-4),(m-5,1),(m-4,2)\} \cup (\{m-3\} \times \{3,7,11,\ldots,n-3\}) \cup \{(m-3,1)\} \cup (\{m-2\} \times \{2,4,6,\ldots,n-4\}) \cup \{(m-2, n)\} \cup (\{m-1\} \times \{1,3,5,\ldots,n-1\}) \cup (\{m\} \times \{2,4,6,\ldots,n\})$; and $\sigma((i,j)) = 0$ otherwise.

Case 5. $m \equiv 1 \bmod 4$, $m \geq 9$ and $n \equiv 2 \bmod 4$, $n \geq 10$.

Subcase 5.1. $m = 9$.

$c((i,j)) = 1$ if $(i,j) \in (\{1\} \times \{3,7,11,\ldots,n-3\}) \cup (\{3\} \times \{1,5,9,\ldots,n-9\}) \cup \{(3, n-1)\} \cup (\{4\} \times \{n-6, n-4, n-2\}) \cup \{(5,1)\} \cup (\{5\} \times \{3,7,11,\ldots,n-7\}) \cup \{(6,n)\} \cup (\{8\} \times \{5,9,13,\ldots,n-9\}) \cup \{(9,n)\}$; $c((i,j)) = 2$ if $(i,j) \in \{(2, n-5)\} \cup (\{5\} \times \{n-5, n-3, n-1\}) \cup (\{7\} \times \{1,5,9,\ldots,n-1\}) \cup (\{9\} \times \{3,7,11,\ldots,n-3\})$; and $c((i,j)) = 0$ otherwise.

Then, $\sigma((i,j)) = 1$ if $(i,j) \in (\{1\} \times \{2,4,6,\ldots,n\}) \cup (\{2\} \times \{1,3,5,\ldots,n-7\}) \cup (\{2\} \times \{n-3, n-1\}) \cup (\{3\} \times \{2,4,6,\ldots,n-4\}) \cup (\{4\} \times \{3,5,7,\ldots,n-9\}) \cup (\{4\} \times \{n-5, n-3, n-1\}) \cup (\{5\} \times \{4,6,8,\ldots,n-6\}) \cup \{(5,n)\} \cup (\{6\} \times \{3,7,11,\ldots,n-7\}) \cup (\{6\} \times \{1, n-5\}) \cup (\{7\} \times \{5,9,13,\ldots,n-9\}) \cup \{8\} \times \{4,6,8,\ldots,n-8\}) \cup \{(8,n)\} \cup (\{9\} \times \{1,3,5,\ldots,n-7\}) \cup (\{9\} \times \{n-1, n-3\})$; $\sigma((i,j)) = 2$ if $(i,j) \in (\{1\} \times \{3,7,11,\ldots,n-3\}) \cup \{(1, n-5),(2, n-6),(2, n-4),(3, n-5)\} \cup (\{3,7\} \times \{n\}) \cup \{(3, n-2)\} \cup (\{4\} \times \{1, n-7\}) \cup (\{5\} \times \{2, n-2, n-4\}) \cup (\{6\} \times \{5,9,13,\ldots,n-9\}) \cup (\{6\} \times \{n-1, n-3\}) \cup (\{7\} \times \{2,4,6,\ldots,n-2\}) \cup (\{8\} \times \{1,3,5,\ldots,n-1\}) \cup (\{9\} \times \{2,4,6,\ldots,n-2\})$; and $\sigma((i,j)) = 0$ otherwise.

Subcase 5.2. $m \geq 13$.

$c((i,j)) = 1$ if $(i,j) \in (\{1,5,9,\ldots,m-8\} \times \{3,7,11,\ldots,n-3\}) \cup (\{3,7,11,\ldots,m-6\} \times \{1,5,9,\ldots,n-1\}) \cup (\{5,9,13,\ldots,m-8\} \times \{n-1\}) \cup (\{m-5\} \times \{n-6, n-4, n-2\}) \cup \{(m-4,1)\} \cup (\{m-4\} \times \{3,7,11,\ldots,n-7\}) \cup \{(m-3, n)\} \cup (\{m-1\} \times \{5,9,13,\ldots,n-9\}) \cup \{(m,n)\}$; $c((i,j)) = 2$ if $(i,j) \in \{(2, n-5)\} \cup (\{m-4\} \times \{n-5, n-3, n-1\}) \cup (\{m-2\} \times \{1,5,9,\ldots,n-1\}) \cup (\{m\} \times \{3,7,11,\ldots,n-3\})$; and $c((i,j)) = 0$ otherwise.

Then, $\sigma((i,j)) = 1$ if $(i,j) \in (\{1\} \times \{2,4,6,\ldots,n\}) \cup (\{3\} \times \{2,4,6,\ldots,n-2\}) \cup (\{5,9,13,\ldots,m-8\} \times \{2,4,6,\ldots,n-4\}) \cup (\{5,9,13,\ldots,m-4\} \times \{n\}) \cup (\{2\} \times \{1,3,5,\ldots,n-1\}) \cup (\{4\} \times \{1,3,5,\ldots,n-3\}) \cup (\{7,11,15,\ldots,m-10\} \times \{2,4,6,\ldots,n-2\}) \cup (\{m-6\} \times \{2,4,6,\ldots,n-8\}) \cup (\{m-5\} \times \{3,5,7,\ldots,n-9\}) \cup (\{m-5\} \times \{n-3, n-1\}) \cup (\{m-4\} \times \{4,6,8,\ldots,n-6\}) \cup (\{m-3\} \times \{3,7,11,\ldots,n-7\}) \cup (\{m-3\} \times \{1, n-5\}) \cup (\{m-2\} \times \{5,9,13,\ldots,n-9\}) \cup (\{m-1\} \times \{4,6,8,\ldots,n-8\}) \cup \{(m-1, n)\} \cup (\{m\} \times \{1,3,5,\ldots,n-7\}) \cup \{6,8,10,\ldots,m-7\} \times \{1,3,5,\ldots,n-3\}) \cup (\{m\} \times \{n-1, n-3\})$; $\sigma((i,j)) = 2$ if $(i,j) \in (\{1\} \times \{3,7,11,\ldots,n-3\}) \cup \{(1, n-5),(2, n-6), (2, n-4), (3, n-5)\} \cup (\{6,10,14,\ldots,m-3\} \times \{n-1\}) \cup (\{3,7,11,\ldots,m-2\} \times \{n\}) \cup (\{4,8,12,\ldots,m-9\} \times \{n-1\}) \cup (\{5,9,13,\ldots,m\} \times \{n-2\}) \cup (\{m-6\} \times \{n-2, n-4, n-6\}) \cup (\{m-5\} \times \{1, n-5, n-7\}) \cup (\{m-4\} \times \{2, n-2, n-4\}) \cup (\{m-3\} \times \{5,9,13,\ldots,n-9\}) \cup (\{m-3\} \times \{n-1, n-3\}) \cup (\{m-2\} \times \{2,4,6,\ldots,n-2\}) \cup (\{m-1\} \times \{1,3,5,\ldots,n-1\}) \cup (\{m\} \times \{2,4,6,\ldots,n-4\})$;and $\sigma((i,j)) = 0$ otherwise. This completes the proof. ■

Theorem 2.4 *If both* $m \geq 6$ *and* $n \geq 6$ *are congruent to* 2 mod 4, *then* $\mathrm{mc}(C_m \square C_n) \leq 3$.

Proof Define $c: V(C_m \square C_n) \to Z_3$ as follows: $c((i,j)) = 1$ if $(i,j) \in (\{1,3,5,\ldots,m-1\} \times \{1,3,5,\ldots,n-1\}) \cup (\{2,4,6,\ldots,m\} \times \{2,4,6,\ldots,n\})$ and $c((i,j)) = 0$ otherwise.

Then, $\sigma((i,j)) = 1$ if $(i,j) \in (\{1,3,5,\ldots,m-1\} \times \{2,4,6,\ldots,n\}) \cup (\{2,4,6,\ldots,m\} \times \{1,3,5,\ldots,n-1\})$; and $\sigma((i,j)) = 0$ otherwise. This completes the proof. ■

We are able to prove the following theorem.

Theorem 2.5 *If* $n \equiv 2 \bmod 4$, *and* $n \geq 6$, *then* $\mathrm{mc}(C_6 \square C_n) = 3$.

3 Conclusion

By the symmetry of m and n in $C_m \square C_n$, the leftover cases are:

$m \equiv 1 \bmod 4$ and $n \equiv 1 \bmod 4$;
$m \equiv 3 \bmod 4$, $m \geq 7$ and $n \equiv 3 \bmod 4$, $n \geq 7$;
$m = 7$ and $n \neq 6$, $n \equiv 2 \bmod 4$; $m = 5$ and $n \equiv 2 \bmod 4$;
$m \equiv 1 \bmod 4$ and $n = 6$; $m \equiv 2 \bmod 4$, $m \geq 10$ and $n \equiv 2 \bmod 4$, $n \geq 10$.
In all the above leftover cases, we conjecture that $\mathrm{mc}(C_m \square C_n) = 3$.

References

1. R. Balakrishnan and K. Ranganathan, "*A textbook of graph theory*", Second Edition, Universitext, Springer, New York, 2012.
2. F. Okamoto, E. Salehi and P. Zhang, "A checkerboard problem and modular colorings of graphs", *Bull. Inst. Combin. Appl.*, 58 (2010), 29–47.
3. F. Okamoto, E. Salehi and P. Zhang, "A solution to the checkerboard problem", *Int. J. Comput. Appl. Math.*, 5 (2010), 447–458.
4. N. Paramaguru and R. Sampathkumar, "Modular chromatic number of $C_m \square P_n$," *Trans. Comb.* 2. No. 2. (2013), 47–72.

Network Analysis of Biserial Queues Linked with a Flowshop Scheduling System

Seema Sharma, Deepak Gupta and Sameer Sharma

Abstract This paper is an attempt to establish a linkage between networks of queues consisting of two parallel biserial servers connected with a common server in series and a multistage flowshop scheduling system having *m* machines. In the queue network, both the arrival and service patterns follow Poisson law. The objective of this paper is to develop an algorithm minimizing the total elapsed time with minimum completion time, average waiting time, and minimum queue length for the proposed queuing–scheduling linkage model. The efficiency of the proposed algorithm is tested by a numerical illustration.

Keywords Biserial servers · Mean queue length · Linkage network · Flowshop scheduling · Elapsed time · Completion time

1 Introduction

Queues or waiting lines arise when the demand for a service facility exceeds the capacity of the server, i.e., customers do not get service immediately upon request, but wait, or the service facilities stand idle. The origin of queuing theory can be traced back to early in the last century when Erlang [1], a Danish engineer, applied this theory extensively to study the behavior of telephone networks. Jackson [2] studied the behavior of a queuing system containing phase-type service. John [3]

S. Sharma (✉) · S. Sharma
D.A.V. College, Jalandhar, Punjab, India
e-mail: seemasharma7788@yahoo.com

S. Sharma
e-mail: samsharma31@gmail.com

D. Gupta
M.M. University, Mullana, Ambala, India
e-mail: guptadeepak2003@yahoo.co.in

G. S. S. Krishnan et al. (eds.), *Computational Intelligence, Cyber Security and Computational Models*, Advances in Intelligent Systems and Computing 246,
DOI: 10.1007/978-81-322-1680-3_37,

has given the formula to find the mean queue *length.* Maggu [4] introduced the concept of bitendom in theory of queues. Vinod et al. [5] discussed the steady-state behavior of a queue model comprised of two systems with biserial channels linked to a common channel. Gupta et al. [6] studied the network queue model comprised of biserial and parallel channels linked with a common server. Sharma et al. [7] studied the analysis of network of biserial queues linked with a common server. Flowshop scheduling with an objective to minimize the makespan is an important task in manufacturing concern. The scheduling problem occurs very commonly whenever there is a choice as to order in which a number of jobs/tasks can be performed. One of the earliest results in flowshop scheduling is an algorithm given by Johnson [8] for scheduling jobs on two or three machines to minimize the total flow time. The literature revealed that a lot of research works have already been conducted in the field of queuing and scheduling theory individually. Only some efforts have been made to establish a linkage between these two fields of optimization. Singh et al. [9] studied the linkage of scheduling system with a serial queue network. Sharma et al. [10] studied the linkage network of queues with a multistage flowshop scheduling system.

In this paper, a linkage between networks of queues consisting of two parallel biserial servers connected with a common server in series and a multistage flowshop scheduling system having m machines is considered in which arrival rate, service rate, and processing time of various jobs/tasks are used to derive various system characteristics. The completion time of jobs/tasks in processing through network of queues (phase I of service) will be the setup time for first machine in flowshop scheduling system (phase II of service). The rest of paper is organized as follows: Sect. 2 describes the mathematical model of the proposed linkage model and its analysis. In Sect. 3, the algorithm is proposed to find various queue characteristics and minimum total elapsed time. In Sect. 4, a numerical illustration to test the efficiency of the proposed algorithm is demonstrated. The paper is concluded in Sect. 5, followed by references.

2 Mathematical Model and Analysis

The entire model is comprised of three servers S_1, S_2, S_3, and a system of machines in series. The servers S_1 and S_2 consist of two biserial service servers. Server S_3 is commonly linked in series with each of two servers S_1 and S_2 for completion of phase I of service demanded either at a subsystem S_1 or at a subsystem S_2. The service time at S_{ij} $(i, j = 1, 2)$ is exponentially distributed. Let mean service rate at S_{ij} $(i, j = 1, 2)$ be $\mu_1, \mu_2, \mu_3, \mu_4$, and μ_5 at S_3, respectively. Queues Q_1, Q_2, Q_3, Q_4, and Q_5 are said to be formed in front of the servers if they are busy. Customers arriving at rate λ_1 after completion of service at S_{11} will go to the network of the servers $S_{11} \to S_{12}$ or $S_{11} \to S_3$ with probabilities p_{12} or p_{15} such that $p_{12} + p_{15} = 1$. Further, customers arriving at rate λ_2 after completion of service at S_{12} will go to the network of the servers $S_{12} \to S_{11}$ or $S_{12} \to S_3$ with probabilities

p_{21} or p_{25} such that $p_{21} + p_{25} = 1$. Similarly, customers arriving at rate λ_3 after completion of service at S_{21} will go to the network of the servers $S_{21} \rightarrow S_{22}$ or $S_{21} \rightarrow S_3$ with probabilities p_{34} or p_{35} such that $p_{34} + p_{35} = 1$ and the customers arriving at rate λ_4 after completion of service at S_{22} will go to the network of the servers $S_{22} \rightarrow S_{21}$ or $S_{22} \rightarrow S_{21}$ with probabilities p_{43} or p_{45} such that $p_{43} + p_{45} = 1$ (Fig. 1).

The completion time (waiting time + service time) of tasks/jobs through queues Q_1, Q_2, Q_3, Q_4, and Q_5 form the setup time for the first machine in phase II of service having m machines in series.

Let P_{n_1,n_2,n_3,n_4,n_5} be the joint probability that there are n_1 units waiting in queue Q_1 in front of S_{11}, n_2 units waiting in queue Q_2 in front of S_{12}, n_3 units waiting in queue Q_3 in front of S_{21}, n_4 units waiting in queue Q_4 in front of S_{22}, and n_5 units waiting in queue Q_5 in front of S_3. In each case, waiting induces a unit in service if any. Also, $n_1, n_2, n_3, n_4, n_5 > 0$. Now, proceeding on the lines of Sharma et al. [7, 10], we get the following standard results: The joint probability function is:

$$
\begin{aligned}
P_{n_1,n_2,n_3,n_4,n_5} &= (1-F_1)^{n_1}(1-F_2)^{n_2}(1-F_2)^{n_3}(1-F_2)^{n_4}(1-F_2)^{n_5}F_1 \cdot F_2 \cdot F_3 \cdot F_4 \cdot F_5 \\
&= (\rho_1)^{n_1}(\rho_2)^{n_2}(\rho_3)^{n_3}(\rho_4)^{n_4}(\rho_5)^{n_5}(1-\rho_1)(1-\rho_2)(1-\rho_3)(1-\rho_4)(1-\rho_5)
\end{aligned}
$$

where $\rho_1 = 1 - F_1, \rho_2 = 1 - F_2, \rho_3 = 1 - F_3, \rho_4 = 1 - F_4$ and $\rho_5 = 1 - F_5$.

The solution in steady state exists if $\rho_1, \rho_2, \rho_3, \rho_4, \rho_5 \leq 1$.

The average number of customers (mean queue length)

$$
L = \sum\nolimits_{n_1=0}^{\infty} \sum\nolimits_{n_2=0}^{\infty} \sum\nolimits_{n_3=0}^{\infty} \sum\nolimits_{n_4=0}^{\infty} \sum\nolimits_{n_5=0}^{\infty} (n_1 + n_2 + n_3 + n_4 + n_5) \cdot P_{n_1,n_2,n_3,n_4,n_5}
$$
$$
= L_1 + L_2 + L_3 + L_4 + L_5
$$

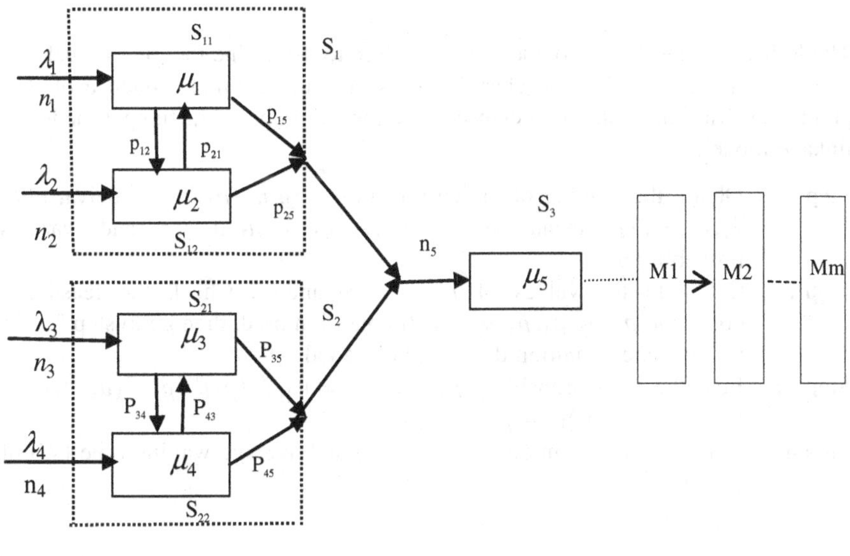

Fig. 1 Linkage network of biserial queue

where $L_1 = \frac{\rho_1}{1-\rho_1}, L_2 = \frac{\rho_2}{1-\rho_2}, L_3 = \frac{\rho_3}{1-\rho_3}, L_4 = \frac{\rho_4}{1-\rho_4}$ and $L_5 = \frac{\rho_5}{1-\rho_5}$.

The average waiting time and the average number of items waiting for service in a service system are important measurements for a manager. Little's law relates these two metrics to the average rate of arrivals to the system, i.e., if $L =$ average number of items in the queuing system, $W =$ average waiting time in the system for an item, and $\lambda =$ average number of items arriving per unit time; then, $L = W\lambda$.

2.1 Theorem

To find the optimal sequence of jobs/tasks to minimize the total elapsed time when a set of n jobs/tasks after completing the phase I of the service entered into the phase II of service consisting of m machines in series with no passing and no preemission is allowed, the following result is considered as derived by Sharma et al. [7, 10].

Statement: Let n jobs J_1, J_2, J_3, …, J_n are processed through m machines M_j ($j = 1, 2, \ldots, m$) in order $M_1 - M_2 - M_3 - \cdots - M_m$ with no passing allowed. Let $A_{i,j}$ represent the processing time of ith job ($i = 1, 2, \ldots, n$) on jth machine ($j = 1, 2, \ldots, m$) such that $\min A_{i,s} \geq \max A_{i,(s+1)}$; $s = 1, 2, \ldots, (m - 2)$, then the optimal schedule minimizing the total elapsed time is given by the following decision rule:

Job J_k proceeds job J_{k+1} if $\min\{G_k, H_{k+1}\} < \min\{G_{k+1}, H_k\}$, where $G_i = A_{i,1} + A_{i,2} + \cdots + A_{i,(m-1)}$ and $H_i = A_{i,2} + A_{i,3} + \cdots + A_{i,m}$.

3 Algorithm

The following algorithm provides the procedure to determine the joint probability function, various other queue characteristics, and the optimal sequence of jobs processing with minimum total elapsed time for the proposed queuing–scheduling linkage model:

Step 1: Obtain the number of the customers n_1, n_2, n_3, n_4; mean arrival rates $\lambda_1, \lambda_2, \lambda_3, \lambda_4$; mean service rates $\mu_1, \mu_2, \mu_3, \mu_4, \mu_5$; and various probabilities.

Step 2: Calculate the values of $\rho_1, \rho_2, \rho_3, \rho_4$ and ρ_5. Check the feasibility condition $\rho_1, \rho_2, \rho_3, \rho_4, \rho_5 \leq 1$. If condition hold, then go to step 3, else steady-state condition does not hold good.

Step 3: Find the joint probability $p_{n_1,n_2,n_3,n_4,n_5} = (\rho_1)^{n_1}(\rho_2)^{n_2}(\rho_3)^{n_3}(\rho_4)^{n_4}(\rho_5)^{n_5}$ $(1-\rho_1)(1-\rho_2)(1-\rho_3)(1-\rho_4)(1-\rho_5)$.

Step 4: Calculate the mean queue length (L) and average waiting time (W) of customers.

Step 5: Find the total completion time (C) of jobs/customers in getting phase I of the service by using the formula

$$C = E(W) + \frac{1}{\mu_1 p_{12} + \mu_1 p_{15} + \mu_2 p_{21} + \mu_2 p_{25} + \mu_3 p_{34} + \mu_3 p_{35} + \mu_4 p_{43} + \mu_4 p_{45} + \mu_5}.$$

Step 6: Define the machines $M_1, M_2, \ldots, M_m$ with processing time $A_1 = A_1 + C$ and $A_2, \ldots, A_m$.

Step 7: Check the consistency of the structural conditions so as to reduce n jobs m machines problem to n jobs 2 machines problem using Theorem 2.1.

Step 8: Apply modified technique as established by Sharma et al. [7, 10] to find the optimal sequence(s) of jobs with minimum total elapsed time.

4 Test of Proposed Algorithm

To test the efficiency of proposed algorithm, the following numerical illustration is carried out. Consider that twelve jobs/customers are processed through the network of queues with the servers S_1, S_2, and S_3. The servers S_1 and S_2 consist of two biserial service servers, and the server S_3 is commonly linked in series with each of two servers. The number of customers, mean arrival rate, mean service rate, and associated probabilities are given in Table 1.

After getting phase I of service in queuing network, jobs/customers are to be processed on the machines M_1, M_2, M_3 and M_4 in series with processing time $A_{i,1}$, $A_{i,2}$, $A_{i,3}$, and $A_{i,4}$, respectively, as given in Table 2.

The objective is to find the mean queue length, the average waiting time of jobs/tasks, and the optimal sequence of jobs/tasks with minimum total elapsed time for the proposed queuing–scheduling linkage model.

Here, we have $\rho_1 = 0.58908, \rho_2 = 0.804598, \rho_3 = 0.583337, \rho_4 = 0.75$ and $\rho_5 = 0.587644$

The joint probability is as follows:

$$\begin{aligned} P_{n_1,n_2,n_3,n_4,n_5,n_6} &= \rho_1^{n_1}\rho_2^{n_2}\rho_3^{n_3}\rho_4^{n_4}\rho_5^{n_5}(1-\rho_1)(1-\rho_2)(1-\rho_3)(1-\rho_4)(1-\rho_5) \\ &= 3.78085 \times 10^{-8}. \end{aligned}$$

Table 1 The detail classification of the queuing model

S. No.	No. of customers	Mean arrival rate	Mean service rate	Probabilities
1	$n_1 = 3$	$\lambda_1 = 2$	$\mu_1 = 12$	$p_{12} = 0.6$, $p_{15} = 0.4$
2	$n_2 = 2$	$\lambda_2 = 3$	$\mu_2 = 9$	$p_{21} = 0.7$, $p_{25} = 0.3$
3	$n_3 = 4$	$\lambda_3 = 5$	$\mu_3 = 15$	$p_{34} = 0.4$, $p_{35} = 0.6$
4	$n_4 = 3$	$\lambda_4 = 4$	$\mu_4 = 10$	$p_{43} = 0.5$, $p_{45} = 0.5$
5	$n_5 = 12$		$\mu_5 = 24$	

Table 2 The machines M_1, M_2, M_3, and M_4 with processing times

Jobs	1	2	3	4	5	6	7	8	9	10	11	12
$M_1(A_{i,1})$	10	12	15	11	10	14	12	10	13	11	14	13
$M_2(A_{i,2})$	9	7	9	9	10	9	10	8	9	9	10	9
$M_3(A_{i,3})$	8	7	6	7	6	4	8	7	6	8	7	6
$M_4(A_{i,4})$	5	4	3	4	4	3	6	3	2	3	5	2

The mean queue length (average no. of customers) $= L = 11.3763$ units
Average waiting time for customers $= E(w) = 0.812593$ units.

The total completion time of jobs/customers when processed through queue network in phase I of service is 0.826879 units. On taking this as the setup time on the first machine M_1 in the phase II of service and using the various steps of the proposed algorithm, the optimal sequence(s) of jobs/customers is $S =$ 7–1–11–5–10–4–2–3–8–12–9–6 with minimum elapsed time of 168.923 units.

5 Conclusion

In this paper, a linkage between networks of queues consisting of two biserial servers connected to a common server with a multistage flowshop scheduling is considered. Various results have been established to find the mean queue length, average waiting time, total completion time, and the optimal sequence of jobs/tasks with minimum total elapsed time. The objective of the paper is of twofold: On the one hand, it optimizes the various queue characteristics such as the mean queue length and total elapsed time of jobs/tasks, and on the other hand, it finds the optimal sequence of jobs/tasks processed with minimum total elapsed time for the proposed queuing–scheduling linkage model. The proposed model finds its application in healthcare centers, administrative setups, supermarkets, office management, communication network, and steel-making plants, and in many other manufacturing setups. The study may further be extended by introducing more complex queuing network linked to a common server, by introducing the concepts of retrial queues, by introducing some of scheduling constraints such as independent setup time and transportation time of jobs/tasks from one machine to another, and by introducing the concept of fuzzy environment in the processing of jobs/tasks.

References

1. A. K. Erlang, "The theory of probabilities and telephone conversations", *Nyt Tidsskrift fur Matematik*, Vol. B, No. 20, pp. 33, 1909.
2. R. R. P. Jackson, "Queuing system with phase type service", *O. R.Quat.*, Vol. 5, pp. 109–120, 1954.
3. D. C. John, "A proof of queuing formula: "$L = \lambda W$"", *Operation Research*, Vol. 13, pp. 400–412, 1965.

4. P. L. Maggu, "Phase type service queue with two channels in Biserial", *J.OP. Res. Soc. Japan*, Vol. 13, No. 1, 1970.
5. K. Vinod, T. P. Singh and R. Kumar, "Steady state behaviour of a queue model comprised of two subsystems with biserial linked with common channel", *Reflection des ERA*, Vol. 1, No. 2, pp. 135–152, 2007.
6. D. Gupta, T.P Singh and R. Kumar, "Analysis of a network queue model comprised of biserial and parallel channel linked with a common server", *Ultra Science*, Vol. 19, No. 2, pp. 407–418, 2007.
7. S. Sharma, D. Gupta and S. Sharma, "Analysis of linkage network of queues with a multistage flowshop scheduling system", International journal of Operation Research, Accepted for publication, 2013.
8. S. M. Johnson, "Optimal two & three stage production schedules with set up times includes", *Nav. Res. Log. Quart.*, Vol. 1, pp. 61–68, 1954.
9. T. P. Singh and V. Kumar, "On linkage of a scheduling system with biserial queue network", *Arya Bhatta Journal of Mathematics & Informatics*, Vol. 1, No. 1, pp. 71–76, 2009.
10. S. Sharma, D. Gupta and S. Sharma, "Analysis of network of biserial queues linked with a common server", International journal of Computing and Mathematics, Accepted for publication, 2013a.

Solving Multi-Objective Linear Fractional Programming Problem—First Order Taylor's Series Approximation Approach

C. Veeramani and M. Sumathi

Abstract In this paper, a method is proposed for solving multi-objective linear fractional programming (MOLFP) problem. Here, the MOLFP problem is transformed into an equivalent multi-objective linear programming (MOLP) problem. Using the first-order Taylor's series approximation, the MOLFP problem is reduced to single-objective linear programming (LP) problem. Finally, the solution of MOLFP problem is obtained by solving the resultant LP problem. The proposed procedure is verified with the existing methods through the numerical examples.

Keywords Linear programming problem · Multi-objective linear programming problem · Taylor's series

1 Introduction

The decision makers, in the sectors such as financial and corporate planning, production planning, marketing and media selection, university planning and student admissions, health care and hospital planning, often face problems to take decisions that optimize department/equity ratio, profit/cost, inventory/sales, actual cost/standard cost, output/employee, student/cost, nurse/patient ratio, etc. The above problems can be solved efficiently through linear fractional programming (LFP) problems.

C. Veeramani (✉) · M. Sumathi
Department of Applied Mathematics and Computational Sciences,
PSG College of Technology, Coimbatore 641004, India
e-mail: veerasworld@yahoo.com

M. Sumathi
e-mail: ramasumathi.psg@gmail.com

G. S. S. Krishnan et al. (eds.), *Computational Intelligence, Cyber Security and Computational Models*, Advances in Intelligent Systems and Computing 246,
DOI: 10.1007/978-81-322-1680-3_38, © Springer India 2014

A general LFP problem can be formulated as follows:

$$\begin{cases} & \operatorname{Max} Z(x) = \frac{c^T x + p}{d^T x + q} \\ \text{subject to} & \\ & Ax \leq b, \\ & x \geq 0 \end{cases} \tag{1}$$

where $x, c^T, d^T \in R^n$, $A \in R^{m \times n}$, $b \in R^m$, and $p, q \in R$.

Charnes and Cooper [2] showed that if the denominator is constant in sign on the feasible region, the LFP problem can be optimized by solving a linear programming problem. Many authors (see [4, 5, 7, 14]) have presented methods for solving LFP problems.

However, in many applications, there are two or more conflicting objective functions, which are relevant, and some compromise must be sought between them. For example, a management problem may require the profit/cost, quality/cost, and other ratios to be maximized and these conflict. Such types of problems can be solved efficiently through multi-objective linear fractional programming (MOLFP) problems. The general form of MOLFP problem can be formulated as follows:

$$\begin{cases} & \operatorname{Max} Z_i(x) = \frac{c_i^T x + p_i}{d_i^T x + q_i} \\ \text{subject to} & \\ & Ax \leq b, \\ & x \geq 0 \end{cases} \tag{2}$$

where $i = 1, 2 \ldots k$, $A \in R^{m \times n}$, $x \in R^n$, $b \in R^m$, $c, d \in R^n$, and $p, q \in R$.

There exist several methodologies to solve MOLFP problem. Kornbluth and Steuer [9] have proposed an algorithm for solving the MOLFP problem for all w-efficient vertices of the feasible region. The solution set concepts of w-efficient subsumes that of s-efficient. Kornbluth and Steuer [8] have discussed a generalized approach for solving a goal program with linear fractional criteria. Nykowski and Zolkiewski [13] have presented a compromise procedure for MOLFP problem. Luhandjula [10] solved MOLFP problem using a fuzzy approach. He used linguistic approach to solve MOLFP problem by introducing linguistic variables to represent linguistic aspirations of the decision maker. Dutta et al. [6] modified the linguistic approach of Luhandjula [10] to solve MOLFP problem. Chakraborty and Gupta [2] have developed fuzzy mathematical programming approach for solving multi-objective LFP problem. Calvete and Gale [1] have proposed a penalty method for solving bilevel linear fractional programming/linear programming problems. In this paper, we propose a method for solving MOLFP problems using first-order Taylor's series approach.

The rest of our work is organized as follows. In Sect. 2, some basic definitions and the concept of solving FLP problem is recalled. In Sect. 3, the method of converting MOLFP problem into MOLP problem is discussed. The procedure of solving MOLFP problem using Taylor's series approach is presented in Sect. 4. In

Sect. 5, the proposed procedure illustrated through a numerical example and compare with the existing methods is given.

2 Linear Fractional Programming Problem

In this section, the general form of LFP problem is discussed. Also, Charnes and Cooper's [2] linear transformation is summarized. The LFP problem can be written as

$$\begin{cases} \text{Max } Z(x) = \frac{c^T x + p}{d^T x + q} = \frac{N(x)}{D(x)} \\ \text{subject to} \\ x \in S = \{x : Ax \leq b, x \geq 0\} \end{cases} \tag{3}$$

For some values of x, $D(x)$ may be equal to zero. To avoid such cases, one requires that either $\{x \geq 0, Ax \leq b \Rightarrow D(x) > 0\}$ or $\{x \geq 0, Ax \leq b \Rightarrow D(x) < 0\}$. For convenience, assume that LFP satisfies the condition that

$$x \geq 0, Ax \leq b \Rightarrow D(x) > 0. \tag{4}$$

Definition 1 The two mathematical programming problem (a) Max $F(x)$, subject to $x \in S$, (b) Max $G(x)$, subject to $x \in U$ will be said to be equivalent iff there is a one to one map f of the feasible set of (a), onto the feasible set of (b), such that $F(x) = G(f(x))$ for all $x \in S$.

Theorem 1 *Assume that no point $(z, 0)$ with $z \geq 0$ is feasible for the following linear programming problem*

$$\begin{cases} \text{Max } c^T z + pt \\ \text{subject to} \\ d^T z + qt = 1 \\ Az - bt = 0 \\ t \geq 0, z \geq 0, z \in R^n, t \in R \end{cases} \tag{5}$$

Now assume that the condition (4), then the LFP (3) us equivalent to linear program problem (5). Consider the two related problems

$$\begin{cases} \text{Max } tN(z/t) \\ \text{subject to} \\ A(z/t) - b \leq 0 \\ tD(z/t) = 1 \\ t > 0, z \geq 0 \end{cases} \tag{6}$$

and

$$\begin{cases} \qquad \text{Max } tN(z/t) \\ \text{subject to} \\ \qquad A(z/t) - b \leq 0 \\ \qquad tD(z/t) \leq 1 \\ \qquad t > 0, z \geq 0 \end{cases} \tag{7}$$

where (6) is obtained from (3) by the transformation $t = 1/D(x), z = tx$ and (7) differs from (6) by replacing the equality constraint $D(z/t) = 1$ by an inequality constraint $tD(z/t) \leq 1$.

Definition 2 The problem (3) will be said to be standard concave–convex programming problem, if $N(x)$ is concave on S with $N(\zeta) \geq 0$ for some $\zeta \in S$ and $D(x)$ is convex and positive on S.

Theorem 2 *Let for some $\zeta \in S$, $N(\zeta) \geq 0$, if (3) reaches a (global) maximum at $x = x^*$, then (7) reaches a (global) maximum at a point $(t, z) = (t^*, z^*)$, where $z^*/t^* = x^*$ and the objective functions at these points are equal.*

Theorem 3 *If* (3) *is a standard concave–convex programming problem which reaches a (global) maximum at a point x^*, then the corresponding transformed problem* (7) *attains the same maximum value at a point (t^*, z^*), where $z^*/t^* = x^*$. Moreover, (7) has a concave objective function and a convex feasible set.*

Suppose in (3), $N(x)$ is concave, $D(x)$ is concave and positive on S and $N(x)$ is negative for each $x \in S$, then

$$\text{Max}_{x \in S} \frac{N(x)}{D(x)} \Leftrightarrow \text{Min}_{x \in S} \frac{-N(x)}{D(x)} \Leftrightarrow \text{Max}_{x \in S} \frac{D(x)}{-N(x)}$$

where $-N(x)$ is convex and positive. Using Theorem (3), the fractional program (3) transformed to the following linear programming problem:

$$\begin{cases} \qquad \text{Max } tD(z/t) \\ \text{subject to} \\ \qquad A(z/t) - b \leq 0 \\ \qquad -tN(z/t) \leq 1 \\ \qquad t > 0, z \geq 0 \end{cases} \tag{8}$$

3 Multi-Objective Linear Fractional Programming Problem

In this section, the general form of MOLFP problem is presented. Furthermore, Chakraborty and Gupta [3] theoretical framework for converting MOLFP problem into MOLP problem is discussed. The MOLFP problem can be written as follows:

$$\begin{cases} \text{Max } Z_i(x) = \frac{c_i^T x + p_i}{d_i^T x + q_i} = \frac{N_i(x)}{D_i(x)} \\ \text{subject to} \\ x \in S = \{x \in R^n : Ax \leq b, x \geq 0\} \end{cases} \tag{9}$$

where $i = 1, 2, \ldots k$, $A \in R^{m \times n}$, $b \in R^n$, $c, d \in R^n$, and $p, q \in R$.

3.1 Theoretical Framework

Let, I be the index set such that $I = \{i : N_i(x) \geq 0 \text{ for some } x \in S\}$ and $I^c = \{i : N_i(x) < 0 \text{ for each } x \in S\}$, where $I \cup I^c = \{1, 2, \ldots k\}$.

Let $D(x)$ be positive on S, where S is non-empty and bounced. For simplicity, let us take the least value of $1/(d_i x + q_i)$ and $1/-(c_i + p_i)$ is t for $i \in I$ and $i \in I^c$, respectively. That is,

$$\cap_{i \in I} \frac{1}{d_i x + q_i} = t \text{ and } \cap_{i \in I^c} \frac{-1}{c_i x + p_i} = t \tag{10}$$

which is equivalent to

$$\frac{1}{d_i x + q_i} \geq t \text{ for } i \in I \text{ and} \frac{-1}{c_i x + p_i} \geq t \text{ for } i \in I^c. \tag{11}$$

Using the transformation $z = tx (t > 0)$, Theorems (2) and (3) and Eq. (11), MOLFP problem (9) may be written as follows:

$$\begin{cases} \text{Max } f_i(z, t) = \{tN_i(z/t), \text{ for } i \in I; tD_i(z/t), \text{ for } i \in I^c\} \\ \text{subject to} \\ tD_i(z/t) \leq 1, \text{ for } i \in I \\ -tN_i(z/t) \leq 1, \text{ for } i \in I^c \\ A(z/t) - b \leq 0, \\ t \geq 0, \ z \geq 0 \end{cases} \tag{12}$$

4 First-order Taylor's Series Approximation for MOLFP Problem

In this section, the method of converting MOLFP problem into LP problem is studied.

In the MOLFP problem, objective functions are transformed into single objective using the first-order Taylor's approximation of MOLFP problem about the solution of the MOLP problem. Here, the obtained Taylor's series polynomial series of the objective functions are equivalent to the fractional objective functions. Then, the MOLFP problem can be reduced into an LP problem. The proposed approach can be explained as follows:

Step 1. Determine $x_i^* = x_{i1}^*, x_{i2}^* \ldots, x_{in}^*$ which is the value(s) that is used to maximize the ith objective function $Z_i(x)$ where $i = 1, 2 \ldots k$.
Step 2. Transform objective function using first-order Taylor's polynomial series.

$$\begin{aligned} Z_i(x) &= Z_i(x_i^*) + (x_1 - x_{i1}^*)\frac{\partial Z_i(x_i^*)}{\partial x_1} + (x_2 - x_{i2}^*)\frac{\partial Z_i(x_i^*)}{\partial x_2} \\ &\quad + (x_n - x_{in}^*)\frac{\partial Z_i(x_i^*)}{\partial x_n} \\ Z_i(x) &= Z_i(x_i^*) + \sum_{j=1}^{n}\left(x_j - x_{ij}^*\right)\frac{\partial z_i(x_i^*)}{\partial x_j} \end{aligned} \tag{13}$$

Step 3. Find satisfactory $x^* = (x_1^*, x_2^* \ldots, x_n^*)$ by solving the reduced problem to a single objective. That is

$$Z(x) = \sum_{i=1}^{k}\left(Z_i(x_i^*) + \sum_{j=1}^{n}\left(x_j - x_{ij}^*\right)\frac{\partial Z_i(x_i^*)}{\partial x_j}\right) \tag{14}$$

By applying above procedure on the problem (9), it is reduced as follows:

$$\begin{cases} & Z = \max \sum_{i=1}^{k}\left(Z_k(x_i^*) + \sum_{j=1}^{n}\left(x_j - x_{ij}^*\right)\frac{\partial Z_k(x_i^*)}{\partial x_j}\right) \\ \text{subject to} & \\ & x \in S = \{x \in R^n : Ax \le b, x \ge 0\} \end{cases}$$

which is a LP problem, can be solved by traditional approach.

The proposed approach for solving MOLFP problem can be summarized in the following manner:

Step 1. Maximize each objective function $Z_i(x)$ subject to the given set of constraints using the method proposed by Charnes and Cooper. Let Z_i^* be the maximum value of $Z_i(x)$ for $i = 1, 2, \ldots, k$.
Step 2. Examine the nature of Z_i^* for all values of $i = 1, 2, \ldots, k$. If $Z_i^* \ge 0$, then $i \in I$, and if $Z_i^* \le 0$, then $i \in I^c$.
Step 3. For $i \in I$, then we may assume the maximum aspiration level is Z_i^* for $i \in I$ and $i \in I^c$, then we may assume the maximum aspiration level is $-1/Z_i^*$.

Step 4. The obtained Z_i^*, for $i = 1, 2, \ldots, k$ and Taylor's series approach, the MOLFP problem can be converted into a single-objective LP problem, which can be solved using classical methods.

5 Numerical Example

The following example studied by Chakraborty and Gupta [3] is considered to demonstrate the effectiveness of the proposed Taylor's series approach.

Example 1 Let us considered a MOLFP problem with two objectives as follows:

$$\begin{cases} \text{Max} \left\{ Z_1(x) = \dfrac{-3x_1 + 2x_2}{x_1 + x_2 + 3} \right. \\ \text{subject to} \quad \left. Z_2(x) = \dfrac{7x_1 + x_2}{5x_1 + 2x_2 + 1} \right\} \\ x_1 - x_2 \geq 1 \\ 2x_1 + 3x_2 \leq 15 \\ x_1 \geq 3 \text{ and } x_1,\ x_2 \geq 0 \end{cases}$$

Here, we observed that $Z_1(x) < 0$ for each x in the feasible region and $Z_2(x) \geq 0$ some x in the feasible region. The above MOLFP problem is equivalent to the following MOLP problem

$$\begin{cases} \text{Max} \ \{f_1(z,t) = z_1 + z_2 + 3t \\ \qquad f_2(z,t) = 7z_1 + z_2\} \\ \text{subject to} \\ 3z_1 - 2z_2 \leq 1 \\ 5z_1 + 2z_2 + t \leq 1 \\ z_1 - z_2 - t \geq 0 \\ 2z_1 + 3z_2 - 15t \leq 0 \\ z_1 - 3t \geq 0 \text{ and } z_1,\ z_2,\ t \geq 0 \end{cases}$$

If the problem is solved for each objective functions one by one, then $Z_1(3.56, 2.56)$ and $Z_2(6.33, 0)$.

Then objective functions of MOLFP problem are transformed using first-order Taylor's polynomial series.

$$Z_1(x) = -2.56x_1 + 0.286x_2 - 0.409$$
$$Z_2(x) = -0.053x_2 + 1.357$$

Hence, the objective function of the MOLFP problem is

$$Z(x) = -2.62x_1 + 0.233x_2 + 0.948$$

Thus, equivalent formulation for the MOLFP problem is obtained as follows:

$$\begin{cases} \text{Max } Z(x) = -2.62x_1 + 0.233x_2 + 0.948 \\ \text{subject to} \\ x_1 - x_2 \geq 1 \\ 2x_1 + 3x_2 \leq 15 \\ x_1 \geq 3 \text{ and } x_1,\ x_2 \geq 0 \end{cases}$$

The problem is solved, and the solution of the problem is $x_1 = 3,\ x_2 = 3$, $Z_1(x) = -0.625$ and $Z_2(x) = 1.15$.

Example 2 Let us considered a MOLFP problem with three objectives as follows:

$$\begin{cases} \text{Max } Z(x) = \left\{ Z_1(x) = \dfrac{-3x_1 + 2x_2}{x_1 + x_2 + 3} \right. \\ Z_2(x) = \dfrac{7x_1 + x_2}{5x_1 + 2x_2 + 1} \\ \left. Z_3(x) = \dfrac{x_1 + 4x_2}{2x_1 + 3x_2 + 2} \right\} \\ \text{subject to} \\ x_1 - x_2 \geq 1 \\ 2x_1 + 3x_2 \leq 15 \\ x_1 + 9x_2 \geq 9 \\ x_1 \geq 3 \text{ and } x_1,\ x_2 \geq 0 \end{cases}$$

Here, we observed that for each x in feasible region, $Z_1 < 0$, $Z_2 \geq 0$, and $Z_3 \geq 0$ for some x in the feasible region. The above MOLFP problem is equivalent to the following MOLP problem

$$\begin{cases} \text{Max } \{f_1(z,t) = z_1 + z_2 + 3t \\ f_2(z,t) = 7z_1 + z_2 \\ f_3(z,t) = z_1 + 4z_2\} \\ \text{subject to} \\ 3z_1 - 2z_2 \leq 1 \\ 5z_1 + 2z_2 + t \leq 1 \\ 2z_1 + 3z_2 + 2t \leq 1 \\ z_1 - z_2 - t \geq 0 \\ 2z_1 + 3z_2 - 15t \leq 0 \\ z_1 - 3t \geq 0 \text{ and } z_1,\ z_2,\ t \geq 0 \end{cases}$$

If the problem is solved for each objective functions one by one, then $Z_1(3.56, 2.56)$, $Z_2(6.33, 0.33)$, and $Z_3(3.5, 2.5)$.

Then, objective functions of MOLFP problem are transformed using first-order Taylor's polynomial series.

$$z_1(x) = -2.62x_1 + 0.286x_2 - 0.409$$
$$z_2(x) = 0.008x_1 - 0.021x_2 + 1.3063$$
$$z_3(x) = -0.039x_1 + 0.094x_2 + 0.7195$$

Hence, the objective of the MOLFP problem is

$$Z(x) = -2.651x_1 + 0.359x_2 + 1.6168$$

Thus, equivalent formulation for the MOLFP problem is obtained as follows:

$$\begin{cases} \text{Max } Z(x) = -2.651x_1 + 0.359x_2 + 1.6168 \\ \text{subject to} \\ x_1 - x_2 \geq 1 \\ 2x_1 + 3x_2 \leq 15 \\ x_1 + 9x_2 \geq 9 \\ x_1 \geq 3 \text{ and } x_1,\ x_2 \geq 0 \end{cases}$$

The problem is solved, and the solution for the above problem is $x_1 = 3, x_2 = 3$, $Z_1(x) = -0.625$, $Z_2(x) = 1.15$, and $Z_3(x) = 0.79$.

The same solution (of the above problems Examples 1 and 2) is obtained by Chakraborty and Gupta [3] using fuzzy mathematical programming technique.

Example 3 The following example considered by Neelam Malhotra and Arora [12] is again used to demonstrate the solution procedures and clarify the effectiveness of the proposed approach. Consider the MOLFP problem with two objectives as follows:

$$\begin{cases} \text{Max } Z(x) = \left\{ Z_1(x) = \dfrac{x_1 + 2x_2}{x_1 + x_2 + 1} \right. \\ \qquad Z_2(x) = \left. \dfrac{2x_1 + x_2}{2x_1 + 3x_2 + 1} \right\} \\ \text{subject to} \\ -x_1 + 2x_2 \leq 3 \\ 2x_1 - x_2 \leq 3 \\ x_1 + x_2 \geq 3 \\ 2 \leq x_1 \leq 25 \\ 1 \leq x_2 \leq 9.5 \end{cases}$$

Here, we observed that for each x in feasible region, $Z_1 \geq 0$ and $Z_2 \geq 0$ for some x in the feasible region. The above MOLFP problem is equivalent to the following MOLP problem

$$
\begin{cases}
\{\text{Max } f_1(z,t) = z_1 + 2z_2 \\
\quad f_2(z,t) = 2z_1 + z_2\} \\
\text{subject to} \\
z_1 + z_2 + t \leq 1 \\
2z_1 + 3z_2 + t \leq 1 \\
-z_1 + zx_2 - 3t \leq 0 \\
2z_1 - z_2 - 3t \leq 0 \\
z_1 + z_2 - 3t \geq 0 \\
z_1 - 2t \geq 0 \\
z_1 - 25t \leq 0 \\
z_2 - t \geq 0 \\
z_2 - 9.5t \leq 0 \\
z_1,\ z_2,\ t \geq 0
\end{cases}
$$

If the problem is solved for each objective functions one by one, then $Z_1(3.07, 3.07)$ and $Z_2(1.92, 1)$.

Then, objective functions of MOLFP problem are transformed using first-order Taylor's polynomial series.

$$z_1(x) = -0.041x_1 + 0.1x_2 + 1.1$$
$$z_2(x) = 0.098x_1 - 0.11x_2 + 0.5388$$

Thus, equivalent formulation for the MOLFP problem is obtained as follows:

$$
\begin{cases}
\text{Max } Z(x) = 0.057x_1 - 0.01x_2 + 1.6388 \\
\text{subject to} \\
-x_1 + 2x_2 \leq 3 \\
2x_1 - x_2 \leq 3 \\
x_1 + x_2 \geq 3 \\
2 \leq x_1 \leq 25 \\
1 \leq x_2 \leq 9.5
\end{cases}
$$

The problem is solved, and the solution of the problem is $x_1 = 3$, $x_2 = 3$, $Z_1(x) = 1.286$, and $Z_2(x) = 0.5625$. The solution obtained by the proposed method is same as the solution obtained by Neelam Malhotra and Arora [12] (using G.P approach) and Mishra [11] (using weighting method).

6 Conclusion

In this paper, a method of solving the MOLFP problems, using Taylor's series approach is discussed. In the proposed method, MOLFP problem is transformed to a MOLP problem and solved by each objective function. The MOLFP problem is converted to a LP problem, using Taylor's series method. The proposed procedure is verified with the existing methods through the numerical examples. The main advantage of the proposed methodology to solve MOLFP problem always yields an efficient solution and reduces the computational complexity.

In our opinion, there are several other points of research and should be studied later on. Some of these points are as follows: A method is extended for solving LFP problems, where the cost of the objective function, the resources, and the technological coefficients are fuzzy in nature. A stochastic approach may be studied for the above problem. A comparison study can be carried out between the fuzzy approach and the stochastic approach for solving problem.

References

1. H.I. Calvete, and C. Gale, "A penalty method for solving bilevel linear fractional programming/linear programming problems", Asia-Pacific Journal of Operational Research 21, 207–224, 2004.
2. A. Charnes, and W.W. Cooper, W W, "Programming with linear fractional functionals", Nav. Res. Logistics Quart. 9, 181–186, 1962.
3. M. Chakraborty, and S. Gupta, "Fuzzy mathematical programming for multi objective linear fractional programming problem", Fuzzy Sets and Systems 125, 335–342, 2002.
4. J.P. Costa, "An interactive method for multiple objective linear fractional programming problems", OR Spectrum, 27, 633–652, 2005.
5. W. Dinkelbach, "On nonlinear fractional programming", Manage. Sci. 13, 492–498, 1967.
6. D. Dutta, R.N. Tiwari, and J.R. Rao, "Multiple objective linear fractional programming—A fuzzy set theoretic approach", Fuzzy Sets and Systems 52, 39–45, 1992.
7. P.C. Gilmore, and R.E. Gomory, "A linear programming approach to the cutting stock problem. Part II", Operational Research, 11, 863–888, 1963.
8. J.S.H. Kornbluth, and R.E. Steuer, "Goal programming with linear fractional criteria", European J. Operational Research, 8, 58–65, 1981.
9. J.S.H. Kornbluth, and R.E. Steuer, "Multiple objective linear fractional programming", Manage. Sci. 27, 1024–1039, 1981.
10. M.K. Luhandjula, "Fuzzy approaches for multiple objective linear fractional optimization", Fuzzy Sets and Systems 13, 11–23, 1984.
11. S. Mishra, "Weighting method for bi-level linear fractional programming problems", European Journal of Operational Research, 183, 296–302, 2007.
12. Neelam Malhotra and S.R. Arora, "An algorithm to solve linear fractional bi-level programming problem via goal programming", OPSEARCH, 37, 1–13, 2000.
13. I. Nykowski, and Z. Zolkiewski, "A compromise procedure for the multiple objective linear fractional programming problem", European Journal of Operational Research. 19, 91–97, 1985.
14. S. Schaible, "Fractional programming I: duality", Manage. Sci. 22, 658–667, 1976.

[illegible] with the existing [illegible] through the numerical examples. The main advantage of the proposed methodology to solve MOLFP problem always yields an efficient solution and reduces the computational complexity.

In our opinion, there are several other points of research and should be studied later. Some of these points are as follows. A method is expected for solving the problems where the objective functions are fractional, [illegible] and the [illegible] constraints [illegible]. [illegible] approach may be applied for the above problems. A comparative study of the interactive [illegible] approach and the stochastic [illegible] should be [illegible].

References

1. [illegible]
2. [illegible]
3. M. [illegible] programming for a [illegible] [illegible] 2003.
4. [illegible] for multiple objective linear fractional programming [illegible]
5. [illegible]
6. [illegible] programming [illegible] 2002.
7. [illegible] A linear programming approach [illegible]
8. [illegible] with [illegible]
9. [illegible]
10. [illegible]
11. [illegible]
12. [illegible]
13. [illegible] multiple [illegible]
14. [illegible]

Voronoi Diagram-Based Geometric Approach for Social Network Analysis

Subbu Surendran, D. Chitraprasad and M. R. Kaimal

Abstract Social network analysis is aimed at analyzing relationships between social network users. Such analysis aims at finding community detection, that is, group of closest people in a network. Usually, graph clustering techniques are used to identify groups. Here, we propose a computational geometric approach to analyze social network. A Voronoi diagram-based clustering algorithm is employed over embedded dataset in the Euclidean vector space to identify groups. Structure-preserving embedding technique is used to embed the social network dataset and learns a low-rank kernel matrix by means of a semi-definite program with linear constraints that captures the connectivity structure of the input graph.

Keywords Social network analysis · Geometric clustering · Voronoi diagram

1 Introduction

Social network analysis is an interesting research area for analyzing the structure and relationships of social network users [1]. Recent works [3, 5, 7, 13] in social network analysis attempt at finding group of closest people in a network (community detection). Usually, visualization techniques are used to analyze such groups in small social networks. However, only a few groups in a social network

S. Surendran (✉)
SCT College of Engineering, Thiruvananthapuram, Kerala, India
e-mail: subusurendran@gmail.com

D. Chitraprasad
TKM College of Engineering, Kollam, Kerala, India
e-mail: dcpvenus@yahoo.com

M. R. Kaimal
Amrita Vishwa Vidyapeetham, Amritapuri Campus, Kollam, Kerala, India
e-mail: mrkaimal@gmail.com

G. S. S. Krishnan et al. (eds.), *Computational Intelligence, Cyber Security and Computational Models*, Advances in Intelligent Systems and Computing 246,
DOI: 10.1007/978-81-322-1680-3_39, © Springer India 2014

are discovered using this approach. Clustering techniques are used for a large social network in identifying more groups and clusters.

A social network is represented as a graph $G = (V, E)$, where V represents vertices (nodes or actors), and E denotes edges (ties or relationship between actors). Most large-scale networks share some common patterns that are not noticeable in small networks. Among all the patterns, the well-known characteristics are as follows: scale-free distribution [8], small-world effect [9], and strong community structure [6].

In a community structure, a group of people tend to interact with each other more than those outside the group. That is, vertices in networks are often found to cluster into tightly knit groups with a high density of within-group edges and a low density of between-group edges [3]. Graph clustering methods are used to find community structure in social networks. In this paper, we propose a geometric approach to find community structure in social networks. Voronoi diagram-based geometric clustering approach is employed here for finding communities from graph. This method can give a polynomial time complexity.

The rest of the paper organized as follows: In Sect. 2, we discuss related works done in this area. Problem statement is specified in Sect. 3. The algorithm is discussed in Sect. 4. Results obtained and complexity analysis of the algorithm are given in Sect. 5.

2 Related Work

Considerable amount of work has been done on community detection in graphs. The Kernighan–Lin algorithm [10] is one of the earliest methods proposed and is still frequently used, often in combination with other techniques. Another popular technique is the spectral bisection method [11], which is based on the properties of the spectrum of the Laplacian matrix. The well-known max-flow min-cut theorem by Ford and Fulkerson [12] is a graph partitioning algorithm. This theorem has been used to determine minimal cuts from maximal flows of graph in clustering algorithms. Flake et al. [23] have used maximum flows to identify communities in the graph of the World Wide Web. The graph may have a hierarchical structure, i.e., may display several levels of grouping of the vertices, with small clusters included within large clusters, which are in turn included in larger clusters, and so on. The most popular hierarchical clustering algorithm is proposed by Newman [13]. A faster version of the Girvan–Newman algorithm has been proposed by Rattigan et al. [14]. Newman and Girvan proposed a function called modularity [3] to evaluate the cluster quality. Based on modularity measure, several community detection algorithms have been proposed [14, 15]. In Vempala et al. [4], some indices measuring the quality of graph clustering are discussed. They define conductance, an index concentrating on the intracluster edges present a clustering algorithm that repeatedly separates the graph. A graph clustering algorithm incorporating the idea of performing a random walk on the graph to identify the more densely connected subgraphs is presented in Van Dongen [16]. The idea of random walks is also used in Harel and Koren [17], but only for clustering geometric data. Ulrik Brandes et al. [2] introduced a method, which combines spectral embedding and decomposition based on minimum spanning trees (MST).

3 Problem Statement and Definitions

Given a social network dataset represented as connected undirected graph $G = (V, E)$; V represents vertices (nodes or actors), and E denotes edges (ties or relationship between actors). Here, $|V| = n$ and $|E| = m$. We wish to find a partition $C = C_1 \cup C_2 \cup \ldots \cup C_k$ of V in k non-overlapping groups $C_1, C_2, \ldots, C_k$. The sets C_i, $i \leq k$ are called communities. The number of intracommunity edges is more compared to the intercommunity edges.

The algorithm overview is shown in Fig. 1. To illustrate the functioning of each step in the proposed approach, we are using well-known Zachary's karate club network dataset [24]. The data were collected from the members of a university karate club by Wayne Zachary in 1977. Each node represents a member of the club, and each edge represents a tie between two members of the club. The network has 34 vertices and 78 undirected edges.

Fig. 1 Algorithm overview

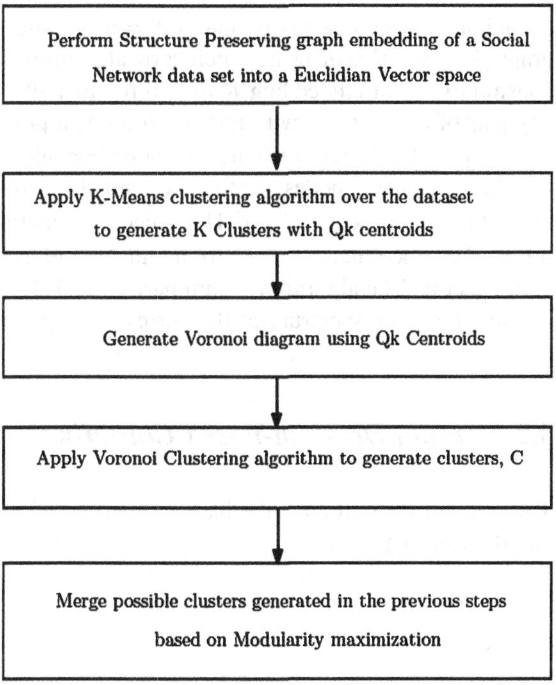

3.1 Structure-Preserving Graph Embedding

Social network dataset is usually represented by graph $G = (V, E)$ with $n = |V|$ nodes and $|E|$ edges. To apply geometric clustering algorithm, we need to embed G into a Euclidean vector space, preserving locality inherent in the graph structure. Traditional graph embedding algorithms such as spectral embedding and spring embedding do not explicitly preserve the structure.

The fundamental optimization underlying traditional spectral embedding method is eigen decomposition. Given a square $n \times n$ matrix K, the matrix can be factorized as $K = \sum_i \lambda_i v_i v_i^T$ where each λ_i is an eigenvalue of K with corresponding eigenvector v_i. We can factorize $K = Y^T Y$ by constructing Y as the stacked eigenvectors of K scaled by the square root of their eigenvalues. This mapping from K to Y forms the basis for the kernel view of principal component analysis (PCA). If all eigenvalues of K are non-negative, K is positive semi-definite, denoted as $K0$. We commonly exploit the useful property that $\mathrm{tr}(K) = \sum_i \lambda_i$.

In this paper, we employed structure-preserving embedding algorithm proposed by Blake and Tony [19]. The idea behind this method is to learn an embedding which preserves the topology of an unweighted undirected input graph defined as an adjacency matrix $A = \mathcal{B}^{n \times n}$. The embedding is represented by a positive semi-definite matrix $K = \Re^{n \times n}$ whose spectral decomposition yields the low-dimensional embedding. This embedding scheme accepts an input graph defined by both a connectivity matrix A and an algorithm ϑ, which accepts as input a kernel matrix K, and outputs a connectivity matrix $\tilde{A} = \vartheta(K)$, such that $\tilde{A}$ is close to the original graph A. Here, ϑ is K-nearest neighbor algorithm. We can define a set of linear constraints on an embedding K to enforce that $\vartheta(K) = A$. The distance between any pair of points *(i, j)*, with respect to a given positive kernel matrix K as $D_{ij} = K_{ij} + K_{jj} - 2K_{ij}$. When ϑ is k-nearest neighbors algorithm, each node is connected to the k other nodes with the smallest distance. This enforced via $D_{ij} > (1 - A_{ij})\max_m(A_{im}D_{im})$. The objective function used is $\max_{K0}$ tr(KA) and limits the trace norm of K to avoid the objective function from growing unboundedly. The algorithm is summarized in Table 1. The embedded karate club dataset and eigen spectrum of the corresponding kernel K is shown in Fig. 2.

3.2 Voronoi Diagram-Based Clustering

This section we introduce the basic properties of Voronoi diagram and Voronoi-based clustering.

Table 1 Structure preserving embedding algorithm

Input:	$A \in \{0,1\}^{N^2}$, connectivity algorithm ϑ, parameters C, § ≥ 0
Step 1:	Solve semi-definite program $\tilde{K} = \arg\max_{K \in k} \mathrm{tr}(KA) - C\S$ s.t. $D_{ij} > (1 - A_{ij})\max_m(A_{im}D_{im}) - \S$ where $D_{ij} = K_{ii} + K_{jj} - 2K_{ij}$
Step 2:	Apply principal component analysis (PCA) to $\tilde{K}$ and use the top eigenvectors as embedding coordinates

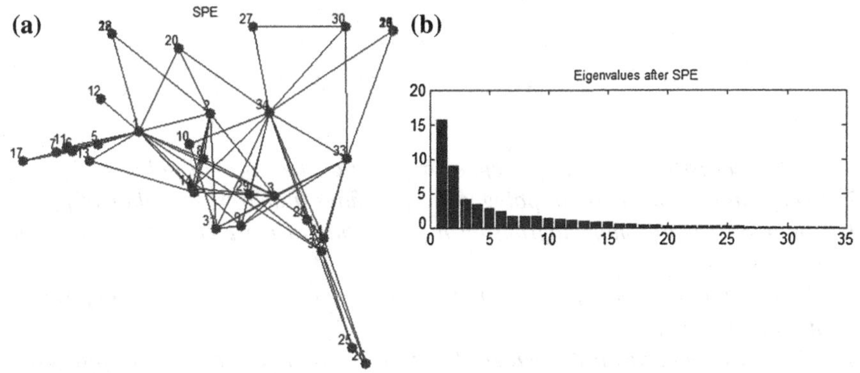

Fig. 2 **a** Result of graph embedding: karate club network embedded into 2D vector space. **b** Eigen spectrum of the kernel matrix of karate club network, K learned using structure-preserving embedding

3.2.1 Voronoi Diagram

Given a set S of n points $p_1, p_2, \ldots, p_n$ in the plane, the Voronoi diagram, denoted as Vor(S), is a subdivision of the plane into Voronoi cells (Fig. 3). The Voronoi cell, denoted as $V(p_i)$ for p_i, to be the set of points q that are closer or as close to p_i than to any other point in S. That is

$$V(p_i) = \{q | \mathrm{dist}(p_i, q)\mathrm{dist}(p_j, q), \quad \forall j \neq i\} \tag{1}$$

where dist is the Euclidean distance function. The Voronoi diagram decomposes the plane into n convex polygonal regions, one for each p_i.The vertices of the diagram are the Voronoi vertices, and the boundaries between two Voronoi cells are referred to as the Voronoi edges. The boundaries of a Voronoi cell $V(p_i)$ is a Voronoi polygon having no more than n-edges. It is one of the most important structures in computational geometry and has been widely used in clustering, learning, and other applications. Voronoi diagrams have some important properties such as [18].

Fig. 3 Voronoi diagram

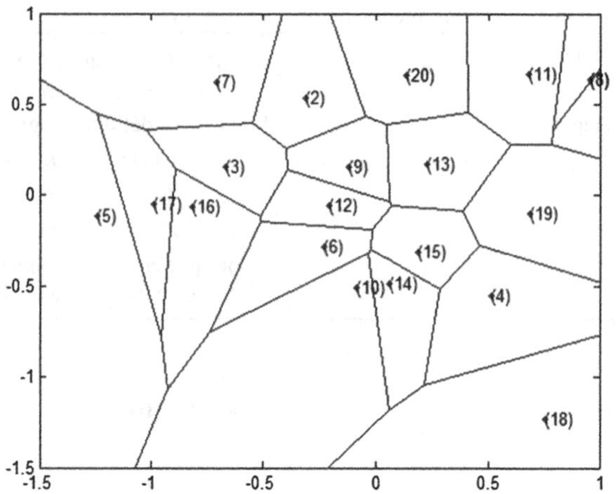

1. *Every nearest neighbor of pi defines an edge of the Voronoi polygon* $V(p_i)$.
2. *Every edge of the Voronoi polygon* $V(p_i)$ *defines a nearest neighbor of* p_i.
3. *For* $n \geq 3$, *a Voronoi diagram on n points has at most* $2n - 5$ *vertices and* $3n - 6$ *edges*
4. *The Voronoi diagram of a set of n points can be constructed in O(n log n) time and this is optimal*
5. *Each Voronoi vertex is the center of a circle touching three or more data points lying in its adjacent Voronoi cells.*

In this paper, we are considering the fifth property to cluster the given dataset.

3.2.2 Clustering Algorithm

Apply *K*-means clustering algorithm over the given dataset and get set of centroids (Q_k), where $C < k << n$, *C* is actual number of clusters, *n* is number of data points, and *k* is the user defined number (number of clusters defined in *K*-means algorithm). Voronoi diagram is created on the set of centroids to get the set of vertices (*P*). The centroids are labeled as unmarked. The vertices are sorted by their *x*-axis coordinates and subcluster_id variable *c* is set to one, and vertices are chosen one at a time. Note that each vertex is the center of the circle touching three or more centroids lying in adjacent Voronoi cells.

For each vertex, this circle is created and if its radius is less than a threshold γ then choose any marked centroids with smallest subcluster id in the circle and mark other centroids (marked or unmarked) to this subcluster_id. If no marked centroid is found, make the unmarked centroids with the subcluster_id, *C* and increment *C* by 1. Repeat the process over all Voronoi vertices. This creates the set of subclusters of centroids. All unmarked are noise centroids. Data points are assigned to the nearest centroids, and this creates the actual clusters or noise. These steps are summarized as follows (Fig. 4):

1. Apply *K*-means algorithm on given dataset (*D*), return set of centroids (*Q*).
2. Create Voronoi diagram on set of centroids (*Q*), return set of vertices (*P*).
3. Set all the centroids (*Q*) as unmarked.
4. Choose vertices (P_i) one by one in increasing order of *x*-axis.
5. Create circle such that the circumference of the circle touches centroids (Qk, $k \geq 3$), where P_i is center of that circle.
6. If the radius (R_i) is less than γ then
    a. If any centroid (*Qk*) is already assigned a subcluster_ id, *C* then assign the unmarked (or readjust the mark) centroids to the smallest subcluster_id, *C*.
7. else
    a. Assign subcluster_id = *c* on unmarked centroids increment *c*;
    b. Set centroids as marked.
8. Continue steps 6–8 till all the vertices (*P*) processed.
9. Return set of subcluster_id. All unmarked centroids are noise centroids.
10. *Assign data points* (*D*) *on nearest centroids.*
11. *Return clusters* (*C*) *and noise* (*N*).

3.2.3 Merge Communities Based on Modularity Maximization

After observing the cluster generated by the Voronoi-based clustering algorithm, communities are found denser and a number of communities are more. With this output, we cannot properly analyze the given social network data. So, we applied a refinement procedure to reduce the number of communities by combining communities generated in the previous step without sacrifices the quality. Here, we are making use of famous community evaluation index *Modularity* proposed by Newman [13].

Modularity measures the strength of a community partition by taking into account the degree distribution. The modularity is the number of edges falling within groups minus the expected number in an equivalent network with edges placed at random. Given a network with *n* vertices and *m* edges, the number of edges between vertices *i* and *j* be A_{ij} (adjacency matrix of *G*), which will normally

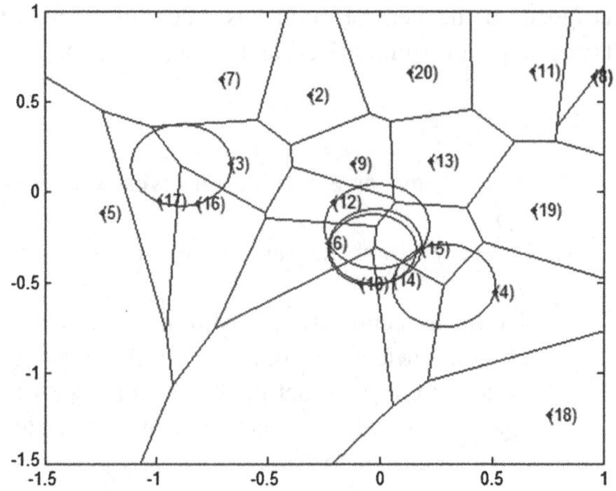

Fig. 4 Grouping Voronoi cells to form clusters

be 0 or 1. The expected number of edges between two nodes with degrees d_i and d_j is $(d_i d_j)/2m$. Thus, the *modularity* Q can be written as follows:

$$Q = \frac{1}{2m} \sum_{l=1}^{k} \sum_{i \in C_l,\, j \in C_l} \left(A_{ij} - \frac{d_i d_j}{2m} \right) \tag{2}$$

The modularity can be either positive or negative, with positive values indicating the possible presence of community structure. Thus, we can search for merging community structures sharing vertices of an edge as per graph, G and merged communities have positive modularity and large compared to individual modularity of clusters before merge. Communities identified in Karate Club dataset is shown in Fig. 5.

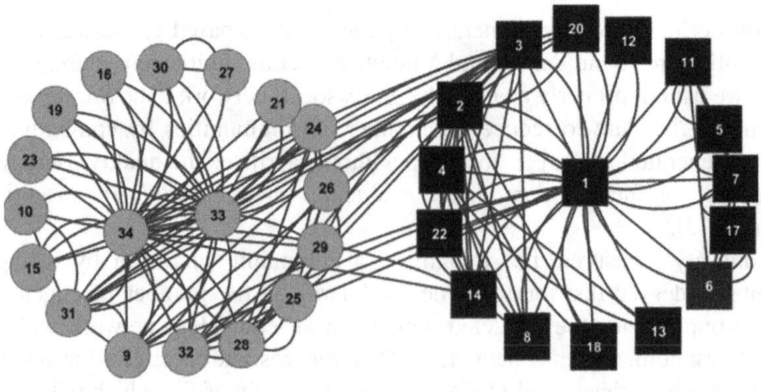

Fig. 5 *Result* two communities are identified from karate club network

4 Results and Discussion

To estimate the goodness of this algorithm, we used the quantitative measure, normalized mutual information, proposed by Andrea Lancichinetti and Santo Fortunato [20]. This measure is borrowed from Shannons information theory. To evaluate the Shannon information content of a partition, one starts by considering the community assignments $\{x_i\}$ and $\{y_i\}$, where x_i and y_i indicate the cluster labels of vertex i in partition X and Y, respectively. Assume that the labels x_i and y_i are values of two random variables X and Y, with joint distribution $P(x,y) = P(X = x, Y = y) = n_{xy}/n$, which implies that $P(X) = P(X = x) = n_x^X/n$ and $P(Y) = P(Y = y) = n_y^Y/n$, where n_x^X, n_y^Y and n_{xy} are the sizes of the clusters labeled by x, y and of their overlap, respectively. The normalized mutual information of two random variables is defined as follows:

$$I_{\text{norm}}(X, Y) = \frac{2I(X, Y)}{H(X) + H(Y)} \tag{3}$$

where $I(X, Y) = \sum_x \sum_y P(x, y) \log \frac{P(x, y)}{P(x)P(y)}$, the mutual information of two random variable and $H(X) = -\sum_x P(x) log P(x)$ is the Shannon entropy of X. The *normalized mutual information*, I_{norm} equals 1 if the partitions are identical, whereas it has an expected value of 0 if the partitions are independent.

4.1 Dataset

To access the goodness of these community detection algorithms, there is a simple network model, the *planted l-partition model*, which is often used in the literature. In this model, one "*plants*" a partition, consisting of a certain number of groups of nodes. Each node has a probability p_{in} of being connected to nodes of its group and a probability p_{out} of being connected to nodes of different groups. As long as $p_{\text{in}} > p_{\text{out}}$, the groups are communities, whereas when $p_{\text{in}} \leq p_{\text{out}}$ the network is essentially a random graph, without community structure. The most popular version of planted *l-partition* model was proposed by Girvan and Newman, GN benchmark. The GN benchmark has two drawbacks: (1) All nodes have same expected degree; (2) all communities have equal size. These features are unrealistic, as complex networks are known to be characterized by heterogeneous distribution of degree and community sizes. To overcome these drawbacks, Andrea et al. [21] have introduced a new class of benchmark graphs, LFR benchmark, that generalize the GN benchmark by introducing power law distributions of degree and community size. Most community detection algorithms perform very well on the GN benchmark, due to the simplicity of its structure. The LFR benchmark, instead, poses a much harder test to algorithms and makes it easier to disclose their limits. Table 2 indicates the result obtained for various LFR benchmark datasets generated using the algorithm proposed by Andrea Lancichinetti and Santo Fortunato [21]. We have

Table 2 Result obtained for various LFR benchmark datasets

No. of vertices	No. of edges	No. of communities present	No. of communities identified	Normalized mutual information I_{norm}
50	362	4	4	0.894071
100	764	6	6	0.891624
500	13,603	7	7	0.967516
1,000	27,696	13	13	0.972413

used the package developed by Lancichinetti and Santo Fortunato available freely in the Web site [22] for generating LFR benchmark dataset.

From these results, it is found that the geometric algorithm proposed here can detect communities with more than 90 % of accuracy. In all the cases, the number of communities identified was same as a number of communities created. It is noticed that a few nodes are misplaced in other groups.

4.2 Computational Complexity

In the first step, we are embedding the dataset $G = (V, E)$ into Euclidean space. This process will take the execution time of $O(N^3 + C^3)$. N is $|V|$, and C represents number of constraints considered for structure-preserving embedding. This is proportional to number of edges in the graph G. For performing k-means clustering, the time complexity will be $O(Nkt)$, where k indicates the number of clusters. We are considering 60 % of Nkt that represents the number of iterations performed during k-means clustering and that will be 3 or 4. Voronoi clustering algorithm requires the running time of $O(k \log k)$. Complexity of final merging algorithm is $O(k^2)$. If we are not considering the running time for graph embedding, complexity of proposed geometric algorithm is $O(N^2)$, $k \sim N$.

5 Conclusion

This work is an initiative to social network data analysis with geometric algorithm. Here, we have employed Voronoi-based clustering algorithm over the embedded social network dataset in the Euclidean space. Since the number of clusters generated by Voronoi-based algorithm is more, we merged together the clusters based on the maximization of popular community measure "modularity." Compared to other graph clustering algorithms, this method is giving better results in community detection. Since Voronoi diagram is used here for grouping cluster members, this algorithm can detect non-globular clusters.

Usually, social network data are growing dynamically. The algorithm proposed here is considering static dataset only. As a future work, we would like to extend the algorithm for analyzing dynamic dataset.

References

1. I-Hsien Ting, "Web Mining Techniques for On-line Social Networks Analysis". Fifth IEEE international conference service systems and service management, 2008, pp. 1–5.
2. Brandes, U., Gaertler, M., and Wagner, D. 2007. Engineering graph clustering: Models and experimental evaluation. Journal of Experimental Algorithmics (JEA).Vol.12, Article 1.1 (2008).
3. Newman, M. E. J. "Fast algorithm for detecting community structure in networks", Physical Review -Series E, Vol 69; Part 6; Part 2, pages 066133 (2004) [5 pages].
4. Vempala S., Kannan R., and Vetta A., "On clusterings good, bad and spectral", In Proceedings of the 41st Annual IEEE Symposium on Foundations of Computer Science (FOCS00). pp. 367–378.
5. M. E. J. Newman and M. Girvan, "Finding and evaluating community structure in networks". Physical Review -Series E Vol. 69, pages 026113 (2004) [15 pages].
6. Charu C Aggarwal, Haixun Wang, "Managing and Mining Graph Data", Advances in Database Systems, Springer Science + Business Media, Chapter 16, pp. 487–513, 2010.
7. Aaron Clauset, M. E. J. Newman, and Cristopher Moore, "Finding community structure in very large networks". Phys. Rev. E Vol. 70, 066111. (2004) [6 pages].
8. MEJ Newman, "Power laws, Pareto distributions and Zipf's law". Contemporary Physics 46:5, 323–351, 2005.
9. J.Travers and S.Milgram, "An experimental study of the small world problem". Sociometry, 32(4):425-443, 1969.
10. B. W. Kernighan and S. Lin "An efficient heuristic procedure for partitioning graphs", Bell System Technical Journal, vol. 49, pp. 291–307 1970.
11. Earl R. Barnes, "An Algorithm for Partitioning the Nodes of a Graph", SIAM. J. on Algebraic and Discrete Methods, 3(4), 541550,1982.
12. Ford L. R., and Fulkerson D. R., "Maximal flow through a network", Can. J. Math 8, 399–404, 1956.
13. M. E. J. Newman, "Modularity and community structure in networks", Proceedings of National Academy of Sciences of the USA, Vol. 103 no. 23 pp. 8577–8582, 2006.
14. Rattigan, M. J., M. Maier, and D. Jensen, "Graph clustering with network structure indices", ICML'07 Proceedings of the 24th international conference on Machine learning, (ACM, New York, NY, USA), pp. 783–790, 2007.
15. Brandes, U., D. Delling, M. Gaertler, R. Gorke, M. Hoefer, Z. Nikolski, and D. Wagner, 2006, URL http://digbib.ubka.unikarlsruhe.de/volltexte/documents/3255.
16. Van Dongen, S. M., "Graph Clustering by Flow Simulation". Ph.D. thesis, University of Utrecht, 2000.
17. David Harel, Yehuda Koren, On clustering using random walks. In Proceedings of the 21st Conference on Foundations of Software Technology and Theoretical Computer S. Lecture Notes in Computer Science, vol. 2245. Springer-Verlag, New York. 1841, 2001.
18. Preparata F. P., M. I. Shamos, "Computational Geometry-An Introduction", Springer-Verlag, August 1985.
19. Blake Shaw, Tony Jebara, "Structure Preserving Embedding", Proceedings of 26th International Conference on Machine Learning, Montreal, Canada, 2009.
20. Andrea Lancichinetti, Santo Fortunato, "Community detection algorithms: A comparative analysis", Physical Review E, vol. 80, Issue 5, id. 056117, 2009.
21. Andrea Lancichinetti, Santo Fortunato, and Filippo Radicchi, "Benchmark graphs for testing community detection algorithms", Physical Review E, 78, 046110, 2008.
22. http://santo.fortunato.googlepages.com/inthepress2.
23. Gary Flake, Steve Lawrence, C. Lee Giles, Frans Coetzee, "Self-Organization and Identification of Web Communities", IEEE Computer, 35(3), 66–71, 2002.
24. W. W. Zachary, "An information flow model for conflict and fission in small groups", Journal of Anthropological Research 33, 452–473 (1977).

Part V
Short Papers

Comparative Analysis of Discretization Methods for Gene Selection of Breast Cancer Gene Expression Data

E. N. Sathishkumar, K. Thangavel and A. Nishama

Abstract DNA microarrays provide an enormous amount of information about genetically conditioned susceptibility to diseases. However, their analysis is uneasy because the number of genes is extremely large with respect to the number of experiments. The problem is that all genes are not essential in gene expression data. Some of the genes may be redundant, and others may be irrelevant and noisy. This research paper studies the gene expression data using rough set theory; it is an intelligent computing method. In this paper, we studied and implemented following discretization methods such as rough discretization (RD), naïve Bayes, max–min, equal width intervals, *K*-means-based discretization, and entropy-based discretization (EBD) for gene selection using rough set quick reduct (QR) for breast cancer gene expression data. Further, the performance of the above algorithms has been evaluated using classification tools available in Weka software and BPN classifier.

Keywords Discretization · Gene expression data · Gene selection · Rough set · Classification

1 Introduction

The DNA microarray technology provides enormous amount of biological information about genetically conditioned susceptibility to diseases. The datasets acquired from microarrays refer to genes via their expression levels. Given

E. N. Sathishkumar (✉) · K. Thangavel · A. Nishama
Department of Computer Science, Periyar University, Salem 636011, India
e-mail: en.sathishkumar@yahoo.in

K. Thangavel
e-mail: drktvelu@yahoo.com

A. Nishama
e-mail: nishama89@gmail.com

G. S. S. Krishnan et al. (eds.), *Computational Intelligence, Cyber Security and Computational Models*, Advances in Intelligent Systems and Computing 246,
DOI: 10.1007/978-81-322-1680-3_40, © Springer India 2014

thousands of gene attributes against hundreds of objects, we face a 'few-objects–many-attributes' problem. Dimensionality reduction in gene expression data can be critical for a number of reasons. Dimensionality reduction is crucial in order to overcome the curse of dimensionality and allow for meaningful data analysis [1]. Rough sets have been used as a feature selection method by many researchers, among them Jensen and Shen [4], Yong Li et al. [6], and Hu et al. [3].

The rough set approach to feature selection consists in selecting a subset of features, which can predict the classes as well as the original set of features. Rough set theory is based on decision tables. According to this, we need to discretize continuous features of a dataset before applying data mining methods based on rough sets. Many discretization methods are frequently used in data mining and knowledge discovery for gene expression data [5]. The main objective of this paper was to compare (experimentally) discretization methods for rough set gene selection.

2 Methods

2.1 Rough Discretization

Rough discretization (RD) is sometimes also called dynamic discretization or roughfication, in dissimilarity to fuzzification known from fuzzy logic [1]. Let the decision system $A = (U, AU\{d\})$ with real-valued conditional attributes. RD is the procedure of creation from A the decision system $A^* = (U \times U,\ A^*U\{d^*\})$, where, for every $x, y, \in U \times U$, we put $d^*(x,y) = d(y)$,and, for every $a \in A$,

$$a^*(x,y) = \text{“}\geq a(x)\text{”} \text{ if and only if } a(y) \geq a(x) \quad (1)$$

$$a^*(x,y) = \text{“}<a(x)\text{”} \text{ if and only if } a(y) < a(x) \quad (2)$$

2.2 Naive Discretization

In naive discretization (ND), a single attribute may be discretized into different intervals. If many attributes have many values, the number of decision rules increases [7].

Step 1: Sort values of the attribute
Step 2: for all adjacent pairs of objects

if $va(i) \neq va(i + 1)$ and $d(i) \neq d(i + 1)$ create a new cut point c

$$c = (va(i) + va(i+1))/2 \quad \text{and} \quad ca = ca\ U\ c \quad (3)$$

2.3 Max–Min Discretization (MMD)

The number of categories the dataset should have must be decided. Then, for each gene *i*, the following two steps are used to discretize the data of the dataset [7].

Step 1: $(\text{Value}_{\max(i)} - \text{Value}_{\min(i)})/N$
Step 2:

$$\text{Value}_{\text{transformed}} = 1 + \text{Round}\left(\left(\text{Value}_{\text{original}} - \text{Value}_{\min(i)}\right)/\text{Step}(1)\right) \quad (4)$$

here, $\text{value}_{\max(i)}$ and $\text{value}_{\min(i)}$ are the maximum value and the minimum value in gene *i*, respectively, and $\text{value}_{\text{original}}$ and $\text{value}_{\text{transformed}}$ are the original value and the transformed value, respectively. N is the number of discrete levels used to discretize the original value, and step (i) is the step size for gene i. After discretizing the dataset, the transformed value of the data is an integer between 1 and $N + 1$.

2.4 K-Means Discretization

K-means discretization (KMD) [6] divides *A(n,:)* into *k* intervals by *K*-means clustering so that adjacent expression values of gene *n* are divided into same interval. In gene expression dataset, each gene attribute is clustered with *K*-means and replaces the attribute values with the cluster membership labels. These labels will be act as discrete values for gene expression dataset.

2.5 Equal Width Discretization

Equal Width Discretization (EWD) [6] divides the number line between $A(n,:)_{\min}$ and $A(n,:)_{\max}$ into k intervals of equal width. Thus, the intervals of gene n have width $w = (A(n,:)_{\max} - A(n,:)_{\min})/k$, with cut points at $A(n,:)_{\min} + w$, $A(n,:)_{\min} + 2w$, …, $A(n,:)_{\min} + (k-1)w$. k is a positive integer and is a user-predefined parameter.

2.6 Entropy-Based Discretization

Entropy-Based Discretization (EBD) was proposed by Fayyad and Irani in 1993. It uses the class information present in the data. The entropy (or the information content) is deliberate on the basis of the class label.

$$\text{Entropy} = -\sum P \log_2 P. \tag{5}$$

Instinctively, it finds the best split so that the bins are as pure as possible, i.e., the majority of the values in a bin correspond to having the same class label. Formally, it is characterized by finding the split with the maximal information gain.

2.7 *Quick Reduct Algorithm*

Rough set theory is a formal mathematical tool that can be applied to reducing the dimensionality of dataset. The rough set attribute reduction method removes redundant input attributes from dataset of discrete values, all the while making sure that no information is lost. The QR algorithm was broadly studied in [2, 5].

3 Experimental Results

3.1 *Dataset*

We study the breast cancer data downloaded from Gene Expression Omnibus (GEO), http://www.ncbi.nlm.nih.gov/projects/geo/gds/gds_browse.cgi?gds=360, analyzed in [1]. It contains 24 core biopsies taken from patients, who are resistant (14 objects) or sensitive (10 objects) to the docetaxel treatment. There are 12,625 gene attributes.

The data studied by rough sets are mainly organized in the form of decision tables. One decision table can be represented as $S = (U, A = C \; U \; D)$, where U is the set of samples, C the condition attribute set, and D the decision attribute set. Above six discretization methods, applied before to gene selection. We apply the QR for finding optimal approximate reducts or genes (Table 1).

Table 1 Selected genes from discretized data

Discretization	QR-selected genes	ID ref
RD	'RPS17,' 'PLD3'	'34592_at,' '36151_at'
ND	'MAPK3'	'1000_at'
MMD	'MAPK3,' 'PCBP1'	'1000_at,' '34305_at'
KMD	'DUSP1,' 'SMAD3'	'1005_at,' 1,433_g_at'
EWD	'KAT2B,' 'RPL3'	'1012_at,' '31722_at'
EBD	'PSG3'	'40857_f_at'

Table 2 Classification accuracy of selected genes

Classifiers	RD	ND	MMD	KMD	EWD	EBD	Without
Naïve Bayes	70.8333	66.6667	83.3330	83.3330	79.1667	91.6667	79.1667
RBF	95.8333	66.6667	79.1667	79.1667	87.5000	91.6667	45.8330
JRip	79.1667	70.8333	87.5000	79.1667	79.1667	95.8333	62.5000
J48	83.3333	66.6667	83.3330	79.1667	75.0000	95.8333	66.6667
BPN	85.7000	62.5000	75.0000	75.0000	80.5000	92.3000	72.5000
Average	82.9733	66.6666	81.6665	79.1666	80.2666	93.4600	65.3332

3.2 Classification Results

The Weka is a well-known package of data mining tools, which provides a variety of known, well-maintained classifying algorithms. This allows us to do experiments with several kinds of classifiers quickly and easily. Some of the classifiers we used in our experiment are naive Bayes, RBF network, JRip, J48, and BPN. Tenfold cross-validation is used in all the above classifiers [2]. Table 2 denotes the RD and EBD used feature selection methods classification accuracy are exceptionally high when compared to all other methods.

4 Conclusion

Choosing a correct discretization method can improve the accuracy of rough set feature selection algorithm. Here, RD, naïve Bayes, max–min, equal width intervals, KMD, and EBD methods are studied and implemented successfully for rough set feature selection method using breast cancer gene expression data. In comparison with other discretization methods, EBD and RD gave enhanced results for rough set feature selection.

Acknowledgments The first author gratefully acknowledges the partial financial assistance under University Research Fellowship, Periyar University, Salem-11, Tamil Nadu, India.

The second author gratefully acknowledges the UGC, New Delhi, for partial financial assistance under UGC-SAP (DRS) Grant No. F.3–50/2011.

References

1. D. Slezak and J Wroblewski, "Rough Discretization of Gene Expression Data" International Conference on Hybrid Information Technology, Nov 9, 2006.
2. E. N. Sathishkumar, K. Thangavel and T. Chandrasekhar,"A Novel Approach for Single Gene Selection Using Clustering and Dimensionality Reduction", International Journal of Scientific & Engineering Research, Volume 4, Issue 5, May-2013. ISSN 2229–5518.
3. Hu, K., Lu, Y and Shi C., "Feature ranking in Rough sets", AI Communications 2003, pp 41–50.

4. Jensen, R. and Shen, Q., "Finding Rough Set Reducts with ant colony optimization". Proceeding of the 2003 UK Workshop on Computing Intelligence 2003, pp. 15–22.
5. T. Chandrasekhar, K. Thangavel and E.N. Sathishkumar, "Verdict Accuracy of Quick Reduct Algorithm using Clustering and Classification Techniques for Gene Expression Data", International Journal of Computer Science Issues, Vol. 9, Issue 1, No 1, January 2012. ISSN (Online): 1694–0814.
6. Yong Li et.al., "Comparative study of discretization methods of microarray data for inferring transcriptional regulatory networks", BMC Bioinformatics 2010, 1471–2105/11/520.
7. Zhi-yong Yan, Cong-fu Xu, Yun-he Pan "Improving naive Bayes classifier by dividing its decision regions", Journal of Zhejiang University-SCIENCE C (Computers & Electronics), Apr. 8, 2011.

Modified Soft Rough set for Multiclass Classification

S. Senthilkumar, H. Hannah Inbarani and S. Udhayakumar

Abstract Rough set theory has been applied to several domains because of its ability to handle imperfect knowledge. Most recent extension of rough set is soft rough set, where parameterized subsets of a universal set are basic building blocks for lower and upper approximations of a subset. In this paper, a new model of soft rough set, which is called modified soft rough set (MSR) where information granules are finer than soft rough sets, is applied for classification of medical data. In this paper, rough-set-based quick reduct approach is applied for selecting relevant features and MSR is applied for multiclass classification problem and the proposed work is compared with bijective soft set (BSS)-based classification, naïve Bayes, and decision table classifier algorithms based on evaluation metrics.

Keywords Soft rough set · Classification · Modified soft rough set · Quick reduct

1 Introduction

Classification and feature reduction are wide areas of research in data mining. Many practical applications of classification involve a large volume of data and/or a large number of features/attributes. Hence, it is necessary to remove irrelevant attributes and only the relevant attributes are used. The new idea is to solve multiclass classification problem with the modified soft rough set (MSR) method.

S. Senthilkumar (✉) · H. H. Inbarani · S. Udhayakumar
Department of Computer Science, Periyar University, Salem 636011, India
e-mail: pkssenthilmca@gmail.com

H. H. Inbarani
e-mail: hhinba@gmail.com

S. Udhayakumar
e-mail: uk2804@gmail.com

G. S. S. Krishnan et al. (eds.), *Computational Intelligence, Cyber Security and Computational Models*, Advances in Intelligent Systems and Computing 246,
DOI: 10.1007/978-81-322-1680-3_41,

In order to solve such a problem in medical diagnosis, attribute reduction, also called feature subset selection, is usually employed as a preprocessing step to select part of the attributes and focus the learning algorithm on relevant information. Rough set has strong ability in data processing and can extract useful rules from them [1]. The main aim of feature selection is to determine a minimal feature subset from a problem domain. A decision rule is a function, which maps an observation to an appropriate action. Deterministic rules correspond to the lower approximation, and non-deterministic rules correspond to the upper approximation [2].

Rough set theory and soft set theory are two different tools to deal with uncertainty. MSR sets satisfy all the basic properties of rough sets and soft sets. In some situations, equivalence relation cannot be defined, which is the basic requirement in rough set theory [3]. In these situations, MSR sets can help us to find approximations of subsets. Our proposed work consists of two parts: Initially, in the preprocessing stage, redundant data are removed and rules are derived from reduced data set. In this study, MSR-based classification is applied for generating decision rules from the reduced data set.

Modified Soft Rough Sets.

Soft set theory [4] deals with uncertainty and vagueness on the one hand, while on the other, it has enough parametrization tools. Feng et al. [5] introduced soft rough sets and comparative analysis of rough sets and soft sets. In order to strengthen the concept of soft rough sets, a new approach called MSR is presented in [6]. The definitions for lower and upper soft rough approximations are given below:

Definition 1 Let (F, A) be a soft set over U, where F is a map $F: A \rightarrow P(U)$. Let $u: U \rightarrow P(A)$ be another map, defined as $u(x) = \{a : x \in F(a)\}$. Then, the pair (U, u) is called MSR approximation space, and for any $X \subseteq U$. Ulower MSR approximation is defined as Table 1.

$$\underline{X}\phi = \{X \in U : \phi(x) \neq \phi(y), \quad \text{for all}\, y \in x^c \tag{1}$$

where $X^c = U - X$ and its upper MSR approximation is defined as follows:

$$\bar{X}\phi = \{X \in U : \phi(x) = \phi(y), \quad \text{for some}\, y \in x \tag{2}$$

If $\underline{X}\phi \neq \bar{X}\phi$, then X is said to be an MSR set.

2 Methodology

The methodology adopted in this work is given in Fig. 1. The MSR-based classification approach is applied for generating rules from the trained data, and rule matching is applied for test data to compute the decision class based on reliability analysis. In this study, the proposed approach is applied for medical diagnosis [7].

Table 1 Sample data for MSR sets

U	E1	E2	E3	D
S1	1	1	0	1
S2	0	1	0	0
S3	0	1	0	1
S4	1	0	0	0
S5	0	0	1	0
S6	1	0	1	1

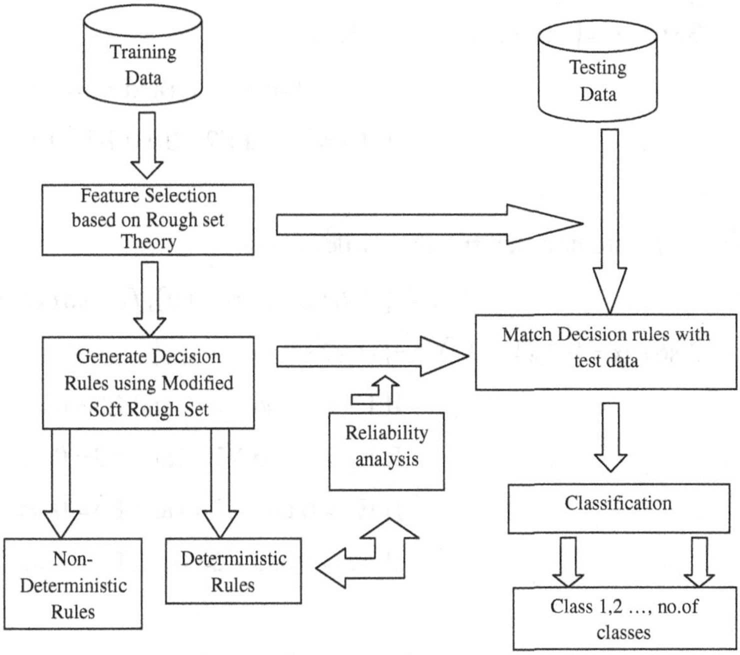

Fig. 1 Overall structure of the proposed work

3 MSR-Based Classification Algorithm

MSR-based classification algorithm is shown in Fig. 2. In this approach, lower and upper soft rough approximations of the given data set based on decision class *X* are constructed. In the second step, AND operation is applied to combine the soft sets. In the third step, deterministic rules are generated based on lower soft rough approximation. In the fourth step, non-deterministic rules are generated based on upper soft rough approximation and support for each non-deterministic rule is computed using step 5 of the algorithm.

The proposed algorithm is explained with example given in Table 1.

Step 1: Construct Modified Soft Rough Set (F, E) from the Dataset given in table 1.

Step 2: Apply AND operation for all conditional attributes (F, E1) ∧ (F, E2) ∧ (F, E3)ϕ(s1) = {e1,e2},ϕ (s2) = {e2} = ϕ (s3),ϕ (s4) = {e1}, ϕ (s5) ={e3},ϕ (s6) = {e1, e3}.

Step 3: Generate deterministic rules by using

$$\underline{X}\phi = \{X \epsilon U: \phi(x) \neq \phi(y), for\ all\ y \epsilon x^c$$

x={s1,s3,s6}, x^c={s2, s4,s5},$\underline{X}\phi$ ={s1,s6}

If E1= 1 and E2= 1and E3= 0 =>d= 1

If E1= 1 and E2= 0and E3= 1 =>d= 1

Step 4: Generate non-deterministic rules using

$$\bar{X}\phi = \{X \epsilon U: \phi(x) = \phi(y), for\ some\ y \epsilon x$$

x={s1,s3,s6}, x^c={s2,s4,s5}, $\bar{X}\phi$={s1,s2,s3,s6}

If E1= 1 and E2= 1and E3= 0 =>d= 1

If E1= 0 and E2= 1and E3= 0 =>d= 0

If E1= 0 and E2= 1and E3= 0 =>d= 1

If E1= 1 and E2= 0and E3= 1 =>d= 1

Step 5: Compute the support value for each non-deterministic rule.

$$\text{Support} = \left(\frac{1}{2} \wedge \frac{1}{2}\right) = 0.5$$

4 Experimental Analysis

Classification is a data mining technique used to predict group membership for data instances. The model is used to classify new objects [7]. In this work, classification accuracy of the proposed approach is compared with three different classifiers BSS classification, naïve Bayes, and decision table using the accuracy

Algorithm: MSR based Classification
Input: Given Dataset with conditional attributes 1, 2, … ,n-1 and the Decision attribute n.
Output: Generated Decision Rules for multiclass values.
Step1: Construct MSR approximation space for the given Dataset
Step2: Apply AND operation for all conditional attributes.
Step3: Generate deterministic rules using

$$\underline{X}\phi = \{X \in U: \phi(x) \neq \phi(y), for\ all\ y \epsilon x^c$$

Step4: Generate non-deterministic rules by using

$$\overline{X}\phi = \{X \epsilon U: \phi(x) = \phi(y), for\ some\ y \in x$$

Step5: Compute the support value for each non-deterministic rule

$support = \frac{support\ (A \wedge B)}{support\ (A)}$ Where *A* is the description on condition attributes and *B* the description on decision attributes.

Fig. 2 MSR-based classification algorithm

Table 2 Performance analysis of classification algorithms for breast cancer and hepatitis

	Breast cancer				Hepatitis			
	MSR sets	Bijective soft set	Decision table	Naïve Bayes	MSR sets	Bijective soft set	Decision table	Naïve Bayes
Precision	0.983	0.972	0.954	0.956	0.848	0.762	0.72	0.747
Recall	0.980	0.961	0.95	0.950	0.735	0.727	0.66	0.717
F-measure	0.981	0.96	0.951	0.953	0.731	0733	0.637	0.71

Table 3 Performance analysis of classification algorithms for Pima Indian diabetes and liver data sets

	Pima Indian diabetes				Liver			
	MSR sets	Bijective soft set	Decision table	Naïve Bayes	MSR sets	Bijective soft set	Decision table	Naïve Bayes
Precision	0.836	0.805	0.758	0.769	0.986	0.981	0.684	0.673
Recall	0.792	0.799	0.766	0.773	0.99	0.976	0.671	0.594
F-measure	0.802	0.796	0.758	0.771	0.988	0.972	0.678	0.59

measures precision, recall, and F-measure. The four medical data sets taken for the experimental analysis of the proposed approach are taken from UCI repository (www.ics.uci.edu/~mlearn/).

From the results provided in Tables 2 and 3, it can be easily concluded that the MSR algorithm is an effective method for medical diagnosis. The proposed classification method is applied to reduced data set, thus reducing the number of rules, which leads to significantly improved classification accuracy and results in a significant difference in a patient's chance for recovery.

5 Conclusion

In this paper, MSR set approach is proposed for engendering deterministic and non-deterministic rules for the multiclass classification of medical data set. Comparison of the proposed approach with familiar BSS classification, decision table, and naïve Bayes classifier algorithms is carried out using performance metrics. The better results show that MSR-based classification is suitable for medical data. Therefore, future treatment decision based on the MSR sets of a medical data is a rational approach toward preventing the outgrowth of metastases.

References

1. C. Velayutham and K. Thangavel, "Unsupervised Quick Reduct Algorithm Using Rough Set Theory", International Journal of electronic science and technology, Vol. 9, pp. 193-201, 2011.
2. L. P. Khoo, S. B. Tor and L. Y. Zhai, "A Rough-Set-Based Approach for Classification and Rule Induction, "International Journal of Advanced Technology, Vol. 15, pp. 438–444, 1999.
3. Z. Pawlak, "Rough sets", International Journal of Computational Information Science, Vol. 11, pp. 341–356, 1982.
4. Molodtsov, "Soft set theory-RST results", Computational Mathematical Application. Vol.37, pp. 19-31, 1999.
5. FengFeng, Xiaoyan Liu, VioletaLeoreanu-Fotea, Young Bae Jun "Soft sets and soft rough sets", Information Sciences, Vol. 181, pp. 1125–1137, 2011.
6. Muhammad Shabir, Muhammad Irfan Ali, TanzeelaShaheen, "Another approach to soft rough sets", Knowledge-Based Systems, Vol. 40, pp. 72–80, 2013.
7. S.Udhaya Kumar, H. Hannah Inbarani, S.Senthilkumar, "Bijective soft set based classification of Medical data", International Conference on PRIME, pp. 517 – 521. 2013.

A Novel Genetic Algorithm for Effective Job Scheduling in Grid Environment

P. Deepan Babu and T. Amudha

Abstract A grid is a set of resources such as CPU, memory, disk, applications, and database distributed over wide area networks and supports large-scale distributed applications. Resources in grid are geographically distributed and linked through Internet, to create virtual supercomputer with vast computing capacity to solve complex problems. Scheduling, resource brokering, and load balancing are the essential functionalities of grid environment. Evolutionary algorithms (EA) operate on a population of potential solutions, applying the principle of survival of the fittest. Genetic algorithms belong to a larger class of EA, which generate solutions to optimization problems using techniques inspired by natural evolution, such as inheritance, mutation, selection, and crossover. This paper proposes a scheduling technique based on genetic algorithm to schedule jobs effectively in a grid. The proposed algorithm is tested with different sizes of preemptive job requests, and analysis of results has shown significant improvement in scheduling performance.

Keywords Grid computing · Evolutionary algorithms · Genetic algorithm · Scheduling

P. Deepan Babu (✉)
Department of IT and CT, VLB Janakiammal College of Arts and Science,
Coimbatore, Tamil Nadu, India
e-mail: pdeepan_13@yahoo.co.in

T. Amudha
Department of Computer Applications, Bharathiar University,
Coimbatore, Tamil Nadu, India
e-mail: amudhaswamynathan@buc.edu.in

G. S. S. Krishnan et al. (eds.), *Computational Intelligence, Cyber Security and Computational Models*, Advances in Intelligent Systems and Computing 246,
DOI: 10.1007/978-81-322-1680-3_42,

1 Introduction

1.1 Grid Computing

Grid computing is an ever-growing area that keeps developing at a constant phase. A grid is a set of resources (such as CPU, memory, disk, applications, and database) distributed over wide area networks and supports large-scale distributed applications [1]. A computational grid is a hardware and software infrastructure that provides dependable, consistent, pervasive, and inexpensive access to high-end computational capabilities [2]. Grid architectures are dynamic in nature, any resource can join or leave the grid, and a resource is disparate and connects different networks. Grid computing achieved various breakthroughs in meteorology, physics, medicine, collaborative, or e-science computing [3] and data-intensive computing. A data grid is a major type of grid [4], used in data-intensive applications, where size of data files reaches terabytes or sometimes petabytes. High-energy physics (HEP) and genetic and earth observation are examples of such applications. Data grid is an integrating architecture that connects a group of geographically distributed resources [5]. Computational grid [6] is developed to solve problems that require processing a large quantity of operations. Many research projects require a lot of CPU time, some requires a lot of memory, and some projects need the ability to communicate in real time. Today, supercomputers are not enough to solve those needs. They do not have the capacity; even if they did, it would not be economically justifiable to use these resources [7]. Computational grids are the solution to all these problems and many more.

1.2 Evolutionary Algorithm

The basic idea behind the evolutionary algorithm (EA) is that given a population of individuals [8], the environmental pressure causes natural selection (survival of the fittest) and this causes a rise in the fitness of the population. EAs are principally a stochastic search and optimization method based on the principles of natural biological evolution.

At each generation of the EA, a new set of approximations is created by the process of selecting individuals according to their level of fitness in the problem domain and reproducing them using variation operators [9]. A GA approach starts with a generation of individual. Individuals are encoded as strings known as chromosome, and a chromosome corresponds to a solution to solve the problem. Each individual is evaluated by the fitness function. Three major operations, selection, crossover, and mutation, [10] are part of the GA based on some key parameters such as fitness function, crossover probability, and mutation probability [11]. All these parameters are used for the optimization of task scheduling [12].

2 Scheduling in Grid

Scheduling is a major component of grid environment because it deals with heterogeneous resources. Scheduler has a collection of resources, selects appropriate resources for job, and allocates the job. Three techniques are used shortest-job-first algorithm, arbitrary scheduling algorithm and proposed genetic-based scheduling algorithm (GSA). Shortest-job-first (SJF) scheduling algorithm associates with each process the length of the process's next processing time. When the resource is available, it assigns the process, which has smallest processing time next. Arbitrary scheduling implies that each individual or unit being entered into a trial has the same chance of receiving each of the possible interventions.

2.1 Proposed Genetic-Based Scheduling Algorithm

The proposed GSA works based on genetic algorithm. GSA allocates the job to resources by applying GA operators namely selection, crossover, and mutation. GSA starts its operation by creation of initial population, selects an individual from population by roulette wheel selection, and applies two-point crossover to allocate the job. Mutation (swap mutation) is rarely applied, when already selected job is chosen again. Table 1 shows the various parameters used in GSA. Shown below is the pseudocode of GSA.

Table 1 Parameter settings for proposed genetic algorithm

Parameter	Type	Values	Description
NOR (No. of requests)	–	Depends on job size	Population size
Selection	Roulette wheel selection	Selects random	Selection probability
Crossover	2-point crossover	0.7	Crossover probability
Mutation	Swap mutation	0.3	Mutation probability
Generations	–	100–1,000	Number of generations in GA

Genetic-Based Scheduling algorithm

```
Scheduling (Number Of Request)
1.  Create InitialPopulation().
2.  Set A :=0 , Z := 0.
3.  Repeat While Limit, do:
        a.  Set Z := Z+1, NodeCount := 0, Count := 0.
        b.  Call Selection().
        c.  Call Crossover().
        d.  Set Sol := Child.
        e.  If (A = 0), then
                i.  Repeat For I = 0 to Limit by 1, do:
                        i.  If (Node[i] = Sol), then
                                1.  Set Alloval[a] := Sol
                                2.  Set A := A+1.
                            Else:
                                Repeat For I =0 to Limit by 1 do:
                                        If (Node[i] = Sol), then Set NodeCount := 1.
                                        [End of If.] [End of For.]
                        [End of If.] [End of For.]
        Else:
                        A. If (NodeCount = 1), then
                              Repeat For J=0 to A by 1 do:
                                  If (Alloval[J] = Sol then  Set Count := 1.
                                  [End of If.] [End of For.]  [End of IF.]
                        B. If (Count = 0), then
                                a.  Set Alloval[A] := Sol.
                                b.  Set A := A+1.
                    Else:
                                i.    Call Mutation().
                                ii.   Set MnodeCount := 0, Mcount := 0.
                                iii.  Set Msol := Child.
                                iv.   Repeat For X = 0 to Size by 1 do:
                                          If (Node[X] = Msol), then  Set MnodeCount := 1.
                                        [End of If.]
                                      [End of For.]
                                v.    If (MnodeCount = 1)
                                          i.  Repeat For Y = 0 to A by 1 do:
                                                  If (Alloval[Y] = Msol), then Set Mcount := 1.
                                              [End of if.]
                                               [End of For.]
                                          ii.  If (Mcount = 0), then
                                                     i.  Set Alloval[A] := Msol.
                                                    ii.  Set A := A + 1.
                                              [End of If.]
                                          [End of If.]
                        [End of Step B IF.]
                [End of Step e IF.]
[End of Step 3 Loop.]
4. Call TurnATime( Alloval, NR, NM, Size).  [Calculates the Fitness of Individual.]
5. Exit
```

```
Algorithm Selection()
1.  Set Pran1 := Random(Limit).
2.  Set Pran1 := Random(Limit).
3.  Set Parent1 := InitialPopulation[Pran1].
4.  Set Parent2 := InitialPopulation[Pran2].
```

```
Algorithm CrossOver()
1.  Set Fv := Random(Limit).
2.  Set Sv := Random(Limit).
3.  Repeat For I=0 to Fv by 1 do:
        Set Child[I] := Parent1[I].  [End of For.]
4.  Repeat For I = Fv to Sv by 1 do:
                  Set Child[I] := Parent2[I].  [End of For.]
5.  Repeat For I = Sv to Col by 1 do:
                  Set Child[I] := Parent1[I].   [End of For.]
```

```
Algorithm Mutation()
1.  Set Fv := Random(MutationPoint).
2.  Set Sv := Random(MutationPoint).
3.  If (Child[Fv] = 0), then Set Child[Fv] := 1.
4.  Else (Child[Fv] = 0), then Set Child[Fv] := 0.
        [End of If.]
5.  If (Child[Sv] = 0), then        Set Child[Sv] := 1.
    Else (Child[Sv] = 0), then      Set Child[Sv] := 0.
      [End of If.]
```

3 Job Request: Preemptive Request

The term 'preemptive' in operating system indicates the job is split into many subjobs, to be processed with multiple resources. Same notion is used in preemptive request; request is split into many subrequests. Every request is independent of each other with no priorities. All the machines and jobs are simultaneously available at initial state. One machine processes an operation at a time, and processing cannot be interrupted. Transportation time of job from one machine to another machine is negligible. Migration of job is not allowed. The preemptive split is identified by the function S_p which has three parameters *N*, *I*, and *R*. *N* defines the number of subrequests, *I* defines the time interval of subrequest, and *R* specifies the resources where the subrequests are processed. Figure 1 illustrates preemptive request; four requests and their subrequests are shown. SR indicates subrequest, and *R* indicates resource at which subrequest executes. The user is restricted to specify the number of subrequests and required resources. The preemptive requests use three constraints to execute its work. Operator precedence constraint indicates that the order of operation of job is fixed, and processing of operations cannot be interrupted. Machine constraint indicates that a single subrequest is executed at a time. Request constraint indicates no overlaps between two subrequests of a request, i.e., if a subrequest finishes at time *T*, the following subrequest must start after that.

4 Implementation and Results

4.1 Environmental Setup

The grid environment is simulated using GridSim with collaborative approach with resource negotiator and resource. Resource negotiator performs two functions using (1) information collector (i.e., resource information) and (2) scheduler (i.e., allocates the job) based on algorithm. The grid environment was simulated with

Fig. 1 Preemptive requests

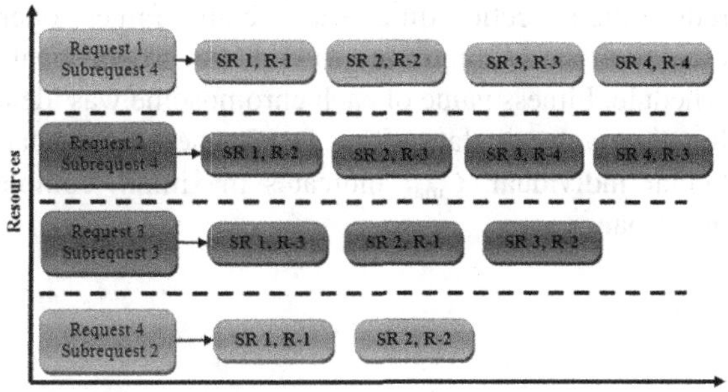

ten resources of heterogeneous capability. Resources are ready at the initial stage. The job information submitted to resource negotiator consists of the number of tasks in the job, time at which task completes execution, and resource requirement. The environment was tested using five types of benchmark jobs with 10, 15, 20, 30, and 50 requests [13]. Each job contains different task and resource requirements.

4.1.1 Completion Time

Completion time indicates the time at which Resources (R) finalize the processing of previous assigned task and already planned task. In Eq. 1, ReadyTime [R] indicates the time when Resource R finished its previously assigned tasks and ExpectedTimeToComplete [T][R] indicates completion time of task T.

$$\begin{aligned}\text{Completion }[R] &= \text{ReadyTime }[R] \\ &+ \sum_{\text{T}\in\text{Tasks}|\text{Schedule}[\text{T}]=\text{R}} \text{ExpectedTimeToComplete }[T][R]\end{aligned} \quad (1)$$

4.1.2 Makespan

Makespan indicates minimum time taken to complete every job.

$$\text{Makespan} = \text{Min}\{\text{Completion }[I], I \in \text{Allocation}\} \quad (2)$$

4.2 *Fitness Evaluation*

$$\text{Fitness}_i = \frac{C_i - M}{C_{\max} - M} \quad (3)$$

Genetic algorithm mimics survival-of-the-fittest principle of nature to make search process. Genetic algorithm targets the total finish time for the required schedule to reduce the execution time of a schedule. Fitness operator is very important as the fitness of the chromosome is directly proportional to the length of associated schedule. Fitness value of each chromosome was identified with makespan. Fitness function was calculated from Eq. 3 where C_i indicates objective function of particular individual, $C_{\max}$ indicates maximum completion time, and M indicates makespan.

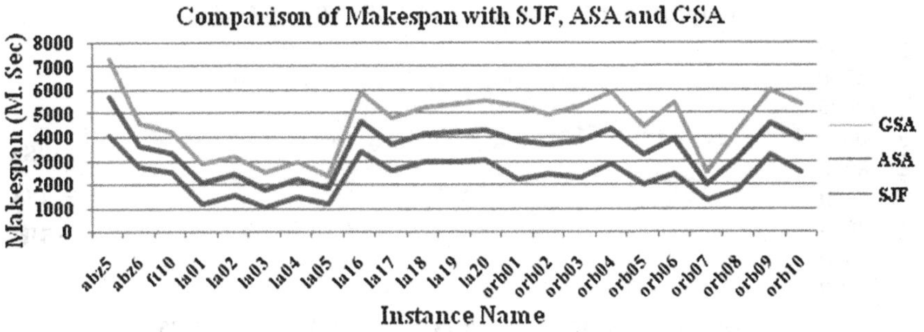

Fig. 2 Comparison chart of optimal makespan—10 request jobs

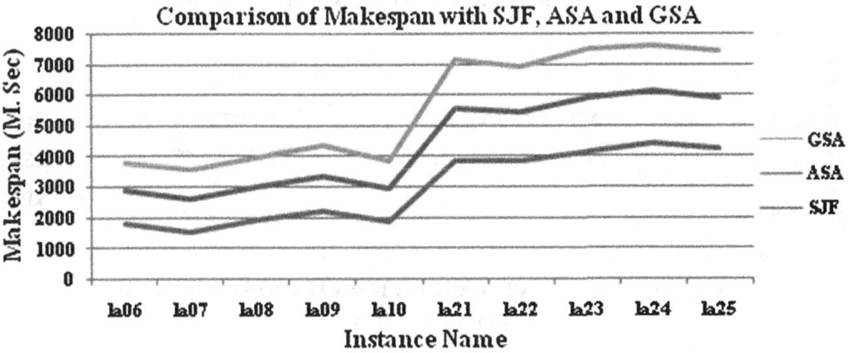

Fig. 3 Comparison chart of optimal makespan—15 request jobs

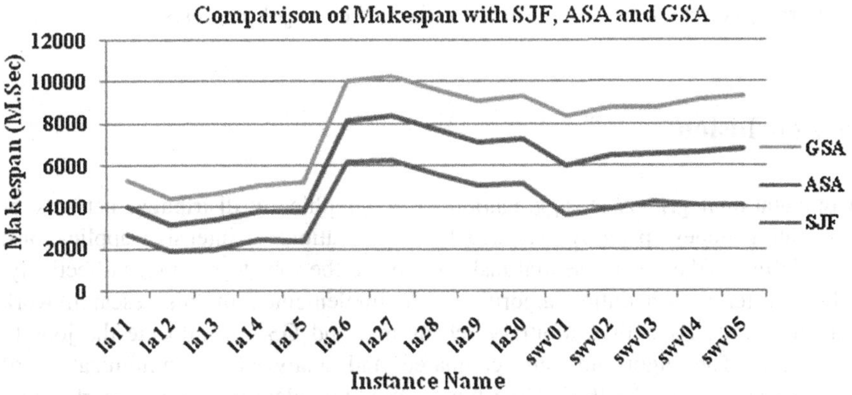

Fig. 4 Comparison chart of optimal makespan—20 request jobs

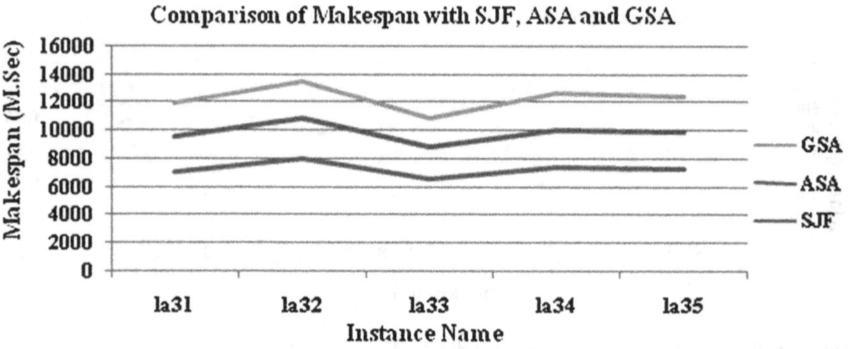

Fig. 5 Comparison chart of optimal makespan—30 request jobs

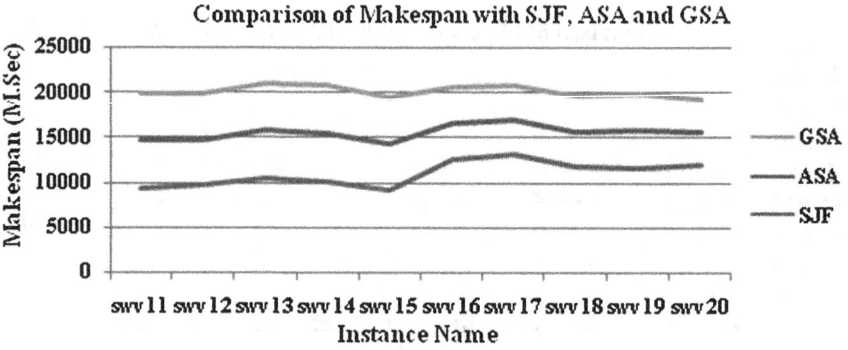

Fig. 6 Comparison chart of optimal makespan—50 request jobs

Figures 2, 3, 4, 5, and 6 show the implementation results of three algorithms SJF, ASA, and GSA. The algorithms were tested with five types of job request. It was identified that GSA has given best makespan as compared to shortest-job-first and arbitrary scheduling algorithms for five different job requests.

5 Conclusion

Computational grid is an aggregation of geographically distributed network of computing nodes specially designed for computationally intensive applications. Scheduling is the most essential task to allocate the jobs to resources effectively. Three different scheduling algorithms are implemented in this research work namely shortest job first, arbitrary scheduling, and GSA to allocate the jobs to resources. The algorithms are compared and analyzed for minimization of makespan. Among the three algorithms, GSA was identified as best performing scheduling technique in the grid environment.

References

1. Zahida Akhtar., "Genetic Load and Time Prediction Technique for Dynamic Load Balancing in Grid Computing", Information Technology Journal, 2007.
2. Joshy Joseph., Craig Fellenstein., "Grid Computing", IBM Press, 2005.
3. Paniagua. C., Xhafa. F., Caballe. S., Daradoumis. T., "A Parallel Grid Based Implementations For Real Time Processing Of Event Log Data In Collaborative Applications", International Journal of Web and Grid Services archive, Vol 6, Issue 2, June 2010.
4. Yaser Nemati., Faramarz Samsami., Mehdi Nikhkhah., "A Novel Data Replication Policy in Data Grid", Australian Journal of Basic and Applied Sciences, 6(7): 339–344, ISSN 1991–8178, 2012.
5. Lizhe Wang., Gregor von Laszewski., Marcel Kunze., Jie Tao., "Provide Virtual Machine Information for Grid Computing", IEEE System Journal, Vol. X, No. X, XXX 2008.
6. Prakash. S, Vidyarthi. D. P., "Load Balancing in Computational Grid Using Genetic Algorithm", Advances in Computing, Scientific & Academic Publishing, 2011.
7. Jia Yu., Rajkumar Buyya., "A Taxonomy of Scientific Workflow Systems for Grid Computing", SIGMOD Record, Vol. 34, No. 3, 2005.
8. Sylvain Cussat-Blanc., Herve Luga., Yves Duthen., "Genetic Algorithms and Grid Computing for Artificial Embryogeny", GECCO, ACM, 2008.
9. Lee Wang., Howard Jay Siegel., Vwani P., Roychowdhury., Anthony A. Maciejewski., "Task Matching and Scheduling in Heterogeneous Computing Environments Using a Genetic-Algorithm-Based Approach", Journal Of Parallel And Distributed Computing, Article No. PC971392, 1997.
10. Rachhpal Singh., "An Optimization of Process Scheduling Based on Heuristic GA", International Journal of Networking & Parallel Computing, Vol 1, Issue 1, September 2012.
11. Tavakkoli Moghaddam. R., Shahsavari Pour. N., Mohammadi Andargoli. H., Abolhasani Ashkezari. M. H., "Duplicate Genetic Algorithm for Scheduling a Bi-Objective Flexible Job Shop Problem", International Journal of Research in Industrial Engineering, Vol 1, Number 2, 2012.
12. http://www.civil.iitb.ac.in/tvm/2701_dga/2701-ga-notes/gadoc/gadoc.html.
13. http://people.brunel.ac.uk/~mastjjb/jeb/info.html.

A Multifactor Biometric Authentication for the Cloud

Shabana Ziyad and A. Kannammal

Abstract The omnipresence of high-speed Internet has paved a promising future for cloud computing. Cloud computing technology deals with providing software, platform, and infrastructure as a service to the clients. This ultimately leads to transmission of the client's confidential data in the cloud. This factor is the motivation to provide a secure biometric authentication framework to the cloud. This paper proposes a multifactor biometric authentication system for cloud computing environment. The biometric features adopted here are palm vein and fingerprint. The idea is to handle the biometric data in a secure fashion by storing the palm vein biometric data in multicomponent smart cards and fingerprint data in the central database of the cloud security server. In order to enhance security, the part of biometric data matching process is performed on the card with Match-on-Card technology and data never leave the smart card.

Keywords Secure cloud · Palm vein · Fingerprint · Multicomponent smart card · Cryptosystem · Authentication

1 Introduction

With the upward movement of more and more data on the Internet, new paradigms such as cloud computing have emerged as one of the milestones in information technology. Cloud computing has matured with heterogeneous group of cloud service providers. In cloud computing environment, the client has permission to

S. Ziyad (✉)
College of Engineering and Computer Sciences, Salman Bin Abdul Aziz University, Alkharj, Riyadh, Kingdom of Saudi Arabia
e-mail: ziyadshabana@yahoo.com

A. Kannammal
Coimbatore Institute of Technology, Coimbatore, Tamilnadu, India
e-mail: kannaphd@gmail.com

G. S. S. Krishnan et al. (eds.), *Computational Intelligence, Cyber Security and Computational Models*, Advances in Intelligent Systems and Computing 246, DOI: 10.1007/978-81-322-1680-3_43,

access only the resources permitted to them; hence, it is essential to authorize the clients with credentials, which will enable them to access the permissible resources in the cloud. With the increase in intrusions such as man-in-the-middle attacks, flooding attacks, phishing, and malware injection attacks, securing the cloud becomes a challenge. Identity theft is the most challenging security issue in cloud computing. As a result, it is imperative for cloud to authenticate every user accessing the cloud. Hence, there is an inevitable need to provide a strong authentication framework for the cloud.

One of the best ways to improve Web-based user authentication systems without compromising its usability and ubiquity is the implementation of biometric technology. Multifactor authentication refers to a compound implementation of two or more classes of authentication factors to authenticate the identity of a person. Identity proofing is the process of verifying user identities (for example, if a user is indeed who the individual claims to be) before provisioning user with resources. This paper analyses factors such as something known to the user, something held by the user, and something inherent only to the user to develop a satisfactory identity proof solution for clients of the cloud.

Two major vulnerabilities that specifically deserve attention in the context of biometric authentication are "*spoof attacks*" at the user interface and "*template database leakage.*" A *spoof attack* involves presenting a counterfeit biometric trait not obtained from a live person. *Template database leakage* refers to a scenario where a legitimate user's biometric template information becomes available to an adversary [1]. This is a dangerous scenario because it makes it easier for the adversary to recover the biometric pattern by simply reverse engineering the template. Hence, there is a need to provide a more secure place to store the template and that is a "*smart card*" [2].

The smart card is a boon to authentication and identity management. Smart cards are accepted as one of the secure and reliable forms of electronic identification. Latest smart card integrated circuits have specific coprocessors for efficient execution of several cryptographic algorithms and a set of peripherals to enable a flexible use of the controller for various kinds of applications.

This paper proposes a multifactor biometric authentication system by storing biometric data on a smart card and later matching that data with the client's biometric data using *Match-on-card* technology. The paper is organized as follows: Sect. 2 surveys the existing authentication systems in cloud computing, Sect. 3 describes the smart card architecture for storing biometric data, Sect. 4 proposes a secure multifactor biometric authentication scheme for cloud, and Sect. 5 describes the conclusion.

2 Survey of Existing Cloud Authentication System

As cloud is emerging as the repository of data, it becomes inevitable to throw light on the existing cloud authentication systems and to enhance the security of this rich repository by providing access to the authorized clients using more than one factor. The objective of this section is to analyze the security architecture of the cloud.

The cloud computing architecture proposed in [3, 4] has variety of servers for providing various types of services to the clients. The foremost security point is communication access point (CAP), which is capable of handling varied communication protocols. It is followed by security access point (SAP), which is the strong authentication server. This is followed by the application access point server (AAP), which is capable of distributing the service requests that arrive in the cloud to various individual servers and finally the standard certificate authority (CA) server which issues certificates to the cloud. Certificate is issued to a client and to the SAP server by certification authority (CA) and stored in an identity management system (IDMS) X.500 compliant directory. The client is then authenticated with the single sign-on authentication, a service provided by the SAP server. The disadvantage of the above authentication model is that if the security server is hacked, then the entire cloud is hacked. The credential text freely moves across the Internet, posing a serious threat to data security.

A secure authentication using single sign-on was proposed in [5]. The study of user inconvenience and administrative overhead due to multiple authentications has led to the proposal of single sign-on. Whenever a user requests service from a server, it authenticates with a central server, providing its credentials. Once authorized, further request to any server for resources is granted by the centralized server on generating username and password. SSO is available in three tiers. Tier1 SSO is based on proven standards, connecting a Web session to the application and authenticating back to a central directory that acts as the single place to enforce policies. Tier2 SSO may be necessary in organizations with many legacy applications that cannot be integrated into more modern user directories, such as Microsoft Active Directory. Tier3 SSO is a quick fix approach, relieving user password burden and IT password resets, but at the cost of optimizing security practices [6]. The disadvantage of single sign-on is that if the centralized server is accessed by unauthorized entities, then the cloud is insecure.

In strong authentication framework proposed in [7], the client enters the user identity number, password, and smart card. After the verification by the local server, the user requests for cloud login. The authentication data are sent to the user via mobile network. Now, these data along with smart card and ID are sent to the server for final authentication. Advantages of this framework are the ability to change password with ease, no free movement of private data of clients in the network, and verification of user's legitimacy before accessing the cloud. The authentication control is toward client side and that resists attacks. The disadvantage of this scheme is that there are a lot of hectic activities in this scheme.

The technique presented in [8] generates password by concatenating passwords at multiple level. Authentication takes place at different levels—organization, team, and user levels. At user level, it authenticates the user's privileges to access the particular cloud resource. Advantage of this technique is that it uses multilevel approach. It is quite difficult to break multilevel security as compared to single level. The disadvantage of this method is that there is a risk of password being hacked. According to technique proposed by [9], authentication is done by considering different dimensions of cloud such as vendor details, consumer details, services, and privileges. Creating a multidimensional password based on user's personal data is little hectic as the user has to remember a set of data which will generate the correct password. The comparison of various authentication techniques proposed with techniques in practice is listed in Table 1.

3 Biometric Smart Card Design

In the past three decades, the joint efforts in the field of material science, VLSI design, and information technology have led to the exploration of new smart cards that are capable of storing biometric features of the user being identified and authenticated. The smart card identity cards are difficult to forge or duplicate. The combination of hardware and software of the smart card has not only made it tamper resistant but also reacts to tampering effects, countering possible attacks. The smart cards are available in three categories—contact cards, contactless cards, and multicomponent cards. Multicomponent cards provide tamper-proof storage of biometric data such as images of iris, finger, face, and signature. ANSI-INCITS has created a series of standards specifying the interchange format for the exchange of biometric data of all the above-said categories [10]. It performs the basic functions of enrollment, verification, and identification and a database interface to allow a biometric service provider to identify and manage high-performance data.

The basic functionality of smart card is divided into three layers—hardware layer, operating system layer, and application layer. Hardware layer is the basic architecture comprising of microcontroller, RAM, ROM, EEPROM, cryptoprocessor, and input/output interface. The operating system layer is characterized by the operating system that efficiently manages smart card system resources. The application layer is characterized by the smart card applications that are stored in an EEPROM and is accessible to the user [11].

The multicomponent cards enable encryption, decryption, and matching process on the card itself, and any communication with the card reader is done securely. Hence, the card securely maintains the biometric information of the clients. This provides enhanced security and privacy to the data. The components of multicomponent smart card include smart card processor NXP SmartMX + cryptoengines, RF interface [ISO14443/13, 56 MHZ], optical display, clock, battery, and button.

4 Proposed System

The advantages and disadvantages of the various existing client authentication models for the cloud have been discussed in the above section. A little bit of research is also done on the smart cards to be more specific, multicomponent smart

Table 1 Comparison of existing authentication methods

Authentication scheme	Defining characters	Advantages	Authentication location	Reliability and security	Disadvantages
Tier1 SSO	Standard-based, cross-domain authentication	Passwords are eliminated. User directory information is safe	Server	Very high reliability and security	Requires participation and coordination between connecting parties
Tier2 SSO	Proprietary, single-domain authentication	It provides password reliefs	Server	Moderate	Highly customized and expensive
Tier3 SSO	Credential replay, cross-domain authentication	The password generation and updates are outside of IT control	Server	Low	Credential texts are freely passed over the Internet
Strong authentication	One-time key is sent to the mobile network of client based on smart card, user identity, and password	Smart card, password, and an extra out-of-band factors provide better security	Server	High	Need for additional hardware and software
Multilevel authentication	Passwords generated and concatenated at multiple levels—cross-domain authentication	Security levels of authentication are improved in cloud environment	Server	Moderate	Difficulty in remembering passwords
Multidimensional authentication model	Passwords generated on multiple parameters of client details in cloud	Risk of hacking password by brute-force attack is greatly reduced	Client/server	High	User has to provide a set of confidential data for password generation

cards storing biometric data. These two concepts are merged together to implement a strong authentication framework for the cloud. There has been a proposal on multimodal systems involving three biometric traits on single sign-on framework [12]. But proposed authentication framework is made secure by combining the concepts of biometrics with cryptography. In the enrollment phase, the smart card is made with a set of unique security parameters for each client. This is verified by certificate authority server by validating some documents given by the client as a proof of his identity. In the registration phase, the biometric data are obtained from the client. The palm vein data are encrypted with the data already stored in smart card. The encrypted biometric palm vein data are stored in the smart card. The fingerprint data are encrypted with data from smart card and stored in the central database.

The functional flow diagram is shown in Fig. 1. Smart card possesses the authentication number and palm vein data along with other authentication details. The palm vein data from the sensor are transformed and matched with transformed palm vein template stored in the smart card. If the matching is favorable, then handshaking with the server takes place at the terminal. Thereby, the ciphertext is sent through a secure channel to the terminal. In the terminal, matching is performed. Based on the result of matching, the access to the resource is permitted or denied. The entire authentication process is divided into three phases: (1) initialization phase, (2) registration phase, and (3) verification phase.

1. Initialization Phase:

The smart card manufacturer generates a random number, which is the authentication number. The manufacturer then generates a large prime number, that is, p and q of equal length. The value of n is computed as $n = pq$. The authentication number along with p and q is stored in the smart card.

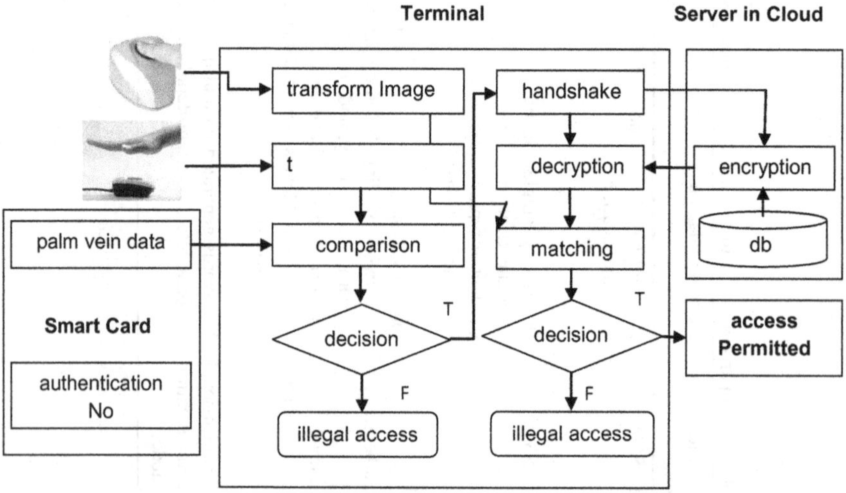

Fig. 1 Function flow diagram of proposed system

2. Registration Phase:

The user registers with the certification authority and receives a smart card from them by providing physical documents to prove their identity. Initially, the user is given a user identity number. The user is allowed to choose a password, and this password is stored in the smart card by some encryption techniques. The user then registers his or her palm vein image. This image is read by the sensor, and the points are selected to form a template. This template is later transformed by any one of the latest transformation techniques and finally stored in the smart card. The fingerprint image of the user is then read by a sensor and maintained in the central database of the server.

3. Verification Phase:

The user inserts the card to the reader and sends login request to the server. The verification phase is performed as follows. The user provides the password PW′ to the terminal. If the password PW′ matches with the password in the smart card, then the user identity number is extracted. In the next step, the user provides the palm vein PV impression via sensor. The transformed template PV already stored in the smart card is now compared with live biometric template also a transformed PV′ using the same transformation techniques. If comparison succeeds, the authentication moves to next step. Otherwise, it stops here. The fingerprint of the user is extracted using the sensor and transformed to compare with the already-registered biometric data in the database. In order to compare the fingerprint biometric data, the RSA algorithm is used for encryption and decryption of biometric data. Terminal calculates $\varphi\ (n)$ as follows:

$$\varphi(n) = \varphi(p)\cdot\varphi(q) = (p-1)(q-1). \tag{1}$$

It chooses integer e such that $1 < e < \varphi\ (n)$ and $gcd\ (e,\ \varphi\ (n)) = 1.e$ and $\varphi\ (n)$ are coprimes. e is released as a public component. The terminal also computes d, the multiplicative inverse of e ($modulo\ \varphi\ (n)$). This is computed using extended Euclidean algorithm. Now, d is the private key component. Terminal sends public key (n, e) to the server. It keeps d, the secret private key.

Fingerprint data FP is taken from the user with a sensor. This is transformed using a proven technique to FP′. The server converts the transformed fingerprint image FP″ stored in the database into a numerical value m such that $0 < m < n$ by using an agreed-upon reversible protocol known as a ***padding scheme***. The ciphertext CFP″ is computed as follows:

$$\mathrm{CFP}'' \equiv m^{e} (\operatorname{mod} n) \tag{2}$$

This ciphertext is sent to the terminal. Terminal can decrypt m from the ciphertext c using private key d using the formula:

$$m \equiv c^{d} (\operatorname{mod} n) \tag{3}$$

The original message *M* is recovered from *m* by ***reverse padding method***. The image is reconstructed from the numeric string *M*. In comparison, if the two images are matched, then the login request is accepted; otherwise, it is rejected (Fig. 2).

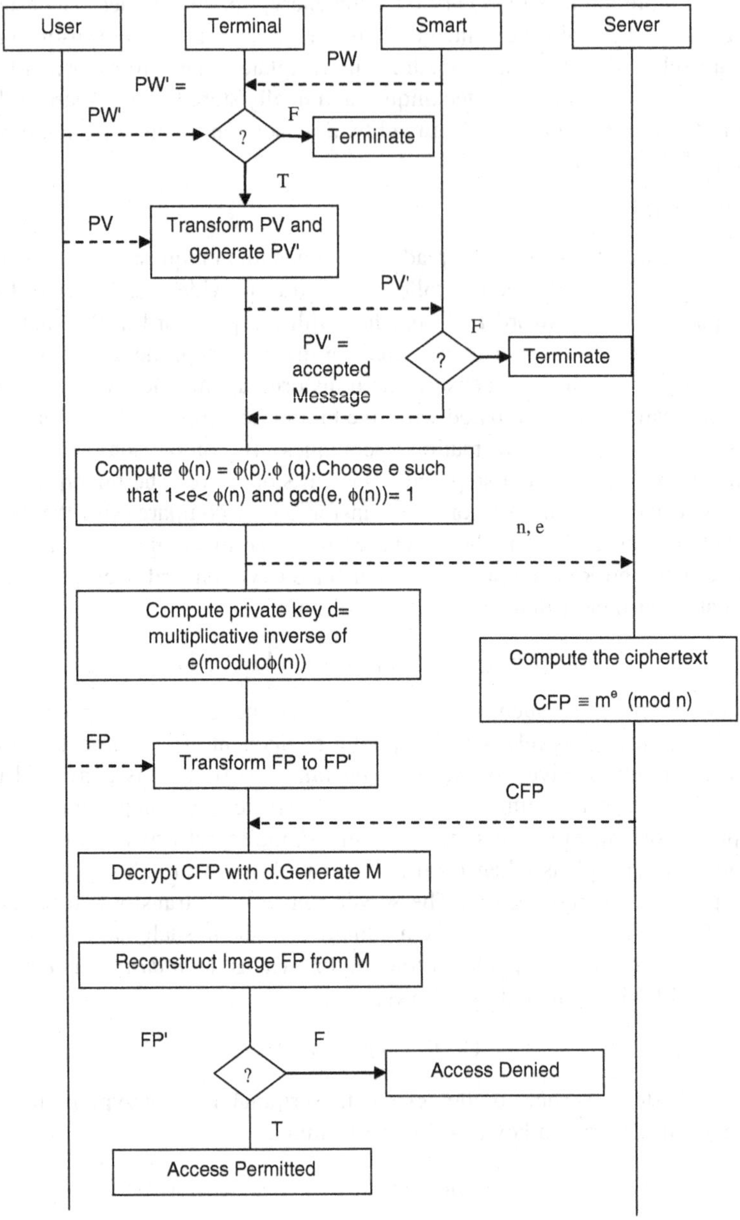

Fig. 2 Sequence diagram for the proposed system

5 Conclusion

This paper surveys the existing authentication systems in the cloud. Further, the merits and demerits of each implemented and proposed method are also analyzed. This paper proposes to merge the proved technology of maintaining the fingerprint data in the database server with the latest Match-on-Card technology to provide a full-fledged authentication system for cloud. Therefore, a proposed multifactor biometric authentication technique considers both the fingerprint and palm vein authentication factors along with traditional smart card, user identity, and user password for authentication in the cloud. This takes the authentication to the three factors—something user knows, something the user has, and something the user is. This proposed technique provides better security compared to the existing authentication techniques.

References

1. A.K. Jain, P. Flynn, and A.A. Ross, "Spoof Detection Schemes", *Handbook of Biometrics*, eds., Springer, 2007, pp. 403–424.
2. Smart cards and Biometrics, A Smart Card Alliance Physical Access Council White Paper March 2011.
3. Security Architecture for Cloud Computing Environments White Paper– February 1, 2011.http://security.setecs.com.
4. Ram Kumar Singh, Aniruddha Bhattacharjya," Security and Privacy Concerns in Cloud Computing", International Journal of Engineering and Innovative Technology (IJEIT) Volume 1, Issue 6, June 2012.
5. Ashish G. Revar and Madhuri D. Bhavsar, "Securing User Authentication Using Single Sign On in Cloud Computing",Institute of Technology, Nirma University, IEEE, December 2011.
6. Ping Identity, "Three Tiers of single Sign On",White Paper.
7. Amlan Jyoti Choudhury, Pardeep Kumar,Mangal Sain, Hyotaek Lim, Hoon Jae-Lee," A Strong User Authentication Framework for Cloud Computing"- Asia -Pacific Services Computing Conference - 2011 IEEE.
8. Dinesha H A,"Multi-level Authentication Technique for Accessing Cloud Services", International Conference on Computing,Communication and Applications (ICCCA), IEEE, 22–24 February 2012, pp 1–4.
9. Dinesha H A and Dr.V.K Agrawal," Multi-dimensional password generation technique for accessing Cloud services", International Journal on Cloud Computing: Services and architecture (IJCCSA),Vol.2, No.3, June 2012.
10. http://www.smartcardbasics.com/smart-card-standards.html.
11. R. Wolfgang and E. Wolfgang, "*Smart Card Handbook*", John Wiley and Sons, 3rd Edition, 2004.
12. B.Prasanalakshmi, A.Kannammal, "Secure Credential Federation for Hybrid Cloud Environment with SAML Enabled Multifactor Authentication using Biometrics", International Journal of Computer Applications (0975–8887) Volume 53– No.18, September 2012.

Accelerating the Performance of DES on GPU and a Visualization Tool in Microsoft Excel Spreadsheet

Pinchu Prabha, O. K. Sikha, M. Suchithra and K. P. Soman

Abstract Graphic processing units (GPU) have attained a greater dimension based on their computational efficiency and flexibility compared to that of classical CPU systems. By utilizing the parallel execution capability of GPU, traditional CPU systems can handle complex computations effectively. In this work, we exploit the parallel structure of GPU and provide an improved parallel implementation for data encryption standard (DES), one of the famous symmetric key cryptosystems. We also developed a visualization tool for DES in Microsoft Excel Spreadsheet which helps the students to understand the primitive operations that constitute the DES cryptosystem clearly. The main objective of this work is to investigate the strength of parallel implementation, on the basis of execution time on GPU as well as on CPU systems.

Keywords DES on GPU · DES visualization tool · DES parallel algorithm

1 Introduction

Protection of secrets from theft is an activity as old as their existence [1]. Humans have sought several ways to protect their secrets from a third party from the ancient time onward. Some practices include protecting the communication channel using guards and safes, and another way was to encode the confidential message so that nobody can read the content even if they tried to get it [1]. This method has evolved by time and become a vast area known as cryptography. Generally, algorithms used for bulk encryption and decryption utilizes single key on sender's side as well as on receiver's side. Symmetric key cryptosystems are broadly classified into two

P. Prabha (✉) · O. K. Sikha · M. Suchithra · K. P. Soman
Centre for Excellence in Computational Engineering and Networking,
Amrita Vishwa Vidyapeetham, Coimbatore, India
e-mail: pinchuprabha@gmail.com

G. S. S. Krishnan et al. (eds.), *Computational Intelligence, Cyber Security and Computational Models*, Advances in Intelligent Systems and Computing 246,
DOI: 10.1007/978-81-322-1680-3_44,

classes: block ciphers and stream ciphers [5]. As the name indicates, block ciphers map n-bit plain text block into n-bit ciphertext block. Stream ciphers, on the other hand, encrypt a message one bit (or more commonly in computing applications, byte) at a time [1]. During nineties, computers have got more attention; hence, private companies started the invention of new methods like data encryption standard (DES) to protect their digital data [1]. High-performance cryptographic accelerator cards have been introduced to improve the speed of cryptosystems [4]. In this work, we have tried to exploit the existing system resources such as graphical processing unit (GPU) to speed up the cryptoactivities. GPUs also known as visual processing units (VPU) are specialized parallel processors that are initially designed to accelerate the real-time 3D rendering problems [1]. Computational flexibility and increased processing power of GPU made them familiar to use for other computational processes [2] especially in cryptography. Several studies have been done about the application GPU parallel programming on the field of cryptography. Gershon Kedem and Yuriko Ishihara published the first paper dealing with cryptography on graphics hardware in August 1999 [3].

2 Cryptography

Cryptography is the science of keeping secrets, which has been evolved centuries back. It is an ancient art used for protecting data and communications from theft and altering. The main aspect of cryptography is to communicate the information between the intended parties by transforming it into an unreadable format so that an eavesdropper cannot interfere with the transmission. Nowadays, the information authenticity and data security have got greater importance since all the applications and services depend on the Internet, a very unsecure channel. Basically, a cryptosystem includes two sections: encryption which changes the appearance of the message at the sender's side and decryption which translates the ciphertext back to plain text at the receiver side. Both encryption and decryption are done using a key. Based on the number of keys used for encryption/decryption, cryptosystems can be classified into two, symmetric key cryptosystems and asymmetric key cryptosystems [8]. Symmetric cryptosystems use the same key for both encryption and decryption, and asymmetric cryptosystems use two different keys for the same. In this work, we are concentrating on the symmetric key cryptosystem especially DES. DES is a block cryptosystem that comes under the category of Feistel ciphers [7]. Generally, a Feistel cryptosystem contains the following components:

1. Ciphertext and plain text spaces are having size 2t bit string and key size will be N.
2. The number of rounds NR which is a positive number.
3. A key scheduling mechanism for generating unique key for each NR round from a single cryptosystem key K $(K^1, K^2, K^{\mathrm{NR}})$.
4. A round function f_{k^i} corresponding to each round key k^i. The input of the string R contains t bits and output also t bits from $f_{k^i}(R)$.

Encryption: We are having a plain text P of size 2t. Split the plaintext into two bit strings each having a size t: $P = (L_0, R_0)$, where L_0 is the left part and R_0 is the right part of the given plain text. We then go through NR rounds as indicated below

$$\begin{aligned}&\text{FOR each round } i = 1 \text{ to NR}\\ &L_i = R_{i-1}\\ &R_i = L_{i-1} \oplus f_{k^i}(R_{i-1})\\ &\text{End for}\end{aligned}$$

After NR rounds, the ciphertext will be $C = (R_{\mathrm{NR}}, L_{\mathrm{NR}})$

Decryption: The ciphertext C is passed through all of the above steps, but the keys are used in a reversed fashion. DES cryptosystem follows the above procedure which is supported by an initial permutation applied before the Feistel system, and an inverse permutation is applied after NR rounds.

3 DES-Parallel Algorithm

In a cryptosystem, data encryption and decryption are complex problems because of the large size of data. Efficiency can be improved through the parallel implementation of decomposed blocks of data based on the GPU-CUDA programming which solves the data processing problem. CUDA programming mainly include three types of functions [9], host function, device function, and global function. Here, global function is used for the parallel implementation. Global function can only access from the CPU with prefix global. Parallel execution implement with help of multiple threads.

Global Function Parameters

1. Grid: This refers the number of thread block included in the kernel.
2. Block: Blocks are organized into a one-dimensional or two-dimensional grid of thread blocks. The number of thread blocks in a grid is usually dictated by the size of the data being processed or the number of processors in the system.
3. Thread: Each thread is identified by threadIdx, which is a 3-component vector, so that a set of threads can form one-dimensional, two-dimensional, or three-dimensional thread block. There is a limit to the number of threads per block, since all threads of a block are expected to reside on the same processor core and must share the limited memory resources of that core. On current GPUs, a thread block may contain up to 512 threads.

3.1 Parallel Implementation

User reads the plain text and key through CPU. GPU memory is allocated for parallel computation using cudaMalloc() function [6]. The input data is then copied into the allocated GPU memory using cudaMemcpy() function. Once the parallel computation is over, the result is copied back to CPU [6]. In our work, parallelism has been exploited in the following steps,

- In order to generate keys for 16 rounds, do the following left shift operation

$$l_i = \begin{cases} 1, & \text{if } i = 1;2;9;16 \\ 2, & \text{Otherwise} \end{cases}$$

Data structure is array. Total number of threads used in this operation depends on the size of the array. Each thread is accessing the each location in array. At time of shifting, each thread picks the value in the next position concurrently.

```
INPUT : left and right half of key after permutation
OUTPUT : shifted array
global void roundkey (char *C, char *D, char *Cn, char *Dn, int r)
{
int i = threadIdx . x;
if (r == (1 | | 2 | | 9 | | 16))
{
Cn [i] = C[(i +1) % 28];
Dn[i] = D[((i +1) % 28)];
}
else
{
Cn [i] = C[(i +2) %28];
Dn[i] = D[((i +2) %28)];
}
}
```

- In DES, each round function includes XOR operation. All the XOR operations have been done in parallel. The number of threads depends on the size of the string arrays to be xored.

```
INPUT : left two character array with same size
OUTPUT : xored array
global void firstxor (char *t , char *k , int *e , char *rx)
{
```

```
int i = threadIdx . x;
char RXor [4 8], Rexp [4 8];
Rexp [i] = t [e [i] - 1];
rx [i] = Rexp [i] ^ k [i];
}
```

- S-box is the heart of DES cryptosystem. DES algorithm includes eight S-boxes. In our work, all the S-boxes are packed together into a 32 × 16 matrix. Each row of the matrix is considered as a block, and each block contains equal number of threads. So, each S-box constitutes four blocks, and each block contains sixteen threads.

```
global void sbox (char * rx, int *s)
{
\\ sbox
int index = threadIdx . x + blockIdx . x * 6;
int i = blockIdx . x;
int row [8] , col [8];
row [i]=2* rx [(i *6)]+rx [(i *6)+5];
col [i]=(8* rx [(i *6)+1])+(4*rx [( i*6)+2])+(2*rx [( i*6)+3])+( rx [( i*6 )+4);
s [i]=sb[(( i*4)+row[i])][ col[i]];
}
```

4 DES Visualization Tool

4.1 Why Excel?

Microsoft Excel Spreadsheet is an excellent and versatile computational platform for the student community to get the practical experience of the concept what they have learned from the classroom without any support of the superiors. This tool helps the students to understand any complex mathematical computations and algorithms in detail, and they can also visualize what is happening on each step. Since it is a user-friendly mathematical platform, the students can practice from their high school onward. This is a very cost-effective tool compared to the other computational platforms.

4.2 Motivation

DES is the most widely used symmetric cryptographic algorithm for providing data security. Since it is an ancient algorithm, most of the people know the theory behind it. But, they do not have the in-depth knowledge about the operations on each step. By using the Excel Visualization tool, students can practically visualize the mathematical operations behind each step. There is no need of an in-depth literacy about the computers and the languages to work in Excel.

4.3 Implementation

In this paper, we describe the Excel visualization tool for DES algorithm which helps the students to understand the primitive operations involved in the algorithm clearly. DES was developed by IBM and standardized in 1977 by National Bureau of Standards (NBS) [7]. Even though the steps involved in the algorithm are simple, it is very difficult to understand it at a stretch. Here, we are using Excel Spreadsheet as the visualization tool. So, even the high school students can work on this platform. This paper provides the visualization tool for 8-bit DES system and 64-bit DES system. Eight-bit DES cryptosystem has been implemented in the Excel sheet using the inbuilt excel functions. Since 64-bit DES system involves many Feistel rounds, we have used Excel VBA tool for building the visualization tool. Simple DES cryptosystem chooses the input as a plain text of 8 bit length, a key of 12 bit length, and produces an 8-bit output cipher. The encryption algorithm mainly involves three sections: an initial permutation, two Feistel rounds, and one final permutation. When we click on the START button, it clears the entire text fields on the window. Now, the user can enter the 8-bit plain text and 12-bit key in the PLAIN TEXT field and KEY fields. On clicking the PLAIN-INITIAL PERMUTATION button, the plain text will split into 4 bit left half and right half, the permuted output will appear on the L_0 and R_0 text field, respectively. Since the simple DES system involves two Feistel rounds, we require two unique keys for each round. When clicking on the PERMUTED KEY GENERATION button, two 8-bit keys are generated after removing the parity bits and will appear on the KEY1 and KEY2 text field. The Feistel round involves many steps such as expand function, XOR operations, S-Box selection, etc. A step-by-step implementation of all these operations has been shown in the visualization window clearly. Here, we are providing the buttons for the steps involved in the Feistel round as EXPAND, XOR, and SBOX, respectively. By clicking the FINAL PERMUTATION button, 8-bit cipher text will appear on the CIPHER field. Similarly, we have developed the Excel visualization tool for 64-bit DES system. This algorithm takes the input as 64-bit plain text and 56-bit key and produces 64-bit cipher text through an initial permutation, 16 Feistel rounds, and a final permutation. Here, we provide 16 round buttons; by clicking on each of the ROUND button, the corresponding output will appear on the window.

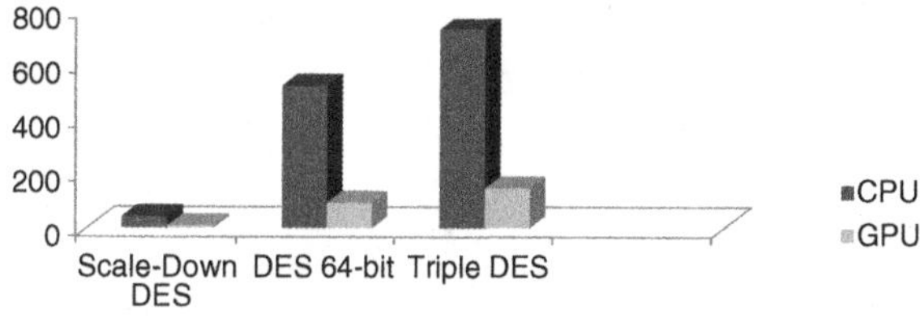

Fig. 1 Execution time CPU versus GPU

5 Conclusion

In this work, we have done a comparative study on the performance of GPU and CPU implementation of DES based on the execution time, and the corresponding performance graph is shown in Fig. 1. The graph compares the execution time needed for scale-down DES, DES, and Triple DES algorithms on CPU as well as on GPU; it shows that the performance of GPU is 60 times faster than CPU. So, with the evolution of GPU, efficient computation of complex algorithms can be done easily. This work also includes Excel visualization tool for 8-bit and 64-bit DES system that will help the students and instructors to understand the DES cryptosystem in an effective manner. All the excel sheets are available on http://nlp.amrita.edu:8080/sisp/wavelet/cwt/des.xlsm.

References

1. Samuel Neves, "Cryptography in GPUs", 2009
2. M. Bobrov1, R. Melton, S. Radziszowski, and M. Lukowiak, "Effects of GPU and CPU Loads on Performance of CUDA Applications"
3. Gershon Kedem, Yuriko Ishihara, "Brute Force Attack on UNIX Passwords with SIMD Computer", Proceedings of the 8th USENIX Security Symposium, August 1999
4. Debra L. Cook John Ioannidis, Angelos D. Keromytis, Angelos D. Keromytis, "Secret Key Cryptography Using Graphics Cards"
5. Alexander Stanoyevitch, "Introduction TO Cryptography With Mathematical Foundations and Computer implementations", 2010
6. Jason Sanders, Edward Kandrot, "CUDA by Example An Introduction To General-Purpose GPU Programming"
7. Jun Tao, Jun Ma, Melissa keranen, "DES visua: A Visualization Tool For the DES cipher", March 26 2011
8. Bruce schneier, "Applied cryptography", 1996
9. Santa Clara, CUDA Programming Guide. NVIDIA Corporation, CA, 2009.

About the Editors

G. Sai Sundara Krishnan is Associate Professor in the Department of Applied Mathematics and Computational Sciences at PSG College of Technology, Coimbatore, India. He has published 15 papers in reputed international journals and conferences and edited two international conference proceedings. Currently, he is supervising nine research scholars pursuing their PhD degrees. His areas of research interest include digital topology and soft sets.

Muthukrishnan Senthil Kumar is Associate Professor, Department of Applied Mathematics and Computational Sciences, PSG College of Technology, Coimbatore, India. He is the recipient of a post doctoral fellowship from the Department of Nanobio Materials and Electronics, (WCU), Gwangju Institute of Science and Technology, Gwangju, Republic of Korea, under the guidance of Prof. K. Sohraby, Curators' Professor, School of Electrical Engineering, University of Missouri-Kansas City. He has over 14 years of teaching and research experience. He has a PhD in retrial queueing models from Anna University, Chennai, India. His achievements are recognized and included in the biographies of Marquis Who's Who in the World 2013. His fields of interest include retrial queueing theory, data communication, and reliability engineering. He serves as a reviewer for several international journals and has also published his research in several international journals of repute.

R. Anitha is Associate Professor in the Department of Applied Mathematics and Computational Sciences at PSG College of Technology, Coimbatore, India. Her research interests include cryptography, computer security, algorithms, and computational models. She has supervised more than 10 PhD scholars. She has published over 40 research papers in refereed journals and has served as an editor for many conferences and workshops. She has served as a PC member for several international conferences such as ICoAC 2011, ICISS 2013, and ADCOM 2013 and organized many research workshops and conferences. She served as the principal investigator (PSG College of Technology) of CDBR-SSE project

G. S. S. Krishnan et al. (eds.), *Computational Intelligence, Cyber Security and Computational Models*, Advances in Intelligent Systems and Computing 246, DOI: 10.1007/978-81-322-1680-3, © Springer India 2014

sponsored by NTRO and is currently the consultant to SETS for a research project sponsored by the Department of Atomic Energy, Government of India. She is a member of several professional bodies including CRSI, ISTE, and ACM.

R. S. Lekshmi is Associate Professor in the Department of Applied Mathematics and Computational Sciences at PSG College of Technology, Coimbatore, India. She has authored 10 publications and guided many students' projects on graph theory and networks. She has organized several research workshops/seminars and was the Program Chair for the International Conference on Mathematical and Computational Models held at PSG College of Technology in 2011.

Anthony Bonato is the Associate Dean, Students and Programs in the Yeates School of Graduate Studies at Ryerson University and Editor-in-Chief of the journal *Internet Mathematics*. He has authored over 90 publications with 40 coauthors on the topics of graph theory and complex networks. In 2011 and 2009, he was awarded Ryerson Faculty SRC Awards for excellence in research. In 2012, he was awarded an inaugural YSGS Outstanding Contribution to Graduate Education Award.

Manuel Graña is currently full professor at the Computer Science and Artificial Intelligence of the University of the Basque Country (UPV/EHU), in the Facultad de Informatica in San Sebastian. His interests include machine learning and pattern recognition, medical image processing and computer-aided diagnosis systems, mobile robot navigation, multi-agent systems with natural inspiration, and social network innovations via computational intelligence. The development of lattice computing approaches in these domains is his principal research endeavor. He is member of MIR Labs, IEEE, and ACM. He has chaired three international conferences (IWANN 2007, HAI 2010, and KES 2012). He has served as an editor of more than 10 books. He has been advisor to more than 20 successful doctoral students. He has coauthored over 100 papers in high-impact research journals.

Author Index

G. S. S. Krishnan et al. (eds.), *Computational Intelligence, Cyber Security and Computational Models*, Advances in Intelligent Systems and Computing 246,
DOI: 10.1007/978-81-322-1680-3, © Springer India 2014